Gmelin Handbook of Inorganic and Organometallic Chemistry

8th Edition

Gmelin Handbook of Inorganic and Organometallic Chemistry

8th Edition

Gmelin Handbuch der Anorganischen Chemie

Achte, völlig neu bearbeitete Auflage

PREPARED AND ISSUED BY — Gmelin-Institut für Anorganische Chemie der Max-Planck-Gesellschaft zur Förderung der Wissenschaften

Director: Ekkehard Fluck

FOUNDED BY — Leopold Gmelin

8TH EDITION — 8th Edition begun under the auspices of the Deutsche Chemische Gesellschaft by R. J. Meyer

CONTINUED BY — E. H. E. Pietsch and A. Kotowski, and by Margot Becke-Goehring

Springer-Verlag
Berlin · Heidelberg · New York · London · Paris · Tokyo · Hong Kong · Barcelona 1991

Gmelin-Institut für Anorganische Chemie
der Max-Planck-Gesellschaft zur Förderung der Wissenschaften

GMELIN HANDBOOK

Dr. J. von Jouanne

Dr. L. Berg, Dr. H. Bergmann, Dr. J. Faust, J. Füssel, Dr. H. Katscher, Dr. R. Keim, Dr. E. Koch, Dipl.-Phys. D. Koschel, Dr. A. Kubny, Dr. P. Merlet, Dr. M. Mirbach, Prof. Dr. W. Petz, Dr. F.A. Schröder, Dr. A. Slawisch, Dr. W. Töpper

Dr. R. Albrecht, Dr. G. Bär, D. Barthel, Dr. N. Baumann, Dr. K. Behrends, Dr. W. Behrendt, D. Benzaid, Dr. R. Bohrer, K.D. Bonn, Dr. U. Busch, Dipl.-Ing. V.A. Chavizon, E. Cloos, Dipl.-Phys. G. Czack, A. Dittmar, Dipl.-Geol. R. Ditz, R. Dowideit, Dipl.-Chem. M. Drößmar, P. Dürr, Dr. H.-J. Fachmann, B. Fischer, Dipl.-Ing. N. Gagel, Dipl.-Phys. D. Gras, Dr. K. Greiner, Dipl.-Bibl. W. Grieser, Dr. I. Haas, Dr. R. Haubold, Dipl.-Min. H. Hein, Dipl.-Phys. C. Heinrich-Sterzel, H.-P. Hente, H.W. Herold, U. Hettwer, Dr. G. Hönes, Dr. W. Hoffmann, G. Horndasch, Dr. W. Huisl, Dr. M. Irmler, B. Jaeger, Dr. R. Jotter, Dipl.-Chem. P. Kämpf, Dr. B. Kalbskopf, Dipl.-Chem. W. Karl, H.-G. Karrenberg, Dipl.-Phys. H. Keller-Rudek, B. Kirchner, Dipl.-Chem. C. Koeppel, R. Kolb, Dr. M. Kotowski, E. Kranz, Dipl.-Chem. I. Kreuzbichler, Dr. V. Kruppa, Dr. W. Kurtz, M. Langer, Dr. B. Ledüc, Dr. A. Leonard, H. Mathis, E. Meinhard, M. Meßer, C. Metz, K. Meyer, E. Mlitzke, Dipl.-Chem. B. Mohsin, Dr. U. Neu-Becker, K. Nöring, Dipl.-Min. U. Nohl, Dr. U. Ohms-Bredemann, I. Rangnow, Dipl.-Phys. H.-J. Richter-Ditten, E. Rudolph, G. Rudolph, Dipl.-Chem. S. Ruprecht, Dr. B. Sarbas, Dr. H. Schäfer, Dr. R. Schemm, Dr. D. Schiöberg, V. Schlicht, Dipl.-Chem. D. Schneider, A. Schwärzel, Dr. B. Schwager, Dr. C. Strametz, G. Strauss, Dr. G. Swoboda, Dr. D. Tille, A. Tuttas, H.-M. Wagner, Dipl.-Phys. J. Wagner, R. Wagner, Dr. E. Warkentin, Dr. C. Weber, Dr. A. Wietelmann, Dr. M. Winter, Dr. B. Wöbke, K. Wolff

GMELIN ONLINE

Dr. R. Deplanque

Dr. P. Kuhn, Dr. G. Olbrich

Dr. R. Baier, Dr. B. Becker, Dipl.-Chem. E. Best, Dr. H.-U. Böhmer, Dipl.-Phys. R. Bost, Dr. A. Brandl, Dr. R. Braun, Dipl.-Chem. R. Durban, R. Hanz, Dr. A. Kirchhoff, Dipl.-Chem. H. Köttelwesch, Dipl.-Ing. W. Korba, Dr. M. Kunz, Dr. L. Leichner, Dipl.-Chem. R. Maass, Dr. A. Nebel, Dipl.-Chem. R. Nohl, Dr. B. Rempfer, Dr. U. Tölle, Dipl.-Ing. H. Vanecek

Volumes published on Platinum-Group Metals

Ir	* Iridium Main Volume — 1939 * Iridium Suppl. Vol. 1 (Metal, Alloys) — 1978 ° Iridium Suppl. Vol. 2 (Compounds) — 1978
Os	* Osmium Main Volume — 1939 Osmium Suppl. Vol. 1 — 1980
Pd	* Palladium 1 (Element) — 1941 * Palladium 2 (Compounds) — 1942
Pt	* Platinum A 1 (History, Occurrence) — 1938 * Platinum A 2 (Occurrence) — 1939 * Platinum A 3 (Preparation of Platinum Metals) — 1939 * Platinum A 4 (Detection and Determination of the Platinum Metals) — 1940 * Platinum A 5 (Alloys of Platinum Metals: Ru, Rh, Pd) — 1949 * Platinum A 6 (Alloys of Platinum Metals: Os, Ir, Pt) — 1951 * Platinum B 1 (Physical Properties of the Metal) — 1939 * Platinum B 2 (Physical Properties of the Metal) — 1939 * Platinum B 3 (Electrochemical Behavior of the Metal) — 1939 * Platinum B 4 (Electrochemical Behavior and Chemical Reactions of the Metal) — 1942 * Platinum C 1 (Compounds up to Platinum and Bismuth) — 1939 * Platinum C 2 (Compounds up to Platinum and Caesium) — 1940 * Platinum C 3 (Compounds up to Platinum and Iridium) — 1940 * Platinum D (Complex Compounds of Platinum with Neutral Ligands) — 1957 Platinum Suppl. Vol. A 1 (Technology of Platinum-Group Metals) — 1986 Platinum Suppl. Vol. A 2 (Isotopes. Atoms. Clusters) — 1989
Rh	* Rhodium Main Volume — 1938 Rhodium Suppl. Vol. A 1 (Metal, Alloys) — 1991 **(present volume)** Rhodium Suppl. Vol. B 1 (Compounds) — 1982 Rhodium Suppl. Vol. B 2 (Coordination Compounds) — 1984 Rhodium Suppl. Vol. B 3 (Coordination Compounds) — 1984
Ru	* Ruthenium Main Volume — 1938 * Ruthenium Suppl. Vol. — 1970

* Completely or ° partially in German

Gmelin Handbook of Inorganic and Organometallic Chemistry

8th Edition

Rh
Rhodium

Supplement Volume A 1

Coordination Compounds with
O- and N-Containing Ligands

With 181 illustrations

AUTHOR G. R. Watts
Bishop Stortford Herts, U.K.

EDITOR Kurt Swars, Gelnhausen,
Bundesrepublik Deutschland

System Number 64

Springer-Verlag
Berlin · Heidelberg · New York · London · Paris · Tokyo ·
Hong Kong · Barcelona 1991

LITERATURE CLOSING DATE: 1988
IN MANY CASES MORE RECENT DATA HAVE BEEN CONSIDERED

Library of Congress Catalog Card Number: Agr 25-1383

ISBN 3-540-93639-4 Springer-Verlag, Berlin · Heidelberg · New York · London · Paris · Tokyo · Hong Kong · Barcelona
ISBN 0-387-93639-4 Springer-Verlag, New York · Heidelberg · Berlin · London · Paris · Tokyo · Hong Kong · Barcelona

Typesetting, printing, and bookbinding: Universitätsdruckerei H. Stürtz AG, Würzburg

Preface

This volume describes the bulk metal rhodium and its alloys with arsenic to ruthenium in accordance with the Gmelin system. The first section, pages 1 through 37, covers in detail the following properties of the metal: Crystallographic properties; self-diffusion; mechanical properties (tensile properties, elastic moduli, hardness, etc.); thermal properties (thermal expansion, vapour pressure, specific heat, thermal conductivity, etc.); electrical and magnetic properties (electric resistivity, superconductivity, Hall effect, thermoelectric force, magnetic susceptibility, etc.); electrochemical behaviour and chemical reactions.

A number of subject matters on metallic rhodium are already contained together with those of other platinum group metals in appropriate chapters of "Platinum" Suppl. Vol. A1, 1986. These include 1) preparation and electrodeposition, 2) zone-refining operations, and 3) chemical vapour deposition, sputtering, and thermal evaporation.

Likewise, "Platinum" Suppl. Vol. 2, 1989, already contains information on the 1) separation and properties of the Rh isotopes, 2) physical properties of the Rh atom and the atomic ions Rh^{+} and Rh^{2+}, 3) the formation of Rh clusters, and 4) the molecular properties of Rh_2.

The second and major portion of this volume, pp. 38 to 273, deals with the physical and chemical properties of rhodium alloys. Phase diagrams of a large number of binary and multicomponent systems are treated in detail as well as the crystallographic and other properties of many intermetallic compounds. There is obviously a great deal of interest in the magnetic properties and possible superconductivity of many Rh-containing intermetallic phases.

Gelnhausen
November 1991

Kurt Swars

Table of Contents

The Metal

General Remarks. Rh, atomic number 45, mean atomic weight 102.9055, is a white, highly reflective metal with a face-centred cubic crystal structure. Despite its structure it is not readily cold-worked, the reasons for this not being fully understood. It is a corrosion-resistant metal and finds applications in thin electrodeposited or sputtered coatings, for crucibles for corrosive fusions, and as a constituent of hard alloys used in tipping pens. The main uses are for catalysts and as an alloying element with platinum in thermocouple and resistance wires.

The preparation of the metal is covered in "Platinum" Suppl. Vol. A1, 1986, pp. 1/41, the electrodeposition on pp. 68/77. However, something must be added on the melting and casting of this material because of the importance these operations have on its purity and hence its properties. The platinum metals as a group have high melting points, and a tendency to form very strongly bonded intermetallic compounds with most of the elements in the refractories usable at high temperature. This is particularly the case under slightly reducing conditions. Also by combination of high melting point and high chill rate usually obtained in the small melts used, entrapment porosity is possible, as is gas dissolution. In many cases, deoxidants such as calcium boride may be added to obtain sound castings, and residual boron in trace quantities may appreciably raise the recrystallization temperature and alter the recrystallization texture. Rhodium itself is now usually melted on a water-cooled hearth under an inert gas, and zone refining operations are often included. Such operations are covered in "Platinum" Suppl. Vol. A1, 1986, pp. 54/57. Chemical vapour deposition, sputtering, and thermal evaporation are also covered in "Platinum" Suppl. Vol. A1, 1986, pp. 43/51.

Properties of the Rh atom have been treated in "Platinum" Suppl. Vol. A2, 1989, on the following pages:

pp. 44/63 properties of Rh isotopes
pp. 141/142 separation of Rh isotopes
pp. 157/164 ground state, ionization energy, optical spectrum of the Rh atom
pp. 262/267 ground state, ionization energy, optical spectrum of the ions Rh^+ and Rh^{2+}
pp. 338/340 properties of the Rh_2 molecule

1 Crystallographic Properties

Lattice Parameter

Room Temperature. Methods of purity testing are given in "Platinum" Suppl. Vol. A1, 1986, pp. 54/7, 60/5. Adequate information on purity measurements, such as residual resistivity, are seldom given when quoting physical properties. Generally, recent measurements are likely to be the most accurate, because of the development of various zone refining techniques. The following figures are quoted in the literature.

year	t in °C	a in Å	Ref.
1933	20	3.7957 ± 0.0003	[1]
1947	26	3.8043	[2]
1953	20	3.7954 kX (3.8053 Å)	[3]
1955	25	3.8031	[4]
1958	26	3.8035	[5]
1959	30	3.803	[6]
1962	20	3.803	[7]
1968	30	3.8034	[8]
1972	26	3.8030	[9]

References:

[1] Owen, E.A.; Yates, E.L. (Philos. Mag. [7] **15** [1933] 472).
[2] Goldschmidt, H.J.; Land, T. (J. Iron Steel Inst. [London] **155** [1947] 221).
[3] Swanson, H.E.; Fuyot, R.K. (NBS-C-539-Vol. 1 [1953] 95 pp.).
[4] Swanson, H.E.; Fuyot, R.K.; Ugrinic, G.M. (NBS-C-539-Vol. 5 [1955] 8/9).
[5] Bale, E.S. (Platinum Metals Rev. **2** [1958] 61/3).
[6] Raub, E.; Beeskow, H.; Menzel D. (Z. Metallk. **7** [1959] 428/31).
[7] Ross, R.G.; Hume-Rothery, W. (J. Less-Common Metals **5** [1963] 258/70).
[8] Singh, H.P. (Acta Crystallogr. A **24** [1968] 469/71).
[9] Schröder, R.M.; Schmitz-Pranghe, N.; Kohlhaas, R. (Z. Metallk. **63** [1972] 12/6).

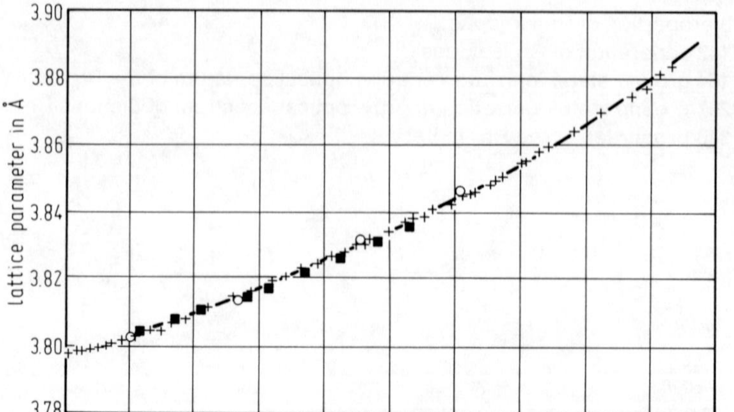

Fig. 1. Lattice parameter of Rh metal between −200 and ~1700 °C.

High Temperatures. There has been controversy over the existence of allotropy in rhodium. An allotropic change at 1030 °C was claimed by [3]. Discontinuity in the thermoelectric power of rhodium was found at 1091 °C by [4] and at 1030 °C by [5]. The changes found by [4] were not regarded as significant by [6, 7] with respect to [5].

More recent high-temperature X-ray work has produced smooth lattice parameter/temperature curves which show no change in the face-centred cubic structure from room temperature up to the melting point [1, 2, 8, 9]; see **Fig. 1** [2].

References:

[1] Ross, R.G.; Hume-Rothery, W. (J. Less-Common Metals **5** [1963] 258/70).
[2] Schröder, R.H.; Schmitz-Pranghe, N.; Kohlhaas, R. (Z. Metallk. **63** [1972] 12/6).
[3] Rudnitskii, A.A.; Polykova, R.S.; Tyurin, I.I. (Izv. Sekt. Platiny No. 29 [1955] 183/9).
[4] Booth, E.T.; Dixon, E.H. (Rev. Sci. Instrum. **9** [1938] 237/41).
[5] Rudnitskii, A.A. (Zh. Neorg. Khim. **1** [1956] 1305/21).
[6] Müller, E.W. (Z. Physik **156** [1959] 399/410).
[7] Vedernikov, M.V. (Advan. Phys. **18** [1969] 337/70).
[8] Bale, E.S. (Platinum Metals Rev. **2** [1958] 61/3).
[9] Raub, E.; Beeskow, H.; Menzel, D. (Z. Metallk. **50** [1959] 428/31).

2 Self-Diffusion

Dislocation Density, Stacking Faults. A figure of 4 to 8×10^{11} lines/cm^2 was reported by [1]. Determinations of stacking faults by [1] and [2] gave <0.008 and 0.004, respectively.

Stacking Fault Energy. Using a method based on the dependence of preferred orientation on stacking fault energy, [3] determined a value of 330 erg/cm^2 whilst [1] obtained a limit of <180 erg/cm^2. Work by [2] resulted in an estimate between 80 and >200 erg/cm^2; a further value of 750 erg/cm^2 is given by [4]. The intrinsic fault energy is given as 190 erg/cm^2, the extrinsic fault energy as 200 erg/cm^2 and the twin fault energy as 112 erg/cm^2 [5].

References:

[1] Dillamore, I.L.; Smallman, R.E.; Roberts, W.T. (Philos. Mag. [8] **9** [1964] 517/26).
[2] Vshikova, N.F.; Noskova, N.I.; Pavlov, V.A. (Fiz. Met. Metalloved. **20** [1965] 480; Phys. Metals Metallog. [USSR] **20** No. 3 [1965] 173).
[3] Seemann, H.J.; Schorr, K. (Phys. Status Solidi **4** [1964] 89/93).
[4] Ahlers, M. (Z. Metallk. **56** [1965] 741).
[5] Papon, A.M.; Simon, J.P.; Guyot, P.; Desjonqueres, M.C. (Philos. Mag. [8] B **39** [1979] 301/2).

Growth of Small Crystals

The growth of small crystals of rhodium has been examined under the 400 keV electron microscope. The crystals were found to grow by coalescence or by the addition of atoms along the surface to give approximately spherical particles [1]. Similar observations revealed the occurrence of a diffuse "cloud" preceding the formation of a new atomic layer [2]. Clusters in sputtered films have been examined with the tunneling electron microscope [3].

References:

[1] Petford-Long, A.K.; Smith, D.J.; Wallenberg, L.R.; Bovin, J.-O. (J. Cryst. Growth **80** [1987] 218/24).
[2] Malm, J.O.; Bovin, J.-O.; Petford-Long, A.K.; Smith, D.J. (J. Cryst. Growth **89** [1988] 165/70).
[3] Green, M.; Richter, M.; Kortright, J.; Carr, R.; Lindau, I. (J. Vac. Tech. A **6** [1988] 428/31).

Surface Self-Diffusion

The following parameters were found for the diffusion equation $D = D_0 \exp(-V_m/kT)$ [1, 2]:

plane	(111)	(311)	(110)	(331)	(100)
D_0 in cm^2/s	2×10^{-4}	2×10^{-3}	3×10^{-1}	1×10^{-2}	1×10^{-3}
V_m in kcal/mol	3.6 ± 0.5	12.4 ± 1.2	13.9 ± 0.8	14.8 ± 0.9	20.2 ± 1.7

References:

[1] Ayrault, G.; Ehrlich, G. (J. Chem. Phys. **57** [1972] 1788/9).
[2] Ayrault, G.; Ehrlich, G. (J. Chem. Phys. **60** [1974] 281/94).

3 Mechanical Properties

Room Temperature Tensile Properties (stress in lb/in^2)

σ_B	δ	0.2% proof	RA in %	E in %	Ref.
102500	7170	9670	22.5	–	[1]a
110100	34000	38500	42.0	–	[1]b
100000	–	–	–	–	[2]
120000 to 130000	–	–	–	30 to 35	[3]
100000	–	–	–	–	[4]
120000 to 130000	–	–	–	–	[5]a
200000 to 230000	–	–	–	–	[5]b
59598	–	9933	20 ± 10	9	[6]
68544	–	–	–	6.5	[7]

σ_B = ultimate tensile strength, δ = limit of proportionality, 0.2% proof = 0.2% proof stress, RA = reduction of area, E = elongation. The materials are described as follows:

[1]a purity as-received; annealed $^1/_4$ h at 900 °C; [1]b same material electron-beam melted, swaged 80% at 900 °C, annealed 15 min at 800 °C; [2] electron-beam melted, tested in air; [3] annealed wire; [5]a annealed wire; [5]b hot-worked wire; [7] annealed in high vacuum.

References:

[1] Holden, F.C.; Douglass, R.W.; Jaffee, R.I. (ASTM Spec. Tech. Publ. No. 272 [1960] 68/79).
[2] Jaffee, R.I.; Maykuth, W.; Douglass, R.W. (Proc. Am. Inst. Mech. Eng. Conf., Detroit 1960).
[3] Metal Progress Data Sheet (Properties of Noble Metals).
[4] Anonymous (Platinum Metals Rev. **16** [1972] 59).
[5] Engelhard Ind. Inc. (Tech. Bull. VI.3 [1965]).
[6] Degussa (Edelmetall-Taschenbuch, Frankfurt a.M. 1967).
[7] Bale, E.S. (Platinum Metals Rev. **2** [1958] 61/3).

Tensile Properties at Various Temperatures. σ_B = ultimate tensile strength, δ = limit of proportionality, RA = reduction of area, E = elongation; stress in lb/in^2. The following table was taken from [1].

t in °C	σ_B	0.2% proof
−196	125165	14223
20	59738	9956
250	69005	12801
500	52626	11379
750	35558	5689
1000	22757	4267
1250	10668	–
1500	5689	–

The following figures are taken from [2].

High-purity rhodium, as-received, annealed 15 min at 900 °C, tested in air [2].

t in °C	σ_B	δ	0.2% proof	RA in %
−196	125000	9210	13300	27.8
25	102500	7170	9670	22.5
250	88500	10800	13780	59.4
500	48800	8380	10900	32.4
750	16950	3410	5820	~90
1000	12320	3500	4930	—

This material was electron-beam melted, swaged 80% at 900 °C, annealed 15 min at 800 °C [2].

t in °C	σ_B	δ	0.2% proof	RA in %
−196	149000	46800	53400	—
25	110100	34000	38500	42
250	91000	37400	41500	55
500	61800	35800	42500	35
750	32600	31500	—	~100
1000	12860	3120	4370	35

The following figures are taken from [3]. The material was annealed for 15 min at 900 °C, and testing was done in vacuum. The material had a grain size of 0.26 mm.

t in °C	σ_B	E in %	0.2% proof	RA in %
750	48700	29	15400	85
1000	16900	61	6900	100
1250	8900	—	3600	100

See also figures in [3, 4]. Some results illustrating the effect of cladding with platinum were published by [5].

References:

[1] Degussa (Edelmetall-Taschenbuch, Frankfurt a. M. 1967).
[2] Holden, F.C.; Douglass, R.W.; Jaffee, R.I. (ASTM Spec. Tech. Publ. No. 272 [1960] 68/79).
[3] Douglass, R.W.; Jaffee, R.I. (Proc. Am. Soc. Testing Mater. **62** [1962] 627/37).
[4] Douglass, R.W.; Jaffee, R.I. (ASTM Reprint No. 65 [1962]).
[5] Reinacher, G. (Metal [Berlin] **17** [1963] 699/705).

Elastic Moduli

Young's Modulus E

E in lb/in² × 10^{-6}	E in kg/mm²	Ref.
46.0	32338*)	[1]
53.3*)	37450	[2]
46.2	32479*)	[3]
54.9*)	38640	[4]
55.0*)	38692	[5]

E in lb/in² × 10^{-6}	E in kg/mm²	Ref.
41.2	28964*)	[6]
50.0	35150*)	[7]
54.9*)	38640	[8]
46.0	32338*)	[9]
55.0	38692	[9]

*) converted figures

A linear decrease of E from 49×10^{-3} to 29×10^{-3} kg/mm² with temperatures rising from 0 to 900 °C was found by [5]; see also [10].

References:

[1] Anonymous (Platinum Metals Rev. **16** [1972] 59).
[2] Köster, W. (Appl. Sci. Res. A **4** [1954] 329/36).
[3] International Nickel (Rhodium 1965).
[4] Degussa (Edelmetall-Taschenbuch, Frankfurt a.M. 1967).
[5] Losinski, M.G.; Fedotow, S.G. (Neue Hütte **3** [1958] 489).
[6] Bale, E.S. (Platinum Metals Rev. **2** [1958] 61/3).
[7] Wise, E.M. (Electrochem. Soc. **97** [1950] 576/646).
[8] Darling, A.S. (Proc. Inst. Mech. Eng. **180** [1965/66] 26/40).
[9] Darling, A.S. (Intern. Metals Rev. **18** [1973] 91/121).
[10] Köster, W. (Z. Metallk. **39** [1948] 1/9).

Other Moduli

Modulus of Rigidity. G = 15300 kg/mm² [1, 3]; 21.6×10^6 lb/in² [2].

Bulk Modulus K. Quoted figures for K: 28010 kg/mm² [3], 27732 kg/mm² [1].

References:

[1] Degussa (Edelmetall-Taschenbuch, Frankfurt a.M. 1967).
[2] Engelhard Ind. Inc. (Data Sheet Metal Progress).
[3] Köster, W. (Appl. Sci. Res. A **4** [1954] 329).

Compressibility $\chi = 1/K$

Most of the experimental work on the compression of materials is due to Bridgman; his work was carried out at pressures of up to 100000 kg/cm² [1, 2, 3].

Work at pressures of up to 400 kbar (409906 kg/cm²) has been carried out; these pressures were created by a shock wave from an explosive device. For values of V/V_0 plotted against pressure see figure in original paper [4]. This experimental work has provided valuable data for attempts to develop appropriate equations of state; in the course of such work, empirical formulae for determining approximations to K or χ sometimes appear. An equation was derived to describe the change in atomic volume with increase in external pressure of the type $(\alpha + \Delta p)\,(\beta + \Delta v) = \alpha\beta$ where α was called the pressure parameter, and β the volume parameter. The pressure parameter was given by $2/\alpha = V_0\, d\chi/dV$ where χ is the compressibility. The parameter $\beta = V_0\, \alpha\chi$ where V_0 is the atomic volume at p = 0. Also, the Grüneisen parameter $\gamma = V/2\beta$ where V is the atomic volume; in principle, the α and β coefficients can be derived from experimentally obtained data [5].

Compressibility was related with pressure by a Taylor series:
$f(p) = -\Delta V/V_0 = ap = bp^2$ where V_0 is the initial volume and p the pressure.
$\beta = -1/V_0(\delta\Delta V/\delta p)_T = a + 2bp$ where $a = \beta_0$ at starting pressure [6].

An empirical expression for β_0 due to [7] is quoted:
$\beta_0 = V^{1.5}/0.003(1+0.0175\,Z)\,T_m)\,10^{-7}$ cm²/kg (V = atomic volume, T_m = melting point, Z = atomic number). This expression gives 2.77×10^6 kg/cm² for rhodium, which agrees with Bridgman. An approximation due to [9] is quoted by [8]:
$\beta = 1/K = 0.0067\,A^{0.75}/D^{1.25}(T_m - 50)$ where A = atomic weight, D = density, T_m = melting point. This gives 2.35×10^6 kg/cm² for rhodium [8, 9]. An equation is derived for the variation of K with temperature: $K(T) = K(T_0)\cdot\{1-2[\gamma(T_0)-1/3]\cdot[\alpha(T)\cdot T-\alpha(T_0)\cdot T_0]\}$ where $\gamma = M/3R\cdot\alpha\cdot K/D$ and M = atomic weight, R = the gas constant, α = volume coefficient of expansion, K = the compression modulus, D = density [10].

An attempt to interpolate between the Thomas-Fermi and the finite strain equations of state is derived by [11]. A relationship between pressure and K is obtained which involves the Bohr radius, the electronic charge, the volume/atom, and the atomic number. [12] developed an interpolation equation to bridge between the experimental results available up to pressures of 1×10^7 (compressions of 2 to 3) and compressions of 10 to 20 where the Thomas-Fermi equation of state is applicable. To represent the cold-pressing of materials in the pressure range of 0 to infinity, an equation of the following form resulted:
$P = \sum_{n=0}^{n=5} b_n r^{-n} \equiv b/r^5(1+\alpha_1 r+\alpha_2 r^{2+\ldots\alpha_5} r^5)$.

Interpolation coefficients are calculated for 50 elements, including rhodium [12].

Murnagan's theory of finite strain is used to develop equations for the effect of pressure on compressibility, and the results are compared with available experimental results. The equations refer to cubic materials, and in simplified form, the following relations are found between pressure and χ_0:

$F = [(V_0/V)^{7/3} - (V_0/V)^{5/3}]$ with $F/P = 2/C_2 = 2\chi_0/3$ = constant, P = pressure, χ = compressibility, V = volume and C_2 is a coefficient.

The constancy of F/P was examined against Bridgman's results; rhodium had an exceptionally high value for $(\delta K_0/\delta P)_0$ of 23; most materials had values between 4 and 7 [13].

References:

[1] Bridgman, P.W. (Proc. Am. Acad. Arts Sci. **68** [1933] 27).
[2] Bridgman, P.W. (Proc. Am. Acad. Arts Sci. **77** [1949] 187/234).
[3] Bridgman, P.W. (Proc. Am. Acad. Arts Sci. **79** [1951] 127/48).
[4] Walsh, J.M.; Rice, M.H.; McQueen, R.G.; Yarger, F.L. (Phys. Rev. [2] **108** [1957] 196/216).
[5] Borelius, G. (Arkiv Fysik **16** [1960] 413/6).
[6] Blanpain, R. (Bull. Classe Sci. Acad. Roy. Belg. [5] **49** [1963] 819/23).
[7] Yoshida, U.; Masui, Y. (Mem. Coll. Sci. Kyoto Imp. Univ. A **24** [1942] 29).
[8] Kempter, C.P. (Phys. Status Solidi **8** [1965] 161/72).
[9] Richards, T.W. (J. Am. Chem. Soc. **37** [1915] 1643).
[10] Schramm, K.-H. (Z. Metallk. **53** [1962] 316/20).

[11] Knopoff, L. (Phys. Rev. [2] **138** [1965] A1445/A1447).
[12] Kalitkin, N.N.; Govorukhina, I.A. (Fiz. Tverd. Tela [Leningrad] **7** [1965] 355/62; Soviet Phys. Solid State **7** [1965] 287/92).
[13] Birch, E. (Phys. Rev. [2] **71** [1947] 809/24).

Poisson's Ratio. The generally quoted figure is 0.26; this was determined by [1]. Methods of determining the transverse contraction and longitudinal elongation under stress are given by [2] (see also figure in original papers). The variation of N during stressing in the elastic range and slightly beyond it was investigated in [3]. **Fig. 2** shows results in the elastic range for rhodium.

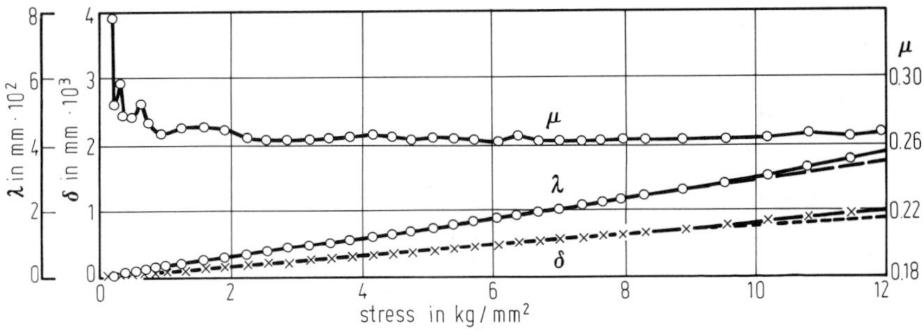

Fig. 2. Elongation λ, transverse contraction δ and Poisson's ratio μ of Rh metal in the elastic range.

References:

[1] Köster, W. (Appl. Sci. Res. A **4** [1954] 329).
[2] Claus, K. (Z. Metallk. **46** [1955] 589/92).
[3] Köster, W.; Scherb, J. (Z. Metallk. **49** [1958] 501/7).

Hardness

Miscellaneous.

condition	H_V in kg/mm^2	Ref.
hot-swaged	180	[1]
annealed*)	190	[1]
annealed	100	[2]
annealed 800 °C	120 to 140	[3]
vacuum-annealed	110	[4]
annealed	120 to 140	[5]
annealed	130	[6]
annealed	100 to 102	[7]
sintered 1300 °C	135	[8]
sintered 1600 °C	128	[8]
cast	139	[9]
annealed 1200 °C	122	[10]

*) electron-beam refined

References:

[1] Calverly, A.; Rhys, D.W. (Nature **183** [1959] 599/600).
[2] Petty, E.R.; O'Neil, H. (Metallurgia **63** [1961] 25/30).
[3] Engelhard Ind. Inc. (Tech. Bull. VI. 3 [1965]).
[4] Bale, E.S. (Platinum Metals Rev. **2** [1958] 61/3).
[5] Degussa (Edelmetall-Taschenbuch, Frankfurt a.M. 1967).
[6] Savitskii, E.; Polyakova, V.N.; Gorina, N.; Roshan, N.R. (Physical Metallurgy of Platinum Metals, MIR Publ., Moscow 1978).

[7] Anonymous (Platinum Metals Rev. **16** [1972] 59).
[8] Lozinskii, M.C.; Fedotov, S.G. (Izv. Akad. Nauk SSSR Otd. Tekh. Nauk **1955** No. 5, pp. 109/13).
[9] Carter, F.E. (Proc. Inst. Met. Div. AIME **78** [1928] 759/85).
[10] Acken, J.S. (Bur. Std. J. Res. [U.S] **12** [1934] 249).

Electroplated Material

Remarks	H_V in kg/mm^2	Ref.
–	549 to 641	[1]
comments on soln. used	800 to 900	[2]
effect of test load	800	[3]
–	800 to 820	[4]
comments on test load	690 to 1095 (Knoop)	[5]
–	594 to 641	[6]
method comments	850 to 998	[7]
–	800 to 1000	[8]

References:

[1] Atkinson, R.H.; Raper, A.R. (Electrodeposit. Tech. Soc. **9** [1934] 77).
[2] Reid, F.H. (Metall. Rev. **8** [1965] 167/211).
[3] Keil, A.; Merkle, E. (Metalloberfläche A **8** [1954] 129/31).
[4] Laister, E.H.; Benham, R.R. (Trans. Inst. Metal Finishing **29** [1953] 1/22).
[5] Wiesner, H.J. (Ann. Tech. Conf. Proc. Am. Electroplat. Soc. **39** [1952] 79/99).
[6] Wernick, S. (Sheet Metal Ind. **22** [1945] 1219/22).
[7] Reid, F.H. (Trans. Inst. Metal Finishing **33** [1955/56] 105).
[8] Lévy, R. (Korrosion **12** [1960] 186/9).

Work-Hardening

Although rhodium almost certainly has a face-centred cubic structure between room temperature and its melting point, it differs from most similarly structured materials in being difficult to cold work; cold reductions in thickness of only 40 to 50% being possible, even when rhodium of very high purity is used. High purity single crystals have withstood reductions of 90% without cracking, but after recrystallization, large cold reductions are not possible [1, 2].

The rate of work-hardening is high; **Fig. 3** shows this [3].

The reasons for the difficulty of cold-working rhodium have not been established; impurities, both metallic and gaseous, probably play a part. However, it has been suggested [4, 5] that high rates of work-hardening are related to a high modulus of rigidity. Also [5] suggested that the flow stress would depend on the force required to push an edge dislocation through the lattice, and that this varied as Gb where b is the Burgers vector. **Fig. 4** appears to support this. It was proposed that the ratio of bulk modulus/modulus of rigidity would give an index of cold-workability; such figures seem to agree with working experience [6].

The usual method of working melted or sintered rhodium is to hot-work at about 1200 °C, gradually reducing the temperature as reduction proceeds. Once a fibrous structure has

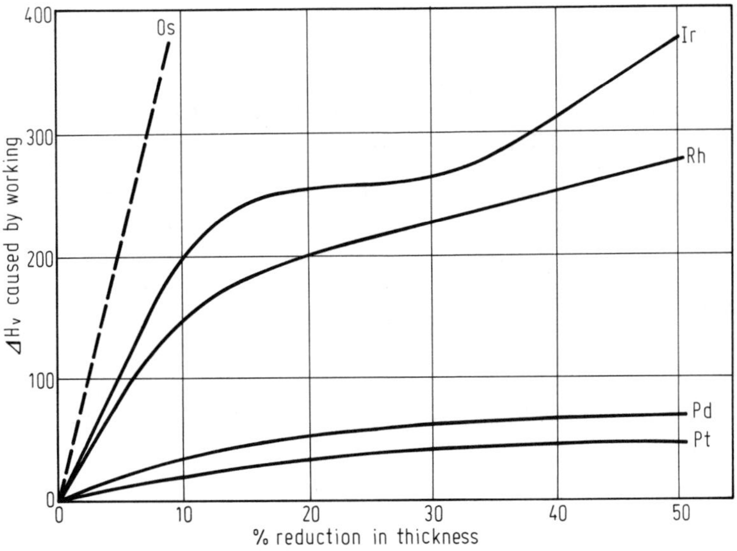

Fig. 3. Rate of work-hardening of Rh metal compared with other metals of the platinum group.

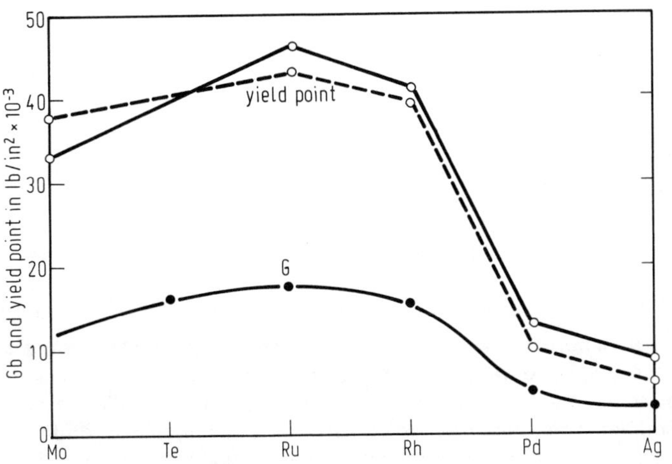

Fig. 4. Product of modulus of rigidity, G, and Burgers' vector, b, for Rh and some other metals.

been obtained, it is possible to cold-work, using stress-relieving anneals at 600 to 800 °C [7]. Complete recrystallization must be avoided, since this severely restricts cold-working. It has been found that cladding with platinum or nickel improves workability and reduces warm-working temperatures [8, 9].

References:

[1] Calverley, A.; Rhys, D.W. (Nature **183** [1959] 599/600).
[2] Bale, E.S. (Platinum Metals Rev. **2** [1958] 61/3).
[3] Office of Naval Research (Contract No. Nonr-2547(00) NRO 39-067).
[4] Pugh, S.F. (Philos. Mag. [7] **45** [1954] 823/43).
[5] Frank, F.C.; Read, W.T. (Phys. Rev. [2] **79** [1950] 722).
[6] Darling, A.S. (Proc. Inst. Mech. Eng. **180** [1965/66] 26/40).
[7] Jaffee, R.I.; Maykuth, D.J.; Douglass, R.W. (Proc. Am. Inst. Mech. Eng. Conf., Detroit 1960).
[8] Reinacher, G. (Metall. [Berlin] **27** [1973] 1/4).
[9] Reinacher, G. (Metall. [Berlin] **14** [1960] 664/8).

Elevated Temperature Hardness

Figures for the change in Vickers hardness H_V with temperature are given by [1] and [2]. Those from [1] are given below.

t in °C	H_V in kg/mm²	t in °C	H_V in kg/mm²
20	127	650	78
100	123	700	67
200	121	750	73
250	113	800	69
300	108	850	67
400	103	900	62
450	96	950	60
500	91	1000	52
550	88	1075	47
600	81		

References:

[1] Losinski, M.G.; Fedotov, S.G. (Izv. Akad. Nauk SSSR Otd. Tekh. Nauk **1956** No. 3, pp. 59/67; Neue Hütte **8** [1958] 489/94).
[2] Degussa (Edelmetall-Taschenbuch, Frankfurt a.M. 1967).

Creep Properties

There are few experimental results on the creep of pure rhodium, and it is difficult to compare those available with each other; from what exists it is clear that rhodium is comparable to molybdenum in strength at high temperatures. Available results are further complicated by differences obtained when testing in air, rather than under inert conditions. In work of a screening nature compression tests at 1000 °C in a vacuum of 2×10^{-2} Torr were used to determine the stress needed to give 1% creep strain in 24 h; a relationship between this stress and atomic number of the elements tested was established [1].

The most comprehensive work was carried out by [2] who made preliminary hot-tensile tests at 750, 1000, and 1250 °C in a vacuum of 1×10^{-6} Torr and in air. These tests showed the importance of oxidation on the results. Stress-rupture tests followed at temperatures of 1000 and 1250 °C, these being done in vacuum [2]. The importance of oxidation during the hot-testing of rhodium was also demonstrated by [3]. Protection was achieved in this case by cladding with a thin skin of platinum. It was shown that whilst bare material gave intercrystalline failures above 500 °C, clad wires gave perfectly ductile fractures. Stress-rupture tests were also carried out on both types of material; however, at high temperatures the platinum coating diffused into the rhodium [3]. Steady-state creep conditions were determined in a vacuum of $<5\times10^{-5}$ Torr. Temperatures between 630 and 1770 °C were used with stresses varying between 0.4 and 46.2 kg/mm² [4]. The activation energy was found to agree well with that for self-diffusion; the latter was calculated as 87 kcal/mol [5], and the activation energy for creep was found to be 91 kcal/mol [4]. **Fig. 5** shows the stress-rupture results of [2] and [3]. **Fig. 6** shows the steady-state creep rate of rhodium, in vacuum, as a function of stress at various temperatures [4].

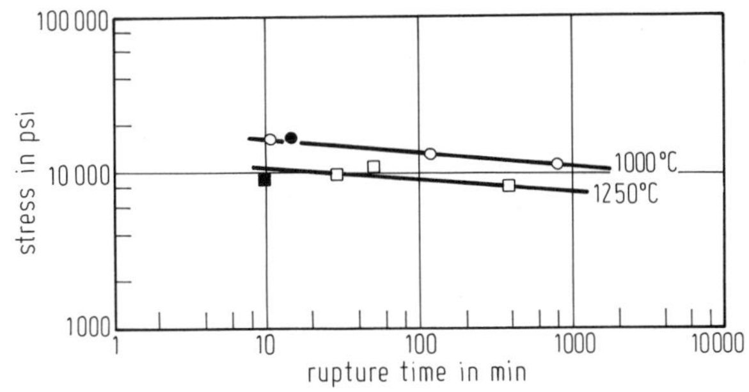

Fig. 5. Stress-rupture data for Rh metal in vacuum.

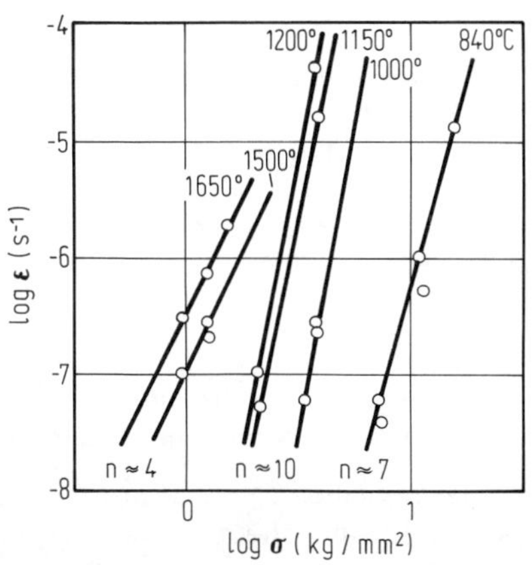

Fig. 6. Steady-state creep rate (ε) of Rh metal as a function of stress (σ).

Activation energy of creep at various temperatures.

σ in kg/mm² . . .	1.0	2.2	4.0	7.4
t in °C	1490 to 1680	1160 to 1300	1000 to 1200	840 to 950
Q_c in kcal/mol . .	89.7	88.9	88.7	99.2

σ = stress, Q_c = activation energy; the standard error of determinations of the activation energy of creep was 5.1 kcal/mol [4].

References:

[1] Allen, N.P.; Carrington, W.E. (J. Inst. Metals [London] **82** [1953/54] 325).
[2] Douglass, R.W.; Jaffee, R.I. (Proc. Am. Soc. Testing Mater. **62** [1962] 627/37).
[3] Reinacher, G. (Metall [Berlin] **17** [1963] 699/705).
[4] Shalayev, V.I.; Tkachenko, I.B.; Pavlov, V.A.; Timofeyev, N.I.; Gushchina, A.V. (Fiz. Met. Metalloved. **29** [1970] 1961/8; Phys. Metals Metallog. [USSR] **29** No. 5 [1970] 170/7).
[5] LeClaire, A.D. (Diffusion in Metals 1969).

Wear

The resistance of rolled rhodium to abrasive wear against a rotating roller was investigated by measuring the volume removed and comparing this with similar tests on osmium, which was given a wear rating of 100. The wear rating A was computed from the expression $A = V_{Os} \times 100/V_{Rh}$. Rhodium was found to have a rating of 4.4 [1]. A test devised to simulate scoring wear conditions similar to those experienced in bearings was used to grade a variety of materials. Rhodium was found to behave very poorly in these tests [2]. A comprehensive testing procedure was devised to assess the wear of electroplated coatings in apparatus designed to simulate service in slip-rings operated against silver-palladium alloy bushes; various plating baths and conditions were used [3].

The abrasive wear of sputtered films on glass substrates was studied in [4]; the degree of oxidation during the early stages of the deposition was found to affect the result [4]. Adhesion and frictional properties of rhodium against diamond, silicon carbide, pyrolitic boron nitride and ferrite in sliding contact have been evaluated by [5, 6].

References:

[1] Winkler, O. (Z. Elektrochem. Angew. Physik. Chem. **49** [1943] 221/8).
[2] Roach, A.E. (Prod. Eng. [New York] **25** [1954] 171/5).
[3] Angus, H.C. (Trans. Inst. Metal Finish. **43** [1965] 135/42).
[4] Cox, R.E.L. (Thin Solid Films **11** [1972] 323/8).
[5] Miyoshi, K.; Buckley, D.H. (Wear **77** [1982] 253/64).
[6] Miyoshi, K.; Buckley, D.H. (ASLE Trans. **27** No. 1 [1984] 15/23).

Velocity of Sound

The bulk sound velocity is given as $v_b = 6.15 \times 10^5$ cm/s. It was derived from the equation $\{E(1-\mu)/\mu(1-\mu-2\mu^2)\}^{1/2}$ where μ is Poisson's ratio and E is the modulus of elasticity [1].

The transverse and longitudinal velocities v_t and v_l are calculated using equations $v_t = (n/\varrho)^{1/2}$; $v_l = [(1-\varrho)\,Y/\varrho(1+\mu)\,(1-2\mu)]^{1/2}$ (n is the modulus of rigidity, Y is Young's modulus, μ is Poisson's ratio, ϱ = density) [2]; $v_t = 3465$ m/s, $v_l = 5604$ m/s; these figures are given with an additional value, the rod wave velocity $v_r = 5066$ m/s [3].

References:

[1] Wehner, G.K. (Phys. Rev. [2] **93** [1954] 633/4).
[2] Narayana, K.L.; Swamy, K.M. (Mater. Sci. Eng. **18** [1975] 157/8).
[3] Swamy, K.M.; Narayana, K.L. (Acustica **54** [1983] 123/5).

Surface Tension

The surface tension at the melting point is $\sigma = 1915$ mJ/m²; the equation given for temperature dependence is σ(t) in mJ/m² = 1915 − 0.664(t − 1966) [1, 2]. $\sigma = 1940$ mJ/m² at the melting point [3], 2000 dyn/cm^2 (numerically equivalent to mJ/m^2) was given by [4]. A figure of $\sigma = 2280 \pm 35$ J/m^2 was determined for solid rhodium just below the melting point [5].

References:

[1] Mit'ko, M.M.; Dubinin, E.L.; Timofeev, A.I.; Chegodaev, A.I. (Soviet Non-Ferrous Metals Res. **6** [1978] 115/7).
[2] Dubinin, E.L.; Timofeev, A.I.; Safonov, S.O.; Chegodaev, A.I. (Soviet Non-Ferrous Metals Res. **4** [1975] 171/2).
[3] Eremenko, V.N. (Ukr. Khim. Zh. [Russ. Ed.] **28** [1962] 427).
[4] Allen, B.C. (Trans. Metall. Soc. AIME **227** [1963] 1175).
[5] Korablev, V.A.; Shkolenko, A.P.; Kabes, A.I. (Fiz. Met. Metalloved. **56** [1983] 409/10).

4 Thermal Properties

Thermal Expansion

range in °C	$\alpha \times 10^{-6}$	Ref.	range in °C	$\alpha \times 10^{-6}$	Ref.
20 to 100	8.3	[1]	20 to 900	10.60	[9]
0 to 100	8.5	[2]	20	8.55	[10]
0 to 100	8.75	[3, 4]	600	11.21	[10]
0 to 300	9.1	[5]	0 to 1500	12.07	[11]
0 to 900	11.1	[6]	0 to 1600	13.15	[12]
23 to 200	8.8	[6]	0 to 1200	10.99	[13]
20 to 1950	12.4	[6]	26 to 1540	14.07	[14]
30	7.89	[7]	0 to −20	8.16	[8]
865	12.80	[7]	−195 to −215	3.06	[8]
30 to 865	10.27	[7]	−270	0.0062	[15]
0 to 40	8.41	[8]	−253	0.25	[16]

A sample of sheet material which had been silver-soldered was used by [16]; this may account for some discrepancies with [8] and [15].

References:

[1] Hidnert, P.; Souder, W. (NBS-C-486 [1950]).
[2] Degussa (Edelmetall-Taschenbuch, Frankfurt a.M. 1967).
[3] Campbell, W.J. (Inform. Circ. U.S. Bur. Mines No. 8107 [1962]).
[4] Pandey, H.D.; Dayal, B. (Phys. Status Solidi **5** [1964] 273).
[5] Leksina, J.E.; Novikova, S.J. (Fiz. Tverd. Tela [Leningrad] **5** [1963] 1094/9; Soviet Phys.-Solid State **5** [1963] 798/801).
[6] Ross, R.G.; Hume-Rothery, W. (J. Less-Common Metals **5** [1963] 258).
[7] Singh, H.P. (Acta Crystallogr. A **24** [1968] 469/71).
[8] Erfling, H.D. (Ann. Physik [5] **34** [1939] 136/60).
[9] Burov, I.V.; Savitskii, E.M.; Tomalin, N.A. (Dokl. Akad. Nauk SSSR **271** [1983] 1370/2).
[10] Pawar, R.R. (Curr. Sci. [India] **37** [1968] 224/5).

[11] Ebert, H. (Physik. Z. **39** [1938] 6/9).
[12] Schröder, R.H.; Schmitz-Pranghe, N.; Kohlhaas, R. (Z. Metallk. **63** [1972] 12/6).
[13] Raub, E.; Beeskow, H.; Menzel, D. (Z. Metallk. **50** [1959] 428/31).
[14] Bale, E.S. (Platinum Metals Rev. **2** [1958] 61).
[15] White, G.K.; Pawlowicz, A.T. (J. Low Temp. Phys. **5/6** [1970] 631/9).
[16] Laquer, H.L. (UCRL-6132 [1960] 7 pp.; N.S.A. **15** [1961] No. 1832).

Grüneisen's Parameter

Much of the essential information on compressibility comes from work by [1] and [2].

The following figures for Grüneisen's parameter γ at room temperature and pressure are quoted in the literature:

γ	2.265	2.29	2.12	2.32
Ref.	[2]	[3]	[4]	[5]

Theoretical values for γ were calculated using equations derived by [7], [8], [9] and [10]; values between 0.37 and 1.2 were obtained [6]. The effect of rising temperature on

γ has been studied by [4]. Calculations from their low-temperature linear expansion coefficient measurements gave γ from -261 °C to $+10$ °C; the authors also deduced γ (electron) and γ (lattice-vibration) components from the T and T^3 terms in their equation for the thermal expansion at temperatures below 10 K. The figures were $\gamma_e = 2.8$ and $\gamma_l = 2.0$ [5].

References:

[1] Bridgman, P.W. (Proc. Am. Acad. Arts Sci. **77** [1949] 187).
[2] Walsh, J.M.; Rice, M.H.; McQueen, R.G.; Yarger, F.L. (Phys. Rev. [2] **108** [1957] 196/216).
[3] Gschneider, K.A. (Solid State Phys. **16** [1964] 275).
[4] Singh, H.P. (Acta Crystallogr. A**24** [1968] 469/71).
[5] White, G.K.; Pawlowicz, A.T. (J. Low Temp. Phys. **2** [1970] 631/9).
[6] Rodionov, K.P. (Fiz. Met. Metalloved. **23** [1967] 1008/12; Phys. Metals Metallog. [USSR] **23** No. 6 [1967] 44/8).
[7] Slater, J.C. (Phys. Rev. [2] **57** [1940] 744).
[8] Dugdale, J.S.; McDonald, D.C. (Phys. Rev. [2] **89** [1953] 832/4).
[9] Gilvarry, J.J. (Phys. Rev. [2] **102** [1956] 331/40).
[10] Druyvesteyn, M.J.; Meyering (Physica **8** [1941] 851).

Vapour Pressure

This has been studied at low pressures by Langmuir rate of evaporation measurements or Knudsen effusion techniques. Such measurements have been made by [1 to 6]. Results are plotted in **Fig. 7** by [1].

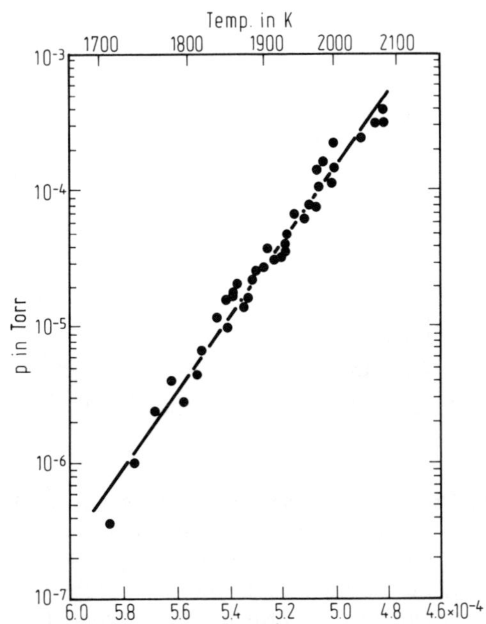

Fig. 7. Vapour pressure of rhodium.

The rate of loss of rhodium at 1800 °C heated in air, argon and vacuum was shown to be less in argon than in vacuum [7]. Similar figures for heating in vacuum between

1300 °C and the melting point are given by [8]. Figures determined for the vapour pressure between 1040 and 3840 °C can be used to calculate rates of evaporation using Langmuir's equation [8, 9]. Equations relating vapour pressure with temperature were derived in several papers; the equation below is given for temperatures between 1709 and 2075 K. $\log p = 6.894 - 27276/T$ (p in atm, T in K) [1].

For the temperature interval 2050 to 2200 K, the following equation is given:

$\log p = 10.28 - 28300/T$ (p in Torr, T in K) [2].

References:

[1] Hampson, R.F.; Walker, R.F. (J. Res. Natl. Bur. Std. A **65** [1961] 289/95).
[2] Panish, M.B.; Reif, J.L. (J. Chem. Phys. **34** [1961] 1915/8).
[3] Dreger, L.H.; Margrave, J.L. (J. Phys. Chem. **65** [1961] 2106/7).
[4] Hasapis, A.A.; Melveger, J.; Panish, M.B.; Rosen, C. (WADD-TR-60-463-Pt. II [1962] 90 pp.; N.S.A. **17** [1963] No. 4933).
[5] Strassmair, H.; Stark, D. (Z. Angew. Physik **23** [1967] 40/4).
[6] Douglas, R.W.; Holden, F.C.; Jaffee, R.I. (High-Temp. Prop. of Refractory Platinum Group Metals, Battelle Mem. Inst. 1959).
[7] Rytvin, E.I.; Malashkin, V.V. (Tr. Inst. Fiz. Metall. Ural. Nauchn. Tsentr Akad. Nauk SSSR No. 28 [1971] 250/3).
[8] Darling, A.S. (Intern. Metals Rev. **18** [1973] 91/122).
[9] Degussa (Edelmetall-Taschenbuch, Frankfurt a.M. 1967).

Boiling Point

date	boiling point	method	Ref.
1938	4773 K	spectroscopic	[1]
1961	4000 ± 100 K	vapour pressure	[2]
1961	3980 K	vapour pressure	[3]
1961	3900 ± 100 K	vapour pressure	[4]
1963	3727 °C	—	[5]
1965	3700 °C	—	[6]
1967	3700 °C	—	[7]
1978	4500 °C	—	[8]

References:

[1] Richardson, D. (Proc. 5th Conf. Spectrosc. and its Applications, J. Wiley 1938).
[2] Hampson, R.F.; Walker, R.F. (J. Res. Natl. Bur. Std. A **65** [1961] 289/95).
[3] Panish, M.B.; Reif, L. (J. Chem. Phys. **34** [1961] 1915/8).
[4] Dreger, L.H.; Margrave, J.L. (J. Phys. Chem. **65** [1961] 2106/7).
[5] Hultgren, R.; Orr, R.L.; Anderson, P.D.; Kelley, K.K. (Selected Values of Thermodynamic Properties of Metals and Alloys, Wiley, New York 1963).
[6] International Nickel (Rhodium 1965).
[7] Degussa (Edelmetall-Taschenbuch, Frankfurt a.M. 1967).
[8] Savitskii, E.; Polyakova, V.N.; Gorina, N.; Roshan, N. (Physical Metallurgy of Platinum Metals, MIR Publ., Moscow 1978).

Melting Point

year	melting point	Ref.	comments
1961	1960 °C	[6]	—
1959	1970 (2400 ± 15 °C)	[9]	at 100 atm
1954	1966 ± 3 °C	[2]	—
1939	1966 ± 3 °C	[1]	—
1960	1966 °C	[4]	—
1960	1970 °C	[5]	see [3]
1967	1960 °C	[8]	—
1978	1960 °C	[9]	—
1965	1960 °C	[7]	—

The effect of pressures up to 100000 atm on the melting point is shown in **Fig. 8.** Reference was made to a shrinkage in volume at 1000 to 1400 °C which necessitated changes in the method of measurement; an exceptionally large expansion in volume was noted on melting. The authors refer to Simon's fusion equation relating melting point with pressure: $(P+a)/a=(T/T_0)^c$.

Values for the constants a and c are given. A plot of c against γ (Grüneisen's constant) gives a functional relationship for sodium, caesium, iron, nickel and platinum, but not for Rh. It is suggested that this may be connected with the previously described shrinkage at 1000 °C and the large volume change on melting [3].

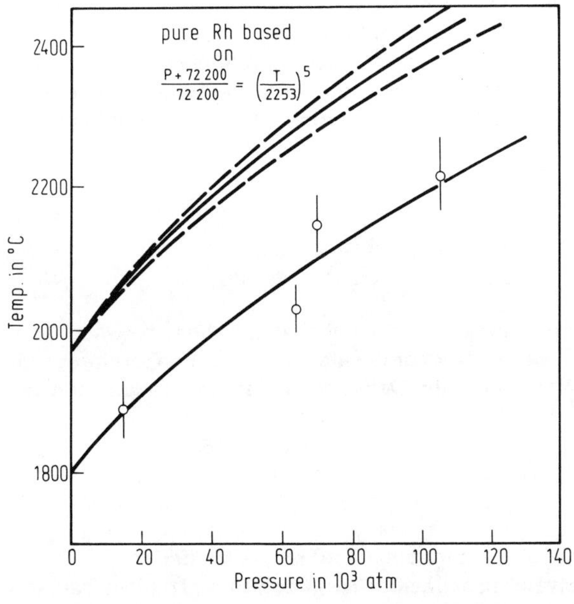

Fig. 8. Fusion curve of rhodium. Lower curve: experimental data; upper curve: Simon's equation for pure Rh using constants from lower curve.

References:

[1] Barber, C.R.; Schofield, F.H. (Proc. Roy. Soc. [London] A **173** [1939] 117/25).
[2] Oriani, R.A.; Jones, T.S. (Rev. Sci. Instrum. **25** [1954] 248/51).
[3] Strong, H.M.; Bundy, F.B. (Phys. Rev. [2] **115** [1959] 278/84).
[4] Shimulis, V.I.; Gryaznov, V.M. (Dokl. Akad. Nauk SSSR **137** [1961] 648/51).
[5] Strong, H.M. (Am. Scientist **48** [1960] 58/79).
[6] Stimson, H.F. (J. Res. Natl. Bur. Std. A **65** [1961] 139/45).
[7] International Nickel (Rhodium 1965).
[8] Degussa (Edelmetall-Taschenbuch, Frankfurt a.M. 1967).
[9] Savitskii, E.; Polyakova, V.N.; Gorina, N.; Roshan, N. (Physical Metallurgy of Platinum Metals, MIR Publ., Moscow 1978).

Specific Heat (C_P in $cal \cdot g^{-1} \cdot {}^{\circ}C^{-1}$)

Near Room Temperature.	atomic heat (C_P)	t in °C	c_P	Ref.
		0	0.059	[1]
		0	0.05893	[2]
		0	0.05893	[3]
	5.87	0	[0.057]	[4]
	6.00	18	[0.0583]	[5]

Above Room Temperature. The figures for c_P usually quoted in reference books originate from [3, 6].

Below Room Temperature. A considerable number of investigators has measured c_P at very low temperatures where the components due to lattice vibrations and electron gas can be separated. In particular, interest has centred on the electronic specific heat because of the information it can provide on the band structure in metals. A particularly good review of the subject is given by [7]. At temperatures $\ll \theta_D$ (Debye temperature), as a first approximation, the electronic specific heat and lattice vibrational specific heat follow the equation $c_P = \gamma T + \alpha T^3$. The coefficient γ is proportional to the density of states of conduction band electrons near the Fermi surface; α is determined by the elastic properties and the density.

Experimental figures for γ:

T in K	—	>10	1.2 to 2.0	1.8 to 4.2	—
γ in mJ/K^2	4.7	4.85	4.9	4.65	4.2
Ref.	[7]	[4]	[8]	[9]	[10]

Experimental figures were given for the atomic heat between 10 and 273 K, and the following expression was derived for the atomic heat at temperatures < 14 K: $C_P = (1/450)^3 + 10^{-3}\,T$, the figure of 450 being the Debye temperature [4]. Values for γ at 300 and 1500 K: 2.24 and 2.40 $J \cdot mol^{-1} \cdot K^{-1}$ [11].

References:

[1] Degussa (Edelmetall-Taschenbuch, Frankfurt a.M. 1967).
[2] Savitskii, E.; Polyakova, V. N.; Gorina, N.; Roshan, N. (Physical Metallurgy of Platinum Metals, MIR Publ., Moscow 1978).
[3] Jaeger, F.M.; Rosenbohm, E. (Proc. Koninkl. Ned. Akad. Wetenschap. **34** [1931] 85/9).

[4] von Clusius, K.; Losa, C.G. (Z. Naturforsch. **10a** [1955] 545/55).
[5] Kelley, K.K. (Bull. U.S. Bur. Mines No. 371 [1934] 42).
[6] Jaeger, F.M.; Rosenbohm, E. (Physica **6** [1939] 1123).
[7] Heiniger, F.; Bucher, E.; Muller, J. (Physik Kondensierten Materie **5** [1966] 243/84).
[8] Wolcott, N.M. (Bull. Inst. Intern. Froid Annexe **1955** 286).
[9] Budworth, D.W.; Hoare, F.E.; Preston, J. (Proc. Roy. Soc. [London] **257** [1960] 250).
[10] Keesom, P.H.; Pearlman, N. (Encycl. Phys. **14** [1956] 282).

[11] White, G.K. (Intern. J. Thermophys. **9** [1988] 839/48).

Debye Temperature. An equation relates Debye temperature θ_D with melting point T_m and the principal quantum number of the period of the element, n (which for rhodium is 5): $\theta_D = 323 \times T_m^{1/2}/(2\,n^2+1)$; the value of 300 K thus obtained was said to agree well with experimentally determined values [1]. The relation $\theta_D = h/k[3\,n\varrho/4\,\pi M]^{1/3}\,v_m$ where h is Planck's constant, k is Boltzmann's constant, N is Avagadro's number, ϱ is the density, and M the atomic weight. The value v_m is an average sound velocity according to the basic assumptions of the Debye theory of specific heat. The relation is attributed to [2], and calculations based on it gave a value of 479 K for rhodium [3].

In a very comprehensive review paper, $\theta_D = 500$ K was quoted [4]; a value of 350 K is given by [5]. At low temperatures (given in parentheses) $\theta_D = 478$ K (1.2 to 2.0) [6], 450 K (~10) [7], and 512 K (1.8 to 4.2) [8].

512 K is quoted by [9], 500 K by [10]. Determinations on specific crystallographic planes of surfaces are given in the table below.

plane	angle	θ_D in K	Ref.
(100)	90°	200	[11]
(111)	90°	210	[11]
(111)	90°	197 ± 12	[12]

Low-energy electron diffraction was used to determine the Debye temperatures of surface atoms during field evaporation, and a series of readings between 100 and 600 K is given below [13].

t in K	100	250	350	430	510	600
θ_D in K	120	80	120	90	100	80

References:

[1] Oscherin, V.N. (Poroshk. Metall. [Kiev] **2** No. 1 [1962] 11/6).
[2] de Launay, J. (Solid State Phys. **12** [1965] 219).
[3] Narayana, K.L.; Swamy, K.M. (Mater. Sci. Eng. **18** [1975] 157/8).
[4] Heiniger, F.; Bucher, E.; Muller, J. (Physik Kondensierten Materie **5** [1966] 243/84).
[5] Gschneider, K.A. (Solid State Phys. **16** [1964] 370).
[6] Wolcott, N.M. (Bull. Inst. Intern. Froid Annexe **1956** 286).
[7] von Clusius, K.; Losa, C.G. (Z. Naturforsch. **10a** [1955] 545).
[8] Budworth, D.W.; Hoare, F.E.; Preston, J. (Proc. Roy Soc. [London] A**57** [1950] 250/62).
[9] Savitskii, E.; Polyakova, V.N.; Gorina, N.; Roshan, N. (Physical Metallurgy of Platinum Metals, MIR Publ., Moscow 1978).
[10] Schröder, R.H. (Z. Metallk. **63** [1972] 12/6).

[11] Castner, D.G.; Somorjai, G.A.; Black, J.E.; Wallis, R.F. (Phys. Rev. [3] B **24** [1981] 1616/23).

[12] Chan, C.-M.; Thiel, P.A.; Yates, J.T.; Weinberg, W.H. (Surf. Sci. **76** [1978] 296).
[13] Ernst, N.; Block, J.H. (Surf. Sci. **91** [1980] L27/L31).

Thermodynamic Data

Standard Entropy. $S_{298}=7.56$ cal/K according to [1].

Latent Heat of Fusion. ΔH_m in kcal/g-atom: 5.2 [2, 4]; 5.15 [3, 5].

An equation relating ΔH_m and melting temperature has been evolved which gave a figure of 4.9059 kcal/g-atom: The relationship is $\Delta H_m=2.3\ (T_m-100)$ cal/g-atom. It is said to give a better approximation than the Richards or Kubaschewski approximations [5].

References:

[1] Fritterer, G.R. (ASM Trans. Quart. **60** [1967] 15/20).
[2] Degussa (Edelmetall-Taschenbuch, Frankfurt a.M. 1967).
[3] Hultgren, R., Orr, R.L.; Anderson, P.D.; Kelley, K.K. (Selected Values of Thermodynamic Properties of Metals and Alloys, New York 1963).
[4] Stull, D.R.; Sinke, G.C. (Thermodynamic Properties of the Elements in Their Standard States, Midland, Mich. 1955).
[5] Xiao, G. (Kexue Tongbao **33** [1988] 896/900).

Latent Heat of Sublimation.

ΔH_{298} in kcal/g-atom	Ref.	ΔH_{298} in kcal/g-atom	Ref.
133.1±0.5	[1]	134.2±0.8	[7]
132.8±0.3	[2]	134.2±2.3	[10]
132.5±2.0	[3]	133	[8]
133	[4]	115	[9]
139	[5]	129 solid	[8]
130.9±0.9	[6]	124 liquid	[8]

References:

[1] Hultgren, R.; Raymond, L.; Anderson, P.D.; Kelley, K.K. (Selected Values of Thermodynamic Properties of Metals and Alloys, New York 1963, 963 pp.)
[2] Panish, M.B.; Reif, L.; (J. Chem. Phys. **34** [1961] 1915/8).
[3] Hampson, R.F.; Walker, R.F. (J. Res. Natl. Bur. Std. A **65** [1961] 289/95).
[4] Degussa (Edelmetall-Taschenbuch, Frankfurt a.M. 1967).
[5] Johnson, B.F.G. (Platinum Metals Rev. **25** [1981] 62).
[6] Strassmair, H.; Stark, D. (Z. Angew. Physik **23** [1967] 40/4).
[7] Dreger, L.H.; Margrave, J.L. (J. Phys. Chem. **65** [1961] 2106/7).
[8] Hasapis, A.A.; Panish, M.B.; Rosen, C. (WADD-TR-60-463-Pt. I [1960] 75 pp.; N.S.A. **15** [1961] No. 5380).
[9] Bichowsky, F.R.; Rossini, F.D. (Thermochemistry of Chemical Substances, Reinhold, New York 1951).
[10] Babeliowsky, T.P.J.H. (Physica **28** [1962] 1160/9).

Properties of positive gas ions are treated in "Platinum" Suppl. Vol. A2, 1989, pp. 262/7.

Thermal Conductivity k

Various units are used in measuring thermal conductivity; the SI units are $J \cdot s^{-1} \cdot m^{-1} \cdot K^{-1}$. The figures reported will be given in the units used by the authors, and conversions given in parentheses.

The following figures at room temperature are quoted in reference books: 0.358 $cal \cdot cm^{-1} \cdot s^{-1} \cdot °C^{-1}$ at 20 °C [1]; 0.210 $cal \cdot cm^{-1} \cdot s^{-1} \cdot °C^{-1}$ at 20 °C [2]; 1.5 $J \cdot cm^{-1} \cdot s^{-1} \cdot °C^{-1}$ (0.358 $cal \cdot cm^{-1} \cdot s^{-1} \cdot °C^{-1}$) [8]; 1.538 $J \cdot cm^{-1} \cdot s^{-1} \cdot °C^{-1}$ (0.367 $cal \cdot cm^{-1} \cdot s^{-1} \cdot °C^{-1}$) [6].

The coefficients are given in an equation for the variation of the thermal resistance with temperature in [3]. In a further paper the same authors report the effect on thermal conductivity of a magnetic field [4].

Measurements at high and low temperatures of k and electric resistance have been prompted to explain anomalies by s-d electron interactions [5, 6].

The Mott model for s-d scattering was used to calculate k and electric resistivity at temperatures between 100 and 1600 K [7]. Results between 80 and 500 K and the variation of the Lorenz number with temperature is shown in [8, 9]. Measurements of k have been made on a 99.995% pure polycrystalline rod at liquid helium temperatures; the effect of a transverse magnetic field was also studied [11]. Measurements on spectroscopically pure material between 1180 and 1573 K gave a straight line relationship shown in **Fig. 9** [12].

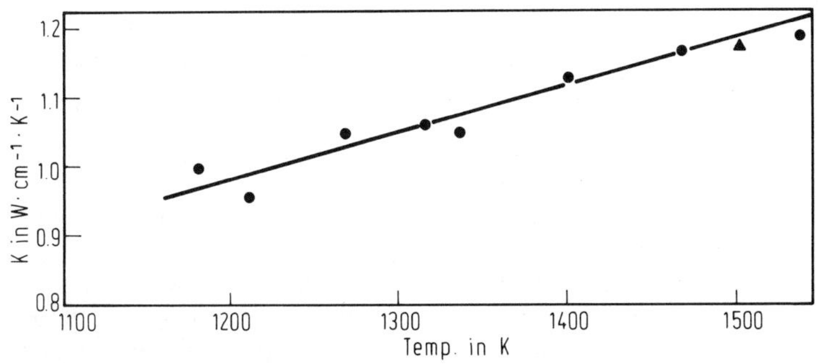

Fig. 9. The temperature variation of the thermal conductivity K of rhodium. ▲ represents the only value of K available in the literature in this temperature range (Sorokin et al. 1969).

A straight line relationship for k with temperatures was shown between 1500 and 2000 K using a least squares analysis of the experimental points [10].

References:

[1] Degussa (Edelmetall-Taschenbuch, Frankfurt a.M. 1967).
[2] Savitskii, E.; Polyakova, V.N.; Gorina, N.; Roshan, N. (Physical Metallurgy of Platinum Metals, MIR Publ., Moscow 1978).
[3] Mendelssohn, K.; Rosenberg, H.M. (Proc. Phys. Soc. [London] A**65** [1952] 388).
[4] Mendelssohn, K.; Rosenberg, H.M. (Proc. Phys. Soc. [London] A**64** [1951] 1057/8).
[5] Kemp, W.R.G.; Klemens, P.G.; Tanish, R.J.I. (Ann. Physik [7] **5** [1959] 35/41).
[6] White, G.K.; Woods, S.B. (Can. J. Phys. **35** [1957] 248/57).
[7] Aisaka, T.; Shimizu, M. (J. Phys. Soc. Japan **28** [1970] 646/54).
[8] Powell, R.W.; Tye, R.P.; Woodman, M.J. (Platinum Metals Rev. **6** [1962] 138/43).
[9] Powell, R.W.; Tye, R.P.; Woodman, M.J. (J. Less-Common Metals **12** [1967] 1/10).
[10] Sorokin, A.G. (Teplofiz. Vys. Temp. **7** [1969] 371/3).
[11] Natarajan, N.S.; Chari, M.B.R. (Indian J. Pure Appl. Phys. **9** [1971] 439/43).
[12] Jain, S.C.; Sharma, B.B.; Reddy, B.K. (J. Phys. D**5** [1972] 155/9).

5 Electrical and Magnetic Properties

Magnetic Susceptibility χ

Room Temperature.

$\chi \times 10^6$	Ref.	$\chi \times 10^6$	Ref.
$+1.1\ cm^3/g$	[1]	1.043 (300 K) cm^3/g	[6]
$+0.9903\ cm^3/g$	[2]	1×10^4 emu/mol	[7]
$+0.9903\ cm^3/g$	[3]	1×10^4 emu/mol	[8]
+0.990 (18 °C) cm^3/g	[4]	5.54×10^{-5} emu/mol	[9]
$+1.11\ cm^3/g$	[5]		

High and Low Temperatures. At 1000 °C, $\chi = +1.80\ cm^3/g$ [4]. Measurements of χ between 80 and 1850 K were obtained by [6 to 10], partly with different results. A later analysis embodied the possible effect of electron-phonon interaction and resulted in a better match with the experimental results of [8], see [11].

The temperature range between 0.37 and 293 K has been studied using magnetic fields of up to 11000 G, with special reference to the effect of small concentrations of iron; the increase of susceptibility over that of pure rhodium was found to be linear [12]. More recent measurements made in the range 0 to 2000 K have been made on pure rhodium [13]. The experimental results taken with other determinations by [14, 15] fit well to a theoretical equation derived by [13]. Three theoretical papers analyse the effect of spin fluctuations on the paramagnetic susceptibility of rhodium [16, 17, 18]. The volume dependence of the magnetism of the metal is analysed by first principles total-energy band calculations using fixed-spin-moment procedure; rhodium was found to be nonmagnetic at zero pressure, but it undergoes first-order transitions to magnetic behaviour at expanded volumes [19].

References:

[1] Honda, K.; Shimizu, Y. (Nature **132** [1933] 565).
[2] Anonymous (Platinum Metals Rev. **16** [1972] 59).
[3] Hoare, F.E.; Walling, J.C. (Proc. Phys. Soc. [London] B **64** [1951] 337/41).
[4] Degussa (Edelmetall-Taschenbuch, Frankfurt a.M. 1967).
[5] Smithells, C.J. (Metals Reference Book, 4th Ed. **2** [1967] 685/9).
[6] Weiss, W.D.; Kohlhaas, R. (Z. Angew. Physik **23** No. 3 [1967] 175/9).
[7] Hoare, F.E.; Matthews, J.C. (Proc. Roy. Soc. [London] A **212** [1952] 137).
[8] Kojima, H.; Tebble, R.S.; Williams, D.E.G. (Proc. Roy. Soc. [London] A **260** [1961] 237/50).
[9] Budworth, D.W.; Hoare, F.E.; Preston, J. (Proc. Roy. Soc. [London] A **257** [1960] 250/62).
[10] Shimitzu, M.; Katsuki, A.J. (J. Phys. Soc. Japan **19** [1964] 1856/61).

[11] Kim, D.J.; Tanaka, C.J. (J. Magn. Magn. Mater. **58** [1986] 254/64).
[12] Knapp, G.S. (J. Appl. Phys. **38** [1967] 1267/8).
[13] Abart, J.; Sänger, W.; Voitländer, J. (J. Magn. Magn. Mater. **28** [1982] 282/4).
[14] Weiss, W.D.; Kohlhaas, R. (Z. Angew. Physik **23** [1967] 175).
[15] Müller, M.; Güntherrodt, H. (private communication 1979).
[16] Shimizu, M.; Kunihara, A. (Phys. Letters A **100** [1984] 218/20).
[17] Trohidou, K.N.; Blackman, J.A.; Cooke, J.F. (Phys. Rev. [3] B **37** [1988] 8154/66).
[18] Shimizu, M. (J. Magn. Magn. Mater. **31/34** [1983] 299/300).
[19] Moruzzi, V.L.; Marcus, P.M. (Phys. Rev. [3] B **39** [1989] 471/87).

De Haas-van Alphen Effect

The pulsed field method was used with fields up to 160 kG. Four fundamental frequencies were identified and labelled α, β, γ and δ. The α had frequencies between 0.2 and 0.6×10^7 and the $\beta < 3 \times 10^7$; the γ were too weak to be disentangled from the harmonics of the β frequencies. The δ was observed near ⟨111⟩ of $\sim 2 \times 10^8$. It was suggested that this was associated with a closed surface centred on Γ [1 to 4].

References:

[1] Coleridge, P.T. (Proc. Roy. Soc. [London] A **295** [1966] 458/75).
[2] Ohlsen, H.; Gustafsson, P.; Nordborg, L. (J. Phys. F **16** [1986] 279/82).
[3] Ketterson, J.B.; Windmiller, L.R.; Hörnfeldt, S. (Phys. Letters A **26** [1968] 115/6).
[4] Carrander, K.; Dronjak, M.; Hörnfeldt, S.P. (J. Phys. Chem. Solids **38** [1977] 289).

Electric Resistivity

Resistivity at 0 °C. The electric resistivity is sensitive to the impurities in the sample examined and improved refining methods tend to result in higher purity in later samples. The available information is therefore presented in chronological order. Some papers give residual resistivity measurements or other impurity checks. The resistivity ϱ is at 0 °C, unless otherwise noted.

year	1934	1952	1954/56	1958	1959	1962	1967	1972
ϱ in μΩ·cm	4.3	4.35	0.0394 [a]	~4.8	4.36 [b]	4.33	4.33	4.3
Ref.	[1]	[2]	[3]	[4]	[5]	[6]	[7]	[8]

[a] in $\Omega/mm^2/M$; [b] theoretical value, see figure in original paper

Resistivity above Room Temperature (temperature coefficient α/°C)

temp. range	0 to 100	0 to 100	0 to 100	0 to 100	0 to 100	0 to 100
$\alpha \times 10^3$	4.57	4.62	4.63	4.63 [a]	4.62	4.57
Ref.	[1]	[3]	[9]	[10]	[8]	[11]

[a] very high purity material

References:

[1] Roeser, W.F.; Wensel, H.T. (J. Res. Natl. Bur. Std. **12** [1934] 519).
[2] Bridgman, P.W. (Proc. Am. Acad. Arts Sci. **79** [1951] 125/79).
[3] Rudnitskii, A.A. (Spravochnik Maschinostroitelya, Vol. 2, Moscow 1956, 560 pp.).
[4] Bale, E.S. (Platinum Metals Rev. **2** [1958] 61/3).
[5] White, G.K.; Woods, S.B. (Phil. Trans. Roy. Soc. [London] A **251** [1959] 273/302).
[6] Powell, R.W.; Tye, R.P.; Woodman, M.J. (Platinum Metals Rev. **6** [1962] 138).
[7] Powell, R.W.; Tye, R.P.; Woodman, M.J. (J. Less Common Metals **12** [1967] 1/10).
[8] Anonymous (Platinum Metals Rev. **16** [1972] 59).
[9] Price, E.G.; Taylor, B. (Nature **195** [1962] 272/3).
[10] Chaston, J.C. (Nature **195** [1962] 793).
[11] Savitskii, E.; Polyakova, V.N.; Gorina, N.; Roshan, N. (Physical Metallurgy of Platinum Metals, MIR Publ., Moscow 1978, p. 42).

Resistivity below 0 °C. Work at low temperatures was carried out to check d-band filling in connection with various theories on the origin of electric resistance. Measurements at 14, 20 and 90 K were made [1]. Calculations of ideal resistivities between 295 and 20 K have been carried out by [2]; figures of 4.78 μΩ·cm at 295 K and 0.0049 μΩ·cm at 25 K were predicted, and experimental determinations at 295 K gave 4.78 and 4.80 μΩ·cm on two samples of widely differing residual resistance [2]. Experimental figures on two high-purity samples with $\varrho_{ice}/\varrho_{He}$ of 180 and 233 μΩ·cm, respectively, gave $\varrho = 0.90$ to 0.92, 2.90 to 2.95 and 4.90 to 4.95 μΩ·cm at 100, 200 and 300 K [3].

The effect of dilute cobalt concentrations on the low-temperature resistance of rhodium between 1.5 and 300 K is given in [4].

General Resistivity. The resistance ratio R_T/R_{295} at temperatures between 295 and 1000 K has been determined and a polynomial expression obtained to fit these results: $R/R_{295} = a_0 + a_1 \cdot T$, where $a_0 = -1.967$, $a_1 = +4.035 \times 10^{-3}$ [5]. The calculation of the phonon-limited resistivity is described by [6].

References:

[1] Potter, H.H. (Proc. Phys. Soc. [London] **53** [1941] 695/705).
[2] White, G.K.; Woods, S.B. (Phil. Trans. Roy. Soc. [London] A **251** [1959] 273/302).
[3] Powell, R.W.; Tye, R.P.; Woodman, M.J. (J. Less-Common Metals **12** [1967] 1/10).
[4] Nagasawa, H. (Phys. Letters A **32** [1970] 271/2).
[5] Garcia, E.Y.; Löffler, D.G. (J. Chem. Eng. Data **30** [1985] 304/5).
[6] Mazin, I.I.; Savitskii, E.M.; Uspenskii, Yu.A. (J. Phys. F **14** [1984] 167/74).

Effect of Pressure. The effect of increasing pressure on the resistivity of rhodium at 30 °C is given by the expression $R/R_0 = -1.764 \times 10^{-6} \cdot p + 9.7 \times 10^{-12} \cdot p^2$ by [1]. Pressures up to 100000 kg/cm² were used. The resistance drops smoothly with increasing pressure.

A theoretical analysis of the variation of the pressure coefficient of electric resistance of rhodium is given in [2].

Effect of Cold-Working. The resistivity change is plotted against extension in tests carried out in liquid nitrogen [3]. Isochronous annealing curves are also given, the annealing being followed by resistance changes, the resistance being measured at −196 °C. The elastic and plastic strain coefficients of resistance have been measured for material in various metallurgical conditions by [4].

References:

[1] Bridgman, P.W. (Proc. Am. Acad. Arts Sci. **81** [1952] 165/251).
[2] Grüneisen, E. (Ann. Physik [5] **40** [1941] 543/52).
[3] van Kuijk, J.G.M. (Physica **30** [1964] 398/400).
[4] Gao, Yi.-Q.; Tong, Li-Z. (Mater. Sci. Eng. **100** [1988] 115/9).

Enthalpy Dependence of Resistivity

The enthalpy dependence of the resistivity of rhodium to temperatures above the melting point was investigated.

Martynyuk, M.M.; Tsapkov, V.I. (Fiz. Met. Metalloved. **37** [1974] 49/54; Phys. Metals Metallog. [USSR] **37** No. 1 [1974] 40/4).

Resistivity of Coatings

Electrodeposited Films. Because of the large differences in the degree of cracking present in such films, which vary with thickness and the conditions used, an absolute figure cannot be expected; a figure of 8.5 $\mu\Omega \cdot cm$ is given by [1].

Sputtered and Evaporated Films. There is a considerable volume of literature covering the effects of heat treatment, gases, type of support, and variation with temperature and thickness; a theoretical analysis of the variations of resistivity and the temperature coefficient with thickness for evaporated films on glass is given in [9].

The following papers are of interest:

subject	Ref.
effect of support	[2, 3]
heat treatment	[4, 5]
effect of gases	[3, 6, 7, 8]
effect of temperature	[1, 2, 3, 7]

References:

[1] Wundt, K. (Oberfläche-Surface **25** [1984] 207/12).
[2] Darmois, E.; Mostovetch, N.; Vodar, B. (Compt. Rend. **228** [1949] 992/3).
[3] Belser, R.B.; Hicklin, W.H. (J. Appl. Phys. **30** [1959] 313/22).
[4] Belser, R.B.; Chester, M.C. (Phys. Rev. [2] **95** [1956] 307/8).
[5] Belser, R.B. (J. Appl. Phys. **28** [1957] 109/16).
[6] Hoffman, D.M.; Coutts, M.D. (J. Vac. Sci. Technol. **13** [1976] 122/6).
[7] Mostovetch, N.; Vodar, B. (Semi-Cond. Mater. Proc. Conf., Reading, Engl., 1950 [1951], pp. 260/81).
[8] Mostovetch, N. (Ann. Phys. [Paris] [12] **8** [1953] 61/125).
[9] Koshy, J. (J. Phys. D**13** [1980] 1339/42).

Superconductivity

The value of T_c is a function of purity, and variations in this may account for differences in reported figures.

In the state of purity available, rhodium was not superconducting above temperatures of 0.1 K [1]. From work on the density of states and characteristic energy losses the absence of superconductivity was explained [2]. $T_c = 0.9$ K is quoted in [3]. Rhodium of the available state of purity was said not to be superconducting at 0.003 K, although the authors deduced that it would be so at 0.2×10^{-3} K [4]; $T_c = 2 \times 10^{-3}$ was found by [5]. A transition temperature of 0.002 K is quoted by [6]; Rh is shown not to be superconducting at 0.086 K by [7]; however, it has been shown that it is at 3.25×10^{-4} K [8, 9]. A good review of the position is given in [10].

References:

[1] Raub, C.J.; Zachariasen, W.H.; Geballe, T.H.; Matthias, B.T. (J. Phys. Chem. Solids **24** [1963] 1093/100).
[2] Claus, H.; Ulmer, K. (Z. Physik **185** [1965] 139/54).
[3] Degussa (Edelmetall-Taschenbuch, Frankfurt a.M. 1967).

[4] Mota, A.C.; Black, W.C.; Brewster, P.M.; Fitzgerald, R.W.; Bishop, J.H. (Phys. Letters **34** [1971] 160).
[5] Ulmer, K. (Solid State Commun. **2** [1964] 327/9).
[6] Savitskii, E.; Polyakova, V.N.; Gorina, N.; Roshan, N. (Physical Metallurgy of Platinum Metals, MIR Publ., Moscow 1978).
[7] Roberts, B.W. (NBS-TN-No. 983 [1978], Tech. Suppl. [1983] 11).
[8] Buchal, Ch.; Pobell, F.; Mueller, R.M.; Kubota, M.; Owers-Bradley, J.R. (Phys. Rev. Letters **50** [1983] 64).
[9] Buchal, Ch.; Welter, J.M. (Platinum Metals Rev. **27** [1983] 170).
[10] Raub, C.J. (Platinum Metals Rev. **28** [1984] 63/74).

Hall Effect, Nernst-Ettingshausen Effect

Hall Effect. The following figures are quoted for the Hall coefficient in cgs units ($\times 10^4$).

t in °C	R_H	resistance ratio	Ref.
18	0.505±0.05	0.1447	[1]
27	0.502	0.0932	[2]
27	0.48	—	[3]
18	0.51	—	[4]

Measurements between 4.2 and 300 K at field strengths between 0 and 22 kG were made by [2]; temperatures between 80 and 900 K and field strengths of 14.5 and 20.2 kG were used by [2, 3]. **Fig. 10** [2] and **Fig. 11** [3] show the results. There is a sign reversal in the Hall potential at 20.4 K with a field of ~18 kG. Determinations using field strengths up to 15 kOe have been made between 77 and 750 K [5].

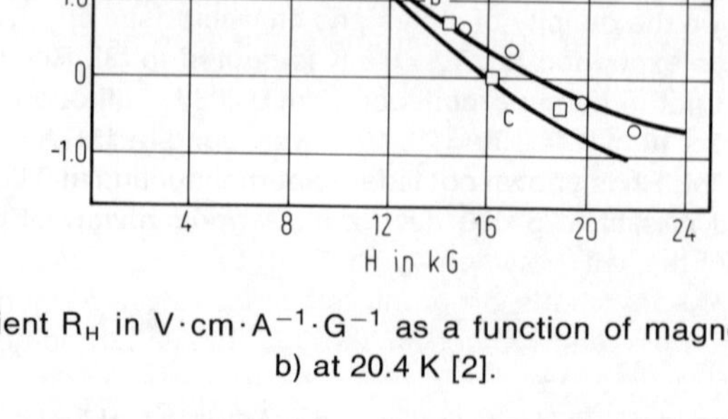

Fig. 10. Hall coefficient R_H in $V \cdot cm \cdot A^{-1} \cdot G^{-1}$ as a function of magnetic field H, a) at 77 K, b) at 20.4 K [2].

Nernst-Ettingshausen Effect Q. This was determined using the same apparatus and specimen as that used for the Hall coefficient determinations by [5].

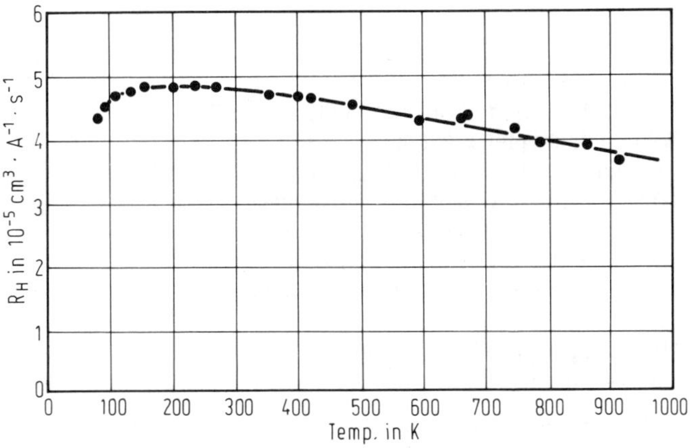

Fig. 11. Dependence of Hall coefficient of Rh on temperature at field strenghts of 14.5 and 20.2 kG [3].

References:

[1] Gehlhoff, P.O.; Justi, E. (Z. Naturforsch. **4a** [1949] 561/2).
[2] Coles, B.R.; Taylor, J.C. (Phys. Chem. Solids **1** [1957] 270/4).
[3] Köster, W. (Z. Metallk. **54** [1963] 619/22).
[4] Degussa (Edelmetall-Taschenbuch, Frankfurt a.M. 1967).
[5] Vasil'yeva, R.P.; Cheremushkina, A.V.; Ivanova, N.N. (Fiz. Met. Metalloved. **35** [1973] 872/5; Phys. Metals Metallog. [USSR] **35** No. 4 [1973] 205/8).

Thermoelectric Force

The thermoelectric properties of spectroscopically pure rhodium are given in [1]. Values of ε (absolute thermoelectric power), σ (Thomson emf), and π (Peltier emf) are given. The thermoelectric force (TEF) is sensitive to trace impurities, particularly at low temperatures and less so at high temperatures; it is also related to electronic structure. Attempts have been made to relate TEF with models based on band theory, so far without convincing agreement with experimental results [2 to 4].

The earlier work on TEF suggested that a discontinuity existed at 1091 °C, and that this was due to an allotropic modification [5]. These results were discounted on statistical grounds by [6], countered in [7]. However, a discontinuity at 1030 °C was reported in a later paper [8]. High-temperature lattice parameter measurements by [9] and [10] seem to make such an allotropic change very unlikely.

Careful measurements between 8 and 1800 K are compared with earlier results by [2, 8, 12]. It will be seen that there is a considerable measure of agreement between the results of [2, 12]. A good analysis of the accuracy of TEF measurements is given in [12]. Further measurements of the TEF between 4 and 300 K have been made using two different purities of material [13]. The results are contrasted with measurements made by [8, 14, 15, 16], and [12]. Some work on changes in the TEF of pure rhodium during annealing at high temperature (1500 °C) are given in [17]. The temperature dependence of the TEF above the Debye temperature has been calculated using the single-site coherent potential approximation; the main experimental features are reproduced [18]. The TEF has been determined experimentally between 77 and 750 K and the results were compared with calculated figures; see illustration in original paper [11].

References:

[1] Savitskii, E.; Polyakova, V.N.; Gorina, N.; Roshan, N. (Physical Metallurgy of Platinum Metals, MIR Publ., Moscow 1978).
[2] Potter, H.H. (Proc. Phys. Soc. [London] **53** [1941] 695/705).
[3] Ricker, T. (Z. Metallk. **54** [1963] 630/40).
[4] Aisaka, T.; Shimizu, M. (J. Phys. Soc. Japan **28** [1970] 646/54).
[5] Booth, E.T.; Dixon, E.H. (Rev. Sci. Instrum. **8** [1937] 381/2).
[6] Wensel, H.T.; Tuckerman, L.B. (Rev. Sci. Instrum. **9** [1938] 237/41).
[7] Booth, E.T.; Dixon, E.H. (Rev. Sci. Instrum. **9** [1938] 242/3).
[8] Rudnitskii, A.A. (Zh. Neorg. Khim. **1** [1956] 1305/21; J. Inorg. Chem. [USSR] **1** No. 6 [1956] 192/208).
[9] Bale, E.S. (Platinum Metals Rev. **2** [1958] 61/3).
[10] Ross, R.G.; Hume-Rothery, W. (J. Less-Common Metals **5** [1963] 258/70).
[11] Vasil'yeva, R.P.; Cheremushkina, A.V.; Ivanova, N.N. (Fiz. Met. Metalloved. **35** [1973] 872/5; Phys. Metals Metallog. [USSR] **35** No. 4 [1973] 205/8).
[12] Vedernikov, M.V. (Advan. Phys. **18** [1969] 337/70).
[13] Huntley, D.J. (Can. J. Phys. **49** [1971] 2610/2).
[14] Walker, C.W.E. (Diss. Simon Fraser Univ., Burnaby, B.C.).
[15] Carter, R.; Davidson, A.; Schroeder, P.A. (J. Phys. Chem. Solids **31** [1970] 2374/7).
[16] Rudnitskii, A.A.; Polykova, R.S.; Tyurin, I.T. (Izv. Sekt. Platiny **29** [1955] 183).
[17] Lapp, G.B.; Maksimova, V.L. (Zh. Neorg. Khim. **2** [1957] 2589/97; Russ. J. Inorg. Chem. **2** No. 11 [1957] 166/79).
[18] Voloshinskii, A.N.; Tsiovkin, Yu.Yu.; Ryzhanova, N.V.; Vishnekov, L.Yu. (Fiz. Met. Metalloved. **67** [1989] 213/20; Phys. Metals Metallog. [USSR] **67** No. 2 [1989] 1/7).

Thermionic Work Function

The Richardson-Dushman equation governs thermionic emission: $j = AT^2e^{-b/t}$ where j is the current density, T the absolute temperature and A and b are constants. A depends on the metal surface and $b = \phi/k$, ϕ being the work function and k the Bolzman's constant.

Values of A and ϕ are given below.

A in A/K^2 . .	33	100	—	—	—	—	—
ϕ in V . . .	4.8	4.9	4.8	4.75	4.8	4.65	5.1 (calculated)
Ref.	[1]	[2]	[3]	[4]	[5]	[6]	[7]

These figures refer to polycrystalline material; it is probable that ϕ is anisotropic with respect to the crystal orientation. No work on single crystals of rhodium could be found, but work by [8] on other platinum metals suggests that this is so.

References:

[1] Wahlin, H.B.; Whitney, L.V. (J. Chem. Phys. **6** [1938] 594).
[2] Weinreich, O.A. (Phys. Rev. [2] **82** [1951] 573).
[3] Degussa (Edelmetall-Taschenbuch, Frankfurt a.M. 1967).
[4] Savitskii, E.; Polyakova, V.N.; Gorina, N.; Roshan, N. (Physical Metallurgy of Platinum Metals, MIR Publ., Moscow 1978).
[5] Kohl, W.H. (Materials and Techniques for Electron Tubes, Reinhold, New York 1960, p. 526).
[6] Michaelson, H.B. (J. Appl. Phys. **21** [1950] 536).
[7] Feibelman, P.J.; Hamann, D.R. (Phys. Rev. [3] B **29** [1984] 6463/7).
[8] Burov, I.V.; Litvak, L.N. (Fiziko-Khimiya Redkikh Metallov [The Physical Chemistry of Rare Metals], Nauka, Moscow 1972, pp. 106/17).

6 Electrochemical Behaviour

Potentials in Aqueous and Acid Solutions

The oxidation potentials of the most commonly encountered compounds are given by [1]. A more complete list is given in [2, 3]. The most important oxidation state is Rh^{3+}. RhO_2 and RhO_3 have been prepared in alkaline solution, but are not stable in acid ones. Rh_2O_3 is soluble in HCl to form complex chlorides and this oxide is also probably soluble in $HClO_4$.

Rh-H_2O. The equilibrium potential/pH diagrams at 25 °C (see figures in original paper) differ in the number of dissolved ions present. Free enthalpy figures for the compounds involved are given as Rh_2O −20, RhO −18, Rh_2O_3 −52.5, RhO_2 −15, Rh^+ 14, Rh^{2+} 28, Rh^{3+} 55, RhO_4 −15 kcal/mol [2].

Acid Solutions. For the reactions $Rh \rightarrow Rh^{3+} + 3e^-$, $Rh \rightarrow Rh^+ + e^-$, $Rh + 6Cl^- \rightarrow RhCl_6 + 3e^-$, and $Rh^+ \rightarrow Rh^{2+} + e^-$, the oxidation potentials E_0 are −0.8, −0.6, −0.44, and −0.6, respectively. **Fig. 12** summarizes the potentials in 1 M acid solution [1].

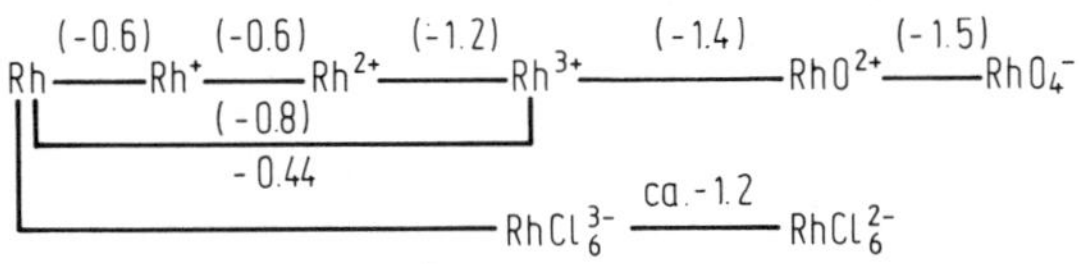

Fig. 12. Oxidation potentials of Rh in 1 M acid solution.

A large number of researches has been carried out on the electrolytic behaviour of rhodium, mainly in 1 M sulphuric acid; the work has mainly been concerned with the anodic and cathodic polarization due to hydrogen and oxygen, the formation of anodic oxide films, and the kinetics of the gas evolution. Later work has tended to use pulsed voltages of varying wave form and also, more detailed examination of oxide film techniques like XPS (X-ray photoelectron spectroscopy) and Auger spectroscopy [4, 5].

Fig. 13 shows a fairly typical current-potential curve in 1 M H_2SO_4 [6].

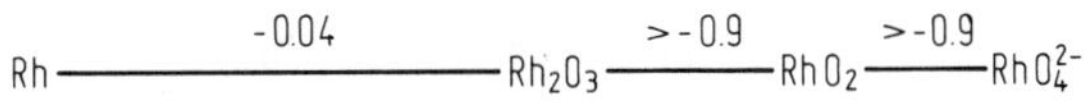

Fig. 13. Typical current-potential curve in 1 M H_2SO_4.

In the text, the significance of various parts of the curve is discussed, features including hydrogen and oxygen adsorption and desorption, oxide formation and dissolution, and the hysteresis in this [6]. Similar work has been carried out in HCl by [7, 8, 9]. The kinetics of oxygen evolution in $HClO_4$ is shown in [10, 11]; [8] has carried out work in $HClO_4$ and HCl. The oxide films were examined with X-rays, electron microscopy and EDA spectroscopy; it was concluded that these were neither Rh_2O_3 nor Rh_2O [11]. Work with HNO_3 and $HClO_4$ solutions is reported, E_0 values of 1.46 and 1.48 V were given for $Rh^{III} \rightarrow Rh^{VI}$ and $Rh^{VI} \rightarrow Rh^{III}$, respectively [12]. Contamination of electrodes can easily occur even after extensive anodic cleaning [4, 5].

Alkaline Solutions. A potential diagram and oxidation states in alkaline solution are shown in a figure by [1]. Work on the electrolytic processes in alkaline solution has been done by [13 to 16]. **Fig. 14**, p. 32 [16] shows current-voltage curves for both acid and alkaline solutions.

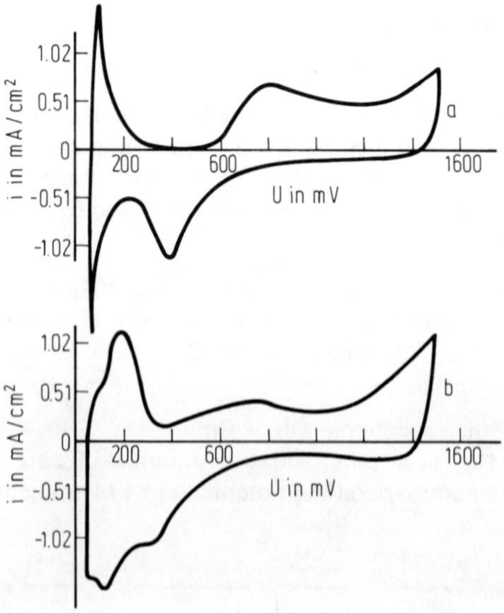

Fig. 14. Current-voltage curves; U = 0.5 V/s in 2.3 M H_2SO_4 (a) and 0.1 M NaOH (b) at 25 °C.

Concerning anodically formed oxide films, a considerable divergence of conclusions is apparent. The following papers are taken in chronological order. The anodic layer was considered to be a chemisorbed one formed by $Rh + H_2O \rightarrow Rh\text{-}OH + H^+ + e^-$; $Rh\text{-}OH \rightarrow Rh\text{-}O + H^+ + e^-$ [16]. Rh_2O_3 or RhO_2 was formed in $HClO_4$; in HCl, Rh_2O_3 + adsorbed ions were found [8]. Rh_2O_3 was thought to be the final layer after a three-stage process: $Rh + H_2O \rightarrow Rh \cdots + H^+ + e^-$; $Rh\text{-}OH \rightarrow Rh\text{-}O + H^+ + e^-$; $Rh\text{-}O \rightarrow Rh_2O_3$ [17].

The anodic polarization was considered to be due to an Rh-O alloy produced by dissolution of the gas in the metal surface [18]. Work by [13] stated that Rh_2O_3 was the final product of anodic treatment, whilst [19] found an Rh-O alloy at the interface. A paper by [20] suggested the reactions involved $Rh \rightarrow Rh_2O \rightarrow Rh_2O_3$. This was clearly observed on oxidation and reduction, although the decomposition of $Rh_2O_3 \rightarrow Rh_2O$ was partly masked by oxygen evolution.

Using an electron microscope with EDAX and X-ray diffraction techniques, the anodically produced film was examined; evidence was produced suggesting that RhO_3 might be an intermediate in reduction of the Rh_2O_3 film [11]. Using X-ray photoelectron spectroscopy, it was found that thick layers of oxide hydroxide were the ultimate product according to $Rh\text{-}H_2O_{ads} \xrightarrow{0.6\,V} Rh\text{-}OH_{ads} \rightarrow Rh(OH)_n \xrightarrow{-1.4\,V} RhOOH$, $n \leqq 3$ [4].

Work on both thermally and electrolytically produced oxidation adlayers using XPS (X-ray photoelectron spectroscopy) studies on such layers resulted in the conclusion that different compounds were produced by the two processes and that this results in different thermal stability. Decomposition of thermally produced films took place in three steps: $Rh_2O_3 \xrightarrow{470\,K} Rh_2O_3, Rh \xrightarrow{\sim 800\,K} RhO_{ads} \xrightarrow{900\,K} Rh$; Rh_2O_3, Rh or "RhO" was found to be stable over a wide temperature range.

Under anodic conditions, anodic polarization in sulphuric acid at 3 to 4 V gave an adlayer of RhOOH with a limiting thickness of ~3 nm. Thermal decomposition took place in two stages: $RhOOH \xrightarrow{420\,K} RhOH_{ads} \xrightarrow{\sim 600\,K} Rh$ [21].

The formation and dissolution of the oxide film is utilised in etching in a 1% HCl-4% NaCl solution employing alternating current [22], and an electrolytic polishing solution using molten $NaNO_3$ and NaCl is given by [23]. The anodic corrosion using superimposed alternating current in HCl solutions resulted in raspberry-red solutions containing Rh^{III} ions [7]. An electrochemical study of the dissolution of rhodium in molten alkali metal chlorides monitored by absorption spectra was carried out by [24]; standard potentials are given.

Corrosion

Theoretical areas of corrosion with and without complex ions are shown in 2 figures (see original paper) [2]. Rhodium concentrations of 10^{-6} to 10^{-7} M are said to inhibit the corrosion of titanium in boiling 2 M HCl. The rhodium ion is found to accelerate that of tin [25, 26, 27]. Additions of 0.5% of rhodium to chromium improved corrosion resistance in boiling acids [28].

References:

[1] Latimer, W.M. (The Oxidation States and Their Potentials in Aqueous Solutions, 2nd Ed., Prentice-Hall, Englewood Cliffs, N.J., 1961).
[2] van Muylder, J.; Pourbaix, M. (Centre Belge Etude Corros. Rappt. Tech. No. 59 [1958] 1/11).
[3] Llopis, J.F.; Tordesillas, I.M. (Encyl. Electrochem. Elem. **6** [1976] 299/329).
[4] Peukert, M. (J. Electroanal. Chem. Interfacial Electrochem. **185** [1985] 379/91).
[5] Petukhova, R.P.; Podlovchenko, B.I.; Gas'kov, A.M. (Elektrokhimiya **21** [1985] 1414/8; Soviet Electrochem. **21** [1985] 1339/43).
[6] Rand, D.A.J. (Proc. Roy. Australian Chem. Inst. **41** [1974] 8/12).
[7] Llopis, J.; Tordesillas, I.M.; Muñiz, M. (Electrochim. Acta **10** [1965] 1045/55).
[8] Llopis, J.; Vazquez, M. (Electrochim. Acta **9** [1964] 1655/63).
[9] Hickling, A.; Salt, F.W. (Trans. Faraday Soc. **36** [1940] 1226/35).
[10] Damjanovic, A.; Dey, A.; Bockris, J.O'M. (J. Electrochem. Soc. **113** [1966] 739/46).

[11] Victori, L.; Marcotegui, J.; Monge, J. (Rev. Iberoam. Corros. Prot. **15** No. 4 [1984] 63/4).
[12] Grube, G.; Autenrieth, H. (Z. Elektrochem. Angew. Physik. Chem. **44** [1938] 296/9).
[13] Khrushcheva, E.I.; Shumilova, N.A.; Tarasevich, M.R. (Elektrokhimiya **2** [1966] 277/82; Soviet Electrochem. **2** [1966] 258/62).
[14] Genshaw, M.A.; Damjanovic, A.; Bockris, J.O'M. (J. Phys. Chem. **71** [1967] 3723/31).
[15] Pentland, N.; Bockris, J.O'M.; Sheldon, E. (J. Electrochem. Soc. **104** [1957] 182/94).
[16] Böld, W.; Breiter, M. (Electrochim. Acta **5** [1961] 169/79).
[17] Shibata, S. (Bull. Chem. Soc. Japan **38** [1965] 1330/7).
[18] Hoare, J.P. (J. Electrochem. Soc. **111** [1964] 232/6).
[19] Hoare, J.P. (Electrochim. Acta **13** [1968] 417/28).
[20] Okamura, T. (Denki Kagaku **37** [1969] 831/5).

[21] Peukert, M. (Surf. Sci. **141** [1984] 500/14).
[22] Raub, E.; Buss, G. (Z. Elektrochem. Angew. Physik. Chem. **46** [1940] 195/202).
[23] Savitskii, E.; Polyakova, V.N.; Gorina, N.; Roshan, N. (Physical Metallurgy of Platinum Metals, MIR Publ., Moscow 1978).
[24] Vasin, B.D.; Ivanov, V.A.; Raspopin, S.P. (Soviet Non-Ferrous Metals Res. **13** [1985] 393/6).
[25] Buck, W.R.; Leidheiser, H. (Nature **181** [1958] 1681/2).
[26] Buck, W.R.; Leidheiser, H. (Z. Elektrochem. **62** [1958] 690/5).
[27] Buck, W.R.; Leidheiser, H. (Nature **204** [1964] 177/8).
[28] Hoar, T.P. (Platinum Metals Rev. **5** [1961] 141/3).

7 Chemical Reactions

Dissolved Gases

Rhodium dissolves large quantities of oxygen in the molten state as well as hydrogen, and this is an important factor in maintaining a high degree of purity in molten material; reaction of most refractories with reducing agents at the high temperatures involved is the reason [1].

The metal will also dissolve both gases at temperatures well below the melting point. That oxygen will diffuse through and dissolve in solid rhodium may be inferred from work on the internal oxidation of alloys quoted by [2]. Field ion microscope studies also clearly show the presence of interstitial oxygen atoms (see figures in original publications) [3, 4]. It has also been suggested that nitrogen may be soluble [5].

No quantitative work has been found on the solubility of oxygen, but several papers have been published on hydrogen. Early work by Sieverts and Jurisch suggested that hydrogen was insoluble in rhodium, and similar work at temperatures between 20 and 450 °C and pressures between 80 and 323 Torr gave the same result [6]. However, a more recent investigation carried out under 1 atm of hydrogen at temperatures between 855 and 1524 °C showed a small solubility. The relative partial molar enthalpy ΔH_H and excess entropy ΔS_H of hydrogen atoms in the solid were calculated as 6.38 ± 0.33 kcal/g-atom H and -11.99 ± 1.31 eu, respectively [7].

Hydrogen determinations have been carried out on rhodium foils produced by electrodeposition at a current density of 10 A/dm². Contents of 0.06% were reported, this falling to 0.015% after annealing [8].

References:

[1] International Nickel (Rhodium 1965, p. 3).
[2] Darling, A.S. (Intern. Metals Rev. **18** [1973] 91/122).
[3] Müller, E.W. (Z. Physik **156** [1959] 399/410).
[4] Müller, E.W. (Struct. Properties Thin Films Proc. Intern. Conf., Bolton Landing, N.Y., 1959).
[5] Caldwell, D.O. (Rev. Sci. Instrum. **23** [1952] 501/2).
[6] Adadurov, I.E.; Pevni, N.I. (Zh. Prikl. Khim. **10** [1937] 1216/9).
[7] McLellan, R.B.; Oates, W.A. (Acta Metall. **21** [1973] 181/5).
[8] Medyanik, V.N.; Kadaner, L.I. (Ukr. Fiz. Zh. **13** [1968] 127/9; Ukr. Phys. J. **13** [1968] 88/9).

Surface Reactions with Oxygen

Although little has been written on dissolved gases in rhodium, a great deal has been written on surface reactions with oxygen. Descriptions of the visible changes which accompany heating the metal in air or oxygen are given in [1 to 5]. Interest in oxidation has of course been stimulated by the increased rate of metal loss at high temperatures under such conditions; for example, when heated at 1000 °C an increase in weight is noted, due to a coating of Rh_2O_3, but at 1400 °C a loss of 33 $mg \cdot cm^{-2} \cdot h^{-1}$ is found which increases fivefold at 1700 °C [6]. The vapour pressure of rhodium oxide plotted against 1/T is shown in [7, 8].

Oxidation studies at temperatures high enough to cause volatilization of oxide coatings have been carried out by [9, 10, 11]; all measured changes in weight to follow the reaction. There is considerable variation in the various results. For example, the activation energies determined from Arrhenius plots were 51 kcal/mol [10] and 29.6 kcal/mol [11], see figures in original papers.

To fully understand the process it is necessary to identify the oxides providing protection at lower temperatures, and the volatile species present at higher temperatures; the problem was complicated by reports that the protective coating was polymorphic [12]. An equilibrium diagram based on work carried out under high pressures of oxygen helps to clarify the situation; see Fig. 4 in "Rhodium" Suppl. Vol. B 1, 1982, p. 9. The stable phase present up to ~670 °C was shown to be RhO_2 and above this temperature, Rh_2O_3 is present in the corundum crystalline form; this transforms to an orthorhombic form at higher temperatures until it decomposes. The decomposition temperature is said to be ~1140 °C at a pressure of 1 atm of oxygen, and ~1030 °C in air at atmospheric pressure [13]. ΔG for Rh_2O_3 has been determined electrochemically by [14], other determinations have been made by [15]. ΔH°_{298} was determined as $-(108000 \pm 4000)$ cal/mol [14].

A paper by [7] showed that the volatile oxide was RhO_2, and that $\Delta G^{\circ} = 45140 \pm 580 - T(4.94 \pm 0.36)$ cal/mol (1200 to 1500 °C), derived from an Arrhenius plot.

Using a different technique, the thermodynamics of the volatile oxide has been investigated. The methods used were "pumping" of oxygen by a filament in a hot cathode ionization gauge, mass spectrometric analyses of gaseous oxidation products flash desorbed from the surface, and mass spectrometric measurements of rates of evaporation of gaseous species effusing from a Langmuir cell [16]. ΔG was determined by a method due to [16, 17]. The growth of the thin "protective" oxide layers has been studied using imaging atom-probe mass spectroscopy and field-ion microscopy. Surface oxides were produced by rhodium field emitter tips in oxygen at pressures between 0.01 and 1.0 Torr at temperatures between 400 and 650 K. Imaging atom-probe analyses indicated that the oxide formed at 1.0 Torr and 500 K and above was Rh_2O_3. The plots of cumulative oxygen atoms against cumulative rhodium atoms carried out at 1.0 Torr of oxygen at 600 K show the growth of the stoichiometric Rh_2O_3 film. At temperatures below 500 K, the oxygen content decreased with decreasing temperature. The activation energy for the early part of the oxidation was determined as 3.1 kcal/mol [18].

References:

[1] Darling, A.S. (Intern. Metals Rev. **18** [1973] 91).
[2] Chaston, J.C. (Platinum Metals Rev. **9** [1965] 51/6).
[3] Chaston, J.C. (Platinum Metals Rev. **13** [1969] 28/9).
[4] Chaston, J.C. (Platinum Metals Rev. **19** [1975] 135/40).
[5] Ismail, M.I. (Platinum Metals Rev. **23** [1979] 22).
[6] Raub, C.J. (Metall **32** [1978] 802/4).
[7] Alcock, C.B.; Hooper, G.W. (Proc. Roy. Soc. [London] A **254** [1960] 551/61).
[8] Alcock, C.B. (Platinum Metals Rev. **5** [1961] 134/9).
[9] Raub, E.; Plate, W. (Z. Metallk. **48** [1957] 529).
[10] Krier, C.A.; Jaffee, R.I. (J. Less-Common Metals **5** [1963] 411/31).

[11] Phillips, W.L. (Trans. Am. Soc. Metals **57** [1964] 33/7).
[12] Wold, A.; Arnott, R.J.; Croft, W.J. (Inorg. Chem. **2** [1963] 1972).
[13] Muller, O.; Roy, R. (J. Less-Common Metals **16** [1968] 129/46).
[14] Kleykamp, H. (Z. Physik. Chem. [N.F.] **67** [1969] 277/83).
[15] Slough, W. (private communication to [1]).
[16] Olivei, A. (J. Less-Common Metals **29** [1972] 11/23).
[17] Franklin, J.E.; Stickney, R.E. (Massachusetts Inst. Technol. Quart. Progr. Rept. No. 101 [1971] 33).
[18] Kellogg, G.L. (Surf. Sci. **171** [1986] 359/76).

Corrosion Resistance

Alkali Melts. Weight losses were measured on 8 cm^2 specimens exposed to melts at a temperature of 410 °C for 1 h; dry and moist oxygen atmospheres were used. There was a marked difference in behaviour between NaOH and KOH melts, attack in the latter being more severe. The moisture content of the covering gas seemed to be unimportant, but oxygen content increased the loss in NaOH melts. In NaOH melts the loss was 20 to 30 mg and in KOH > 10000 mg [1]. Tests in fused salts and NaOH have been carried out by [2]. In this case, weight losses were expressed in $mg \cdot dm^{-2} \cdot d^{-1}$ based on a 1 h exposure. The results for NaOH are lower than those reported by [1]; they are shown below [2].

reacting agent	NaOH	Na_2O_2	Na_2CO_3	$NaNO_3$	KNO_3	NaCN	KCN	$KHSO_4$
t in °C	350	350	920	350	350	700	700	440
weight loss in $mg \cdot dm^{-2} \cdot d^{-1}$	180	336	−48	120	36	25200	11000	1320

Chloride Melts. Reactions between rhodium and alkaline metal chlorides are reported by [3, 4]. The equilibrium potentials of the metal in molten lead chloride and sodium/caesium eutectic with and without additions of $ZrCl_4$ as a complexing agent are given by [5].

Various Chemicals. (exposed at 20 °C unless otherwise stated) [6].

agent		conditions	rating	
Br_2		moist	A	
Cl_2		dry	A	
Cl_2		moist	A	
F_2		—	?	
I_2		moist	B	
H_2S		—	A	
HBr	60%	100 °C	A	
HCl	36%	—	A	
HCl	36%	100 °C	A	
HF	40%	—	A	
H_2SO_4			A	
H_2SO_4			B	100 °C
HNO_3		—	A	
HNO_3		100 °C	A	
aqua regia		—	A	
aqua regia		100 °C	A	
organic acids (moist)		100 °C	A	
NaOCl solution		100 °C	B	

A = no appreciable solution; B = some attack, but not enough to preclude use.

The rhodium-chlorine system has been studied over the temperature range 700 to 1500 °C. $RhCl_2$ and $RhCl_3$ are the important gaseous species [7]. Chlorination and the thermal transformations of the trichloride were examined by thermographic, thermogravimetric, chemical and X-ray methods. It was found that rhodium reacts with chlorine, hydrogen chloride and carbon tetrachloride above 500 °C. RhCl decomposes in a stream of hydrogen to give the metal at 105 to 140 °C; heated in air it decomposes to Rh_2O_3 at 440 to 500 °C [8].

References:

[1] Lux, H.; Betz, E. (Z. Anorg. Allgem. Chem. **310** [1961] 305/19).
[2] Atkinson, R.H.; Raper, A.R.; Middleton, A.B. (J. Iron Steel Inst. [London] **155** [1947] 213/34).
[3] Alexandrov, E.P.; et al. (Izv. Vysshikh Uchebn. Zaved. Tsvetn. Metall. **1981** No. 5, pp. 86/9).
[4] Vasin, B.D.; Ivanov, V.A.; Raspopin, S.P. (Izv. Vysshikh Uchebn. Zaved. Tsvetn. Metall. **1984** No. 2, pp. 77/9).
[5] Vasin, B.D.; Ivanov, V.A.; Raspopin, S.P. (Soviet Non-Ferrous Metals Res. **13** [1985] 393/6).
[6] Uhlig, H.H. (Corrosion Handbook, Wiley, New York 1947).
[7] Bell, W.E.; Tagami, M.; Merten, U. (J. Phys. Chem. **66** [1962] 490/4).
[8] Ivashentsev, Ya.I.; Timonova, R.I. (Zh. Neorg. Khim. **11** [1966] 2189/92; Russ. J. Inorg. Chem. **11** [1966] 1173/5).

Alloys of Rhodium

1 Alloys with Arsenic

The Rh-As System

Phase Diagram. No phase diagram has been determined. However, a number of intermediate compounds have been identified, and some structures and lattice parameters determined. The existence of RhAs, $RhAs_2$, Rh_2As and $RhAs_3$ was established by [1]. A polymorphic transformation in Rh_2As which took place between 650 and 700 °C was found during an investigation of the system at temperatures between 300 and 1000 °C; no liquid phases were found up to 1000 °C at the rhodium-rich end, and none up to 900 °C at the arsenic-rich end. A new phase was found at $Rh_{1.7}As$ and a homogeneous field between $Rh_{1.3}As$ and $Rh_{1.6}As$ [2]. Investigations by X-ray diffraction and density determinations on the compound $RhAs_2$ confirmed the approximate $CoAs_2$-type structure [3]. Lattice parameter measurements were made on Rh_2As with material quenched from above and below the transformation temperature and also, the magnetic susceptibility was measured with rising and decreasing temperature on the α and β forms; the latter measurements suggested that the transformation takes place between 870 and 890 K [4]. The compound $RhAs_3$ has been examined by [5].

Crystallography. The following table gives the lattice parameter determinations obtained by various investigators.

compound	type	a	b in Å	c	β	Ref.
Rh_2As	C1	5.674 a)	—	—		[1]
RhAs	B31 (MnP)	5.62 b)	3.58 b)	6.00 b)		[1]
RhAs	MnP	5.645	3.595	6.061	—	[8]
$RhAs_3$	Do2	8.453 a)	—	—	—	[1]
$RhAs_2$	$CoSb_2$	6.06	6.08	6.15	—	[9]
$RhAs_2$	$CoAs_2$?	complicated pattern			—	[1]
$RhAs_2$	$CoSb_2$	6.059	6.080	6.149	114.74° d)	[8]
α-Rh_2As	C1	5.678 d)	—	—	—	[2]
β-Rh_2As	C23	5.89 b)	3.89 b)	7.32 b)	—	[2]
α-Rh_2As	—	5.675(3)	—	—	—	[4]
β-Rh_2As	—	5.910(2)	3.917(1)	7.367(1)	—	[4]
α-Rh_2As	anti-CaF_2	5.675	—	—	—	[8]
β-Rh_2As	Co_2P	5.904	3.918	7.352	—	[8]
$RhAs_2$	RhP_2	6.041 c)	6.082 d)	6.126 c)	114.333° d)	[2]
$RhAs_2$	$CoSb_2$	6.0629(4)	6.0816(5)	6.1498(4)	114.707(6)° d)	[3]
$RhAs_2$	$CoSb_2$	[5.1416(4)]	[6.0816(5)]	3.2943(2)	90.898(2)° d)	[3]
$RhAs_3$	$CoAs_3$	8.4507(3)	—	—	—	[5]
$RhAs_3$	$CoAs_3$	8.453	—	—	—	[8]
Rh_9As_7	hex. SS	27.118	—	3.512	—	[8]
$Rh_{12}As_7$	~$Cr_{12}P$	79.297(1)	—	3.657(1)	—	[6]
$Rh_{12}As_7$	$P6_3/m$	9.315(7)	—	3.659(4)	—	[7]
$Rh_{1.5}As$	$P6_3/m$	9.219(2)	—	3.532(1)	—	[8]

SS = superstructure

In the above table the accuracies are indicated thus:

[a] ±0.001 Å, [b] ±0.01 Å, [c] ±0.005 Å, [d] ±0.002 Å. In the case of references [3 to 5], the last figures in parentheses are standard deviations. The figures in brackets represent the parameters of the compound interpreted as a marcasite structure for $RhAs_2$ [3].

The compound $Rh_{1.7}As$ and the homogeneous field between $Rh_{1.3}As$ and $Rh_{1.6}As$ first found by [2] has been the subject of several structural investigations. $Rh_{1.7}As$ was found to be hexagonal and closely similar in structure to $Cr_{12}P_7$, but with some slight differences. **Fig. 15** shows the projection of a proposed structure on the (001) plane [7]. Agreement on the similarity with $Cr_{12}P_7$ of this compound was found and an explanation proposed in which the arsenic atom statistically occupying the 4e position is disordered along the 6-fold axis (see figure in original publication) [6]. The lower rhodium end of the homogeneous range, i.e. $Rh_{1.5}As$ to $Rh_{1.3}As$, has been investigated, and the former compound found to have a hexagonal unit cell and a defective $Cr_{12}P_7$-type structure; the latter compound was found to have a superstructure-type lattice of hexagonal form and a = 27.118, c = 3.512 Å. The miscibility gap found to exist between $Rh_{1.5}As$ and $Rh_{1.7}As$ is explained in terms of the location of the arsenic atom on the c axis [8].

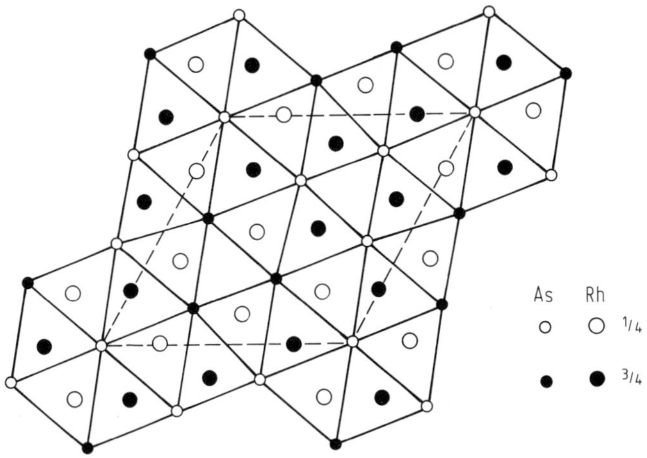

Fig. 15. Projection of a proposed structure for $Rh_{1.7}As$ on the (001) plane [7].

References:

[1] Heyding, R.D.; Calvert, L.D. (Can. J. Chem. **39** [1961] 955/7).
[2] Quesnel, J.C.; Heyding, R.D. (Can. J. Chem. **40** [1962] 814/8).
[3] Kjekshus, A. (Acta Chem. Scand. **25** [1971] 411/22).
[4] Kjekshus, A.; Skaug, K.E. (Acta Chem. Scand. **26** [1972] 2554/6).
[5] Kjekshus, A.; Rakke, T. (Acta Chem. Scand. **28** [1974] 99/103).
[6] Pivan, J.Y.; Guérin, R.; Sergent, M. (J. Less-Common Metals **107** [1985] 249/58).
[7] Lambert-Andron, B.; Dhahri, E.; Chaudoüet, P.; Madar, R. (J. Less-Common Metals **108** [1985] 353/8).
[8] Pivan, J.Y.; Guérin, R. (J. Less-Common Metals **144** [1988] 31/9).
[9] Hulliger, F. (Phys. Letters **4** [1963] 282/3).

Other Physical Properties

Density. The following figures were obtained on $RhAs_2$ by pycnometric and X-ray methods: 8.09 and 8.15 g cm^3, respectively [1].

Superconductivity. The compound $Rh_{12}As_7$ was found to be superconducting with a transition temperature T_c of 0.6 K [2]. RhAs and $Rh_{1.4}As$ were also found to be superconductors with $T_c=0.58$ and 0.56 K, respectively [3].

Semiconduction. $RhAs_3$ was found to be a diamagnetic semiconductor [4].

Electric Resistivity. The resistivity ϱ of $RhAs_3$ plotted against temperature T (in K); $10^3/T$ is shown in a figure [4].

Magnetic Susceptibility. The table below gives the available figures at room temperature.

compound	$\chi\times10^6$ in emu/g	Ref.
$RhAs_3$	-0.219 ± 0.002	[4]
$RhAs_2$	-0.498 [a)]	[5]
$RhAs_2$	-0.412 [b)]	[1]
α-Rh_2As	-0.378 [c)]	[6]
β-Rh_2As	0.131 [c)]	[6]

a) figure estimated from a graph and converted from emu/mol;

b) calculated at room temperature from the author's equation $\chi(T)=-0.400-0.00004\ T$, 80 °C<T<850 °C;

c) calculated at room temperature from authors' equations:
α-Rh_2As $\chi(T)=-0.348+0.000101\ T$; $90<T<880$ K
β-Rh_2As $\chi(T)=0.1285+0.000010\ T$; $90<T<\sim600$ K and $T>880$ K.

Figures in the original papers show the variation of the susceptibilities of $RhAs_2$ and $RhAs_3$ with temperature; in the former χ is expressed in emu/mol and in the latter in emu/g [4, 5].

Fig. 16 shows the effect of increasing and decreasing temperature on the reciprocal of the susceptibility [6].

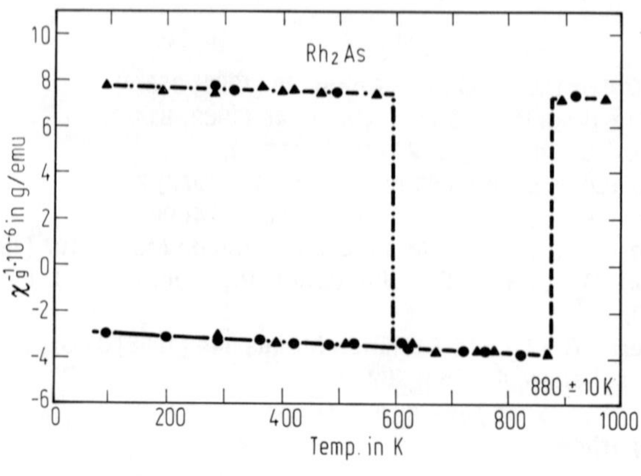

Fig. 16. Reciprocal of the magnetic susceptibility of α- and β-Rh_2As as a function of increasing (▲) and decreasing (●) temperature.

References:

[1] Kjekshus, A. (Acta Chem. Scand. **25** [1971] 411/22).
[2] Raub, C.J.; Zachariasen, W.H.; Geballe, T.H.; Matthias, B.T. (J. Phys. Chem. Solids **24** [1963] 1093).
[3] Raub, C.J. (Platinum Metals Rev. **28** [1984] 63/75).
[4] Pleass, C.M.; Heyding, R.D. (Can. J. Chem. **40** [1962] 590/600).
[5] Bennett, S.L.; Heyding, R.D. (Can. J. Chem. **44** [1966] 3017/30).
[6] Kjekshus, A.; Skaug, K.E. (Acta Chem. Scand. **26** [1972] 2554/6).

2 Alloys with Antimony

2.1 The Rh–Sb System

Phase Diagram. The system up to 50 at% antimony was investigated using microscopical, X-ray, hardness, density and thermal analysis measurements; the compounds $RhSb_3$, $RhSb_2$ and RhSb were identified together with a eutectic horizontal at ~610 °C, and peritectic reactions at ~900 and 1000 to 1100 °C [1]. A later paper covered the whole composition range and the diagram is shown in **Fig. 17** [2]. Rh_2Sb and RhSb were found to melt congruently at ~1440 and 1310 °C, whilst $RhSb_2$ and $RhSb_3$ form peritectically at 1100 and 900 °C, respectively. The maximum solubility of antimony in rhodium at 1150 °C is 8 at%, whilst the solubility of rhodium in antimony is very low, probably <3 at%; the γ phase, Rh_3Sb_2, exists at high temperature and decomposes eutectoidally at ~1000 °C [2].

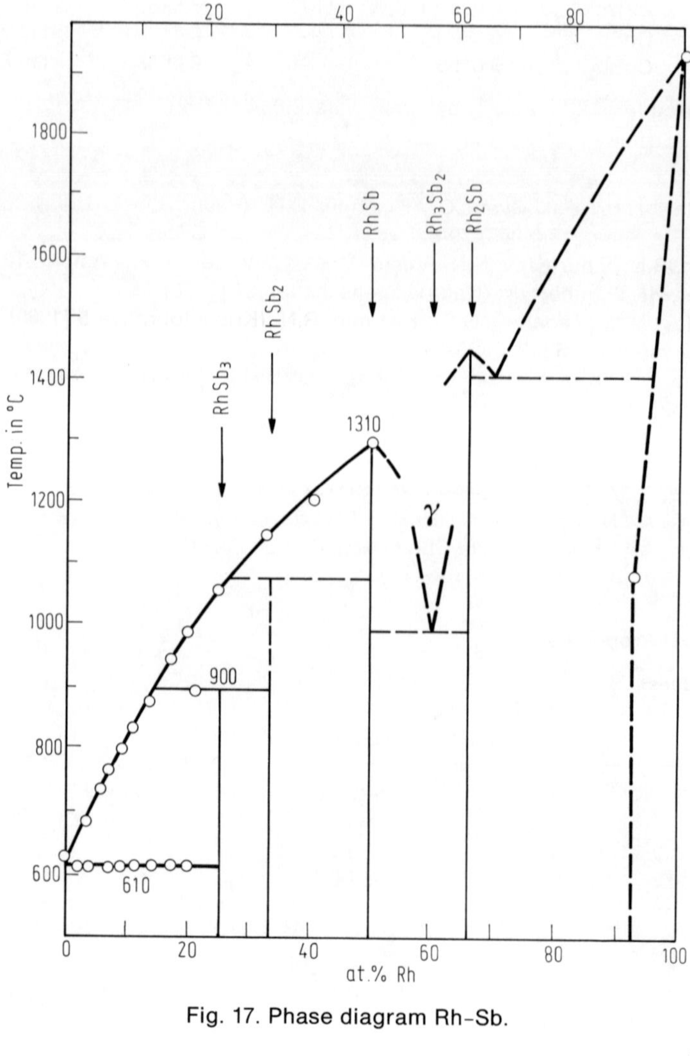

Fig. 17. Phase diagram Rh–Sb.

Crystallography. The following table summarizes the information available.

compound	type	a	b in Å	c	β	Ref.
RhSb	B31 (MnP)	6.333 a)	5.952 a)	3.876 a)	–	[3]
$RhSb_3$	$CoAs_3$	9.230 a)	–	–	–	[1]
$RhSb_3$		9.229	–	–	–	[4]
$RhSb_3$	$CoAs_3$	9.2533_5	–	–	–	[9]
$RhSb_3$	$CoAs_3$	9.2322_6	–	–	–	[10]
$RhSb_2$	$IrSb_2$	6.6 ± 0.2	6.4 ± 0.2	6.7 ± 0.2	$117 \pm 1°$ c)	[1]
$RhSb_2$	P21/c	6.57	6.52	6.66	116.08° c)	[4]
$RhSb_2$	P21/c	6.57 ± 0.02	6.52 ± 0.02	6.66 ± 0.02	116.9° c)	[5]
$RhSb_2$	P21/c	6.57	6.52	6.66	116.0° c)	[2]
$RhSb_2$	P21/c	6.604 ± 0.005	6.557 b)	6.668 b)	116.08° c)	[6]
$RhSb_2$	$CoSb_2$	6.06	6.08	6.15	114.7° c)	[7]
$RhSb_2$	$CoSb_2$	6.6156_4	6.5596_4	6.6858_3	116.8216°	[8]

a) converted from kX units; b) accuracy also ± 0.005; c) ± 0.002 Å

References:

[1] Zhuravlev, N.N.; Pak Gvan, O.; Kuz'min, R.N. (Vestn. Mosk. Univ. Ser. Mat. Mekh. Astron. Fiz. Khim. **13** [1958] 79/82).
[2] Kuz'min, R.N.; Zhuravlev, N.N. (Vestn. Mosk. Univ. Ser. III Fiz. Astron. **18** [1963] 9/14).
[3] Pfisterer, H.; Schubert, K. (Naturwissenschaften **37** [1950] 112/3).
[4] Zhuravlev, N.N.; Zhdanov, G.S.; Kuz'min, R.N. (Kristallografiya **5** [1960] 553/62; Soviet Phys.-Crystallogr. **5** [1960] 532/9).
[5] Zhdanov, G.S.; Kuz'min, R.N. (Kristallografiya **6** [1961] 872/81; Soviet Phys.-Crystallogr. **6** [1961/62] 704/11).
[6] Quesnel, J.C.; Heyding, R.D. (Can. J. Chem. **40** [1962] 814/8).
[7] Hulliger, F. (Phys. Letters **4** [1963] 282/3).
[8] Kjekshus, A. (Acta Chem. Scand. **25** [1971] 411/22).
[9] Kjekshus, A.; Nicholson, D.G.; Rakke, T. (Acta Chem. Scand. **27** [1973] 1315/20).
[10] Kjekshus, A.; Rakke, T. (Acta Chem. Scand. A **28** [1974] 99/103).

Other Physical Properties

Density, Hardness

at% Rh	compound	D in g/cm^3	H_v in kg/mm^2	Ref.
25	$RhSb_3$	7.5 (7.96)	300 (200 to 350 °C)	[1]
33.3	$RhSb_2$	8.9 (9.0)	650 (550 to 710 °C)	[1]
33.3	$RhSb_2$	8.9 (9.14)	–	[2]
33.3	$RhSb_2$	8.85 (8.89)	–	[3]
33.3	$RhSb_2$	9.0 (9.0)	–	[4]
50	RhSb	(10.27)	350 (320 to 420 °C)	[1]

Magnetic Susceptibility. $RhSb_2$ was found to have a magnetic susceptibility $\chi = -0.36 \times 10^6$ emu/g in the temperature range 80 to 750 K [3].

References:

[1] Zhuravlev, N.N.; Pak Gvan, O.; Kuz'min, R.N. (Vestn. Mosk. Univ. Ser. Mat. Mekh. Astron. Fiz. Khim. **13** [1958] 79/82).
[2] Zhdanov, G.S.; Kuz'min, R.N. (Kristallografiya **6** [1961] 872/81; Soviet Phys.-Crystallogr. **6** [1962] 704/11).
[3] Kjekshus, A. (Acta Chem. Scand. **25** [1971] 411/22).
[4] Zhuravlev, N.N.; Zhdanov, G.S.; Kuz'min, R.N. (Kristallografiya **5** [1960] 553/625; Soviet Phys.-Crystallogr. **5** [1960] 532/9).

2.2 Rh-Sb-As Alloy

RhSbAs has a $CoSb_2$-type structure in connection with the superconducting properties of such compounds; the lattice parameters were found to be a=6.36, b=6.28, c=6.41 Å, β=115.4°. The compound was found to have p-type conduction and a thermoelectric force of ~100 μV/°C at room temperature; the room temperature resistivity was of the order of 10 to 100 Ω·cm. Diffuse reflectance measurements indicated an energy gap of the order of 1 eV. The magnetic susceptibility at room temperature was -75×10^6 cgs units/mol and the same at liquid nitrogen temperature. Some of the structural implications are discussed.

Hulliger, F. (Phys. Letters **4** [1963] 282/3).

3 Alloys with Bismuth

The Rh-Bi System

Phase Diagram

The original diagram determined for this system showed the existence of the compounds $RhBi_4$, $RhBi_2$ and RhBi with suggestions from the thermal analysis results that there might be polymorphic changes in $RhBi_4$ and $RhBi_2$ [1]. The discovery that $RhBi_4$ and $RhBi_2$ were superconducting with critical temperatures of 2.75 and 2.2 K, respectively, stimulated interest in the system [2]. Further investigations particularly on Bi_4Rh showed the value of T_c to be variable and sometimes non-superconducting, resulting in the suggestion that Bi_4Rh existed in three modifications α, β and γ [3]. A following paper proposed that the β form was actually Bi_3Rh with an $NiBi_3$-type structure [4]. Determinations using an accurate high-temperature X-ray camera showed that Bi_4Rh was stable up to the melting point [5]. A paper covering the whole system followed [6]. This work was largely confirmed in a later paper by [7]. The diagram shown in **Fig. 18** has been derived from material in the above papers by [8]; the eutectic is placed at 99.3 wt% bismuth (1.4 at% rhodium) taken from

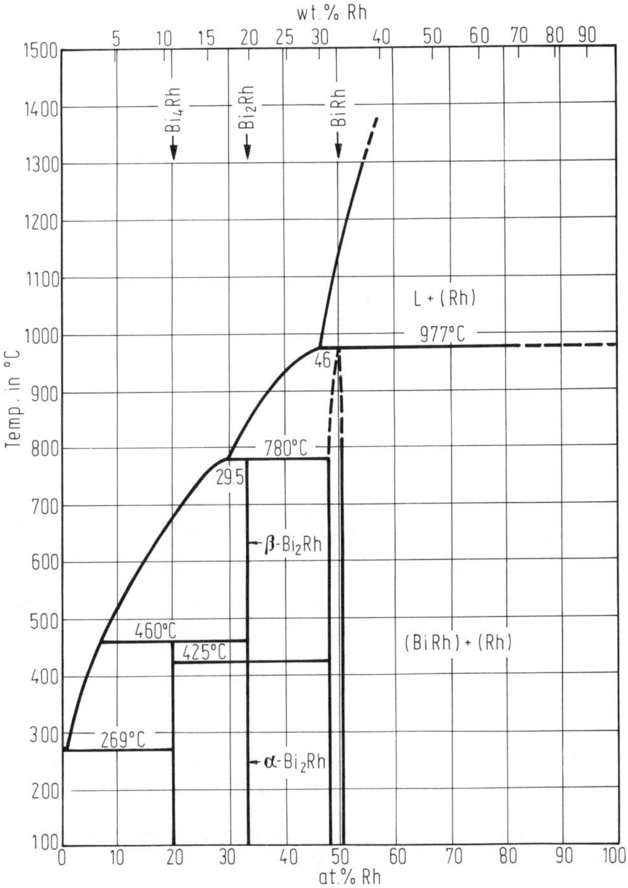

Fig. 18. Phase diagram Rh-Bi.

[9]. A determination of fission products in a liquid-metal nuclear fuel shows a linear increase of the rhodium solubility in liquid bismuth from ~0.6 to 10.2 at% between 283 and 636 °C [10].

References:

[1] Rode, E.I. (Izv. Inst. Izuch. Platiny Drugigkh Blagorodn. Metal. No. 21 [1929]).
[2] Alekseevskii, N. (Zh. Eksperim. Teor. Fiz. **18** [1948] 101/2).
[3] Zhuravlev, N.N.; Zhdanov, G.S. (Izv. Akad. Nauk SSSR Ser. Fiz. **20** [1956] 708/19; Bull. Acad. Sci. USSR Phys. Ser. **20** [1956] 645/9).
[4] Zhuravlev, N.N.; Zhdanov, G.S.; Alekseevskii, N.E. (Vestn. Mosk. Univ. Ser. Mekh. Astron. Fiz. Khim. **14** [1959] 117/27).
[5] Ross, R.G.; Hume-Rothery, W. (J. Less-Common Metals **1** [1959] 304/8).
[6] Ross, R.G.; Hume-Rothery, W. (J. Less-Common Metals **4** [1962] 454/9).
[7] Kuz'min, R.N.; Zhuravlev, N.N.; Zhdanov, G.S. (Zh. Neorg. Khim. **8** [1963] 1906/14; Russ. J. Inorg. Chem. **8** [1963] 991/7).
[8] Shunk, F.A. (Constitution of Binary Alloys, 2nd Suppl., McGraw-Hill, New York 1969, p. 132).
[9] Zhuravlev, N.N.; Zhdanov, G.S. (Zh. Eksperim. Teor. Fiz. **28** [1955] 228/36; Soviet Phys.-JETP **1** [1955] 91/9).
[10] Schweitzer, D.G.; Weeks, J.R. (ASM Trans. Quart. **54** [1961] 185/200).

Crystallography

The table below gives the lattice parameters of the various compounds found.

compound	type	a	b (in Å)	c	β	Ref.
RhBi	NiAs	4.06 a)	—	4.96 a)	—	[1]
α-Bi_2Rh	—	5.9	6.8	7.2	—	[2]
β-Bi_2Rh	—	16.2	7.0	10.5	92.5° c)	[3]
β-Bi_2Rh	a)	7.59±0.01	4.01±0.1	5.53±0.01	—	[13]
BiRh	NiAs	4.094	—	5.663	—	[4]
BiRh	NiAs	4.075	—	5.669	—	[4]
α-Bi_4Rh	b)	14.928	—	—	—	[5]
β-Bi_4Rh	b)	11.4	9.0	4.2	—	[6]
α-Bi_2Rh	b)	6.7	6.8	6.9	117±2° c)	[6]
β-Bi_4Rh	Bi_3Ni c)	11.52	9.03	4.24	—	[7]
β-Bi_3Rh	Bi_3Ni	11.522 b)	9.027	4.24	—	[8]
α-Bi_2Rh	b)	6.96 b)	6.83	7.01	118.2° c)	[9]
RhBi	NiAs	4.0894 a)	—	5.6642	—	[10]
α-Bi_2Rh	$CoSb_2$	6.9207 b)	6.7945	6.9613	117.735° c)	[11]
α-Bi_2Rh	a)	5.9413 b)	6.7945	3.5887	90.379° c)	[11]
$RhBi_4$	bcc	14.9274 b)	—	—	—	[12]

a) in [1]: a, c converted from kX figures ±3 kX; b) in [5]: space group O_h^{10}-Ia3d; b) in [6]: β-Bi_4Rh crystallizes in the rhombic system and α-Bi_2Rh in the monoclinic system.

c) in [7] refers to the change in the authors' nomenclature; what was previously considered to be β-Bi_4Rh was now believed to be Bi_3Rh, isomorphous with Bi_3Ni; b) in [8]: in this paper the figures are actually given as a=9.027±0.006, b=4.24±0.02, c=11.522±0.008 Å; b) in [9] the structure is reported as arsenopyrite-type, space group $P2_1/c$; a) in

[10]: indicates measurement at 20 °C; [b] in [11]: the last decimal places were (4), (4) and (6), β (6).

[a] In [11]: indicates that the parameters are based on a pseudo-marcasite cell structure, last decimal places were (5), (4) and (2), β (2).

The data for the α-, β- and Bi_3Rh structures have been given for the sake of completeness, although their existence under equilibrium conditions has been disproved.

References:

[1] Pfisterer, H.; Schubert, K. (Naturwissenschaften **37** [1950] 112/3).
[2] Zhuralev, N.N.; Zhdanov, G.S. (Zh. Eksperim. Teor. Fiz. **25** [1953] 485).
[3] Glagoleva, V.P.; Zhdanov, G.S. (Tr. Inst. Kristallogr. Akad. Nauk SSSR No. 9 [1953] 311) from Zhdanov, G.S. (Tr. Inst. Kristallogr. Akad. Nauk SSSR No. 10 [1954] 99/116).
[4] Glagoleva, V.P.; Zhdanov, G.S. (Zh. Eksperim. Teor. Fiz. **25** [1953] 248/54).
[5] Glagoleva, V.P.; Zhdanov, G.S. (Zh. Eksperim. Teor. Fiz. **30** [1956] 248/51).
[6] Zhdanov, G.S.; Zhuralev, N.N.; Kuz'min, R.N. (Zh. Neorg. Khim. **3** [1958] 287/97).
[7] Zhuralev, N.N.; Zhdanov, G.S.; Kuz'min, R.N. (Kristallografiya **5** [1960] 553/62; Soviet Phys.-Crystallogr. **5** [1960] 532/9).
[8] Kuz'min, R.N.; Zhdanov, G.S. (Kristallografiya **5** [1960] 869/76; Soviet Phys.-Crystallogr. **6** [1961/62] 830/5).
[9] Zhdanov, G.S.; Kuz'min, R.N. (Kristallografiya **6** [1961] 872/81; Soviet Phys.-Crystallogr. **6** {1961/62] 704/11).
[10] Ross, R.G.; Hume-Rothery, W. (J. Less-Common Metals **1** [1959] 304/8).
[11] Kjekshus, A. (Acta Chem. Scand. **25** [1971] 411/22).
[12] Ross, R.G.; Hume-Rothery, W. (J. Less-Common Metals **4** [1962] 454/9).

Mechanical and Thermal Properties

Density. The densities of RhBi, β-$RhBi_2$, and $RhBi_3$ all lie on a straight line parallel to the straight line for the densities of annealed Bi, $RhBi_4$ and α-$RhBi_2$. At the composition $RhBi_2$ there is a jump in density corresponding to the different modification of the compound; see **Fig. 19**.

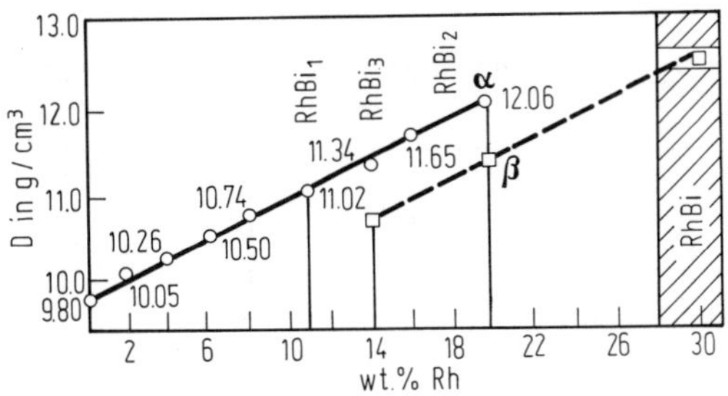

Fig. 19. Density of Rh-Bi alloys.

The actual figures are given in the tables on p. 48; determinations were carried out on powders after annealing at 250 °C for 140 h by immersion in CCl_4 at 20 °C [3].

wt% Rh	phases	D in g/cm^3
0	Bi	9.8
2	Bi + $RhBi_4$	10.05
4	Bi + $RhBi_4$	10.26
6	Bi + $RhBi_4$	10.50
8	Bi + $RhBi_4$	10.74
11	$RhBi_4$	11.02
14	$RhBi_3$	11.34
16	$RhBi_4$ + β-$RhBi_2$	11.65
20	α-$RhBi_2$	12.06
20	β-$RhBi_2$	11.2
~31	RhBi	12.5

In the following table both X-ray and pycnometric figures are given.

compound	$D_{x\text{-}ray}$	D_{pyc}	Ref.
α-Bi_2Rh	11.94	11.86	[4]
α-Bi_2Rh	12.2	12	[1]
β-Bi_2Rh	11.4	11.6	[1]
α-Bi_4Rh	11.0	11.24	[1]
β-Bi_4Rh *)	10.7	–	[1]
γ-Bi_4Rh *)	10.7	–	[1]
BiRh	12.5	12.65	[1]
α-$RhBi_2$	11.79	–	[5]
β-Bi_3Rh *)	11.0	10.7	[6]

*) metastable phase

Hardness. The table below gives the microhardness results obtained on single crystals of the following compounds using a 10 g load except in the case of γ-Bi_4Rh where 5 g was used; the β and γ forms of Bi_4Rh have not been found by later investigators; results are included for reference [1, 2].

compound	mean hardness	range of hardness	Ref.
	in kg/mm^2		
α-Bi_2Rh	230	210 to 260	[1]
β-Bi_2Rh	45	40 to 50	[1]
BiRh	410	370 to 450	[1]
α-Bi_4Rh	105	95 to 115	[1]
α-Bi_4Rh	105	90 to 120	[2]
β-Bi_4Rh	65	60 to 70	[1]
γ-Bi_4Rh	40	30 to 55	[1]

Thermal Expansion. This is shown for the compound Bi_4Rh at temperatures up to ~160 °C in **Fig. 20** due to [7].

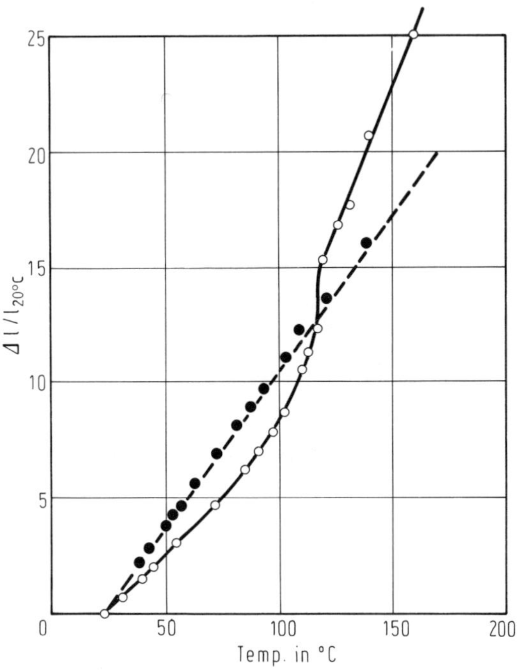

Fig. 20. Thermal expansion of Bi_4Rh.

References:

[1] Zhuralev, N.N.; Zhdanov, G.S.; Glagoleva, V.P. (Zh. Eksperim. Teor. Fiz. **28** [1955] 228/35, Suppl., pp. 235/6).

[2] Alekseevskii, N.E.; Zhdanov, G.S.; Zhuralev, N.N. (Zh. Eksperim. Teor. Fiz. **28** [1955] 237/40; Soviet Phys.-JETP **1** [1955] 99/102).

[3] Kuz'min, R.N.; Zhuralev, N.N.; Zhdanov, G.S. (Zh. Neorg. Khim. **8** [1963] 1906/14; Russ. J. Inorg. Chem. **8** [1963] 991/7).

[4] Kjekshus, A. (Acta Chem. Scand. **25** [1971] 411/22).

[5] Zhdanov, G.S.; Kuz'min, R.N. (Kristallografiya **6** [1961] 872/81; Soviet Phys.-Crystallogr. **6** [1962] 704/11).

[6] Kuz'min, R.N.; Zhdanov, G.S. (Kristallografiya **5** [1960] 869/76; Soviet Phys.-Crystallogr. **6** [1961/62] 830/5).

[7] Alekseevskii, N.E.; Brandt, N.B.; Kostina, T.I. (Izv. Akad. Nauk SSSR Ser. Fiz. **16** [1952] 233/63).

Electrical and Magnetic Properties

Electric Resistivity. This is shown plotted against temperature for the compounds RhBi and $RhBi_4$ in **Fig. 21**, p. 50, and **Fig. 22**, p. 50 [1].

Magnetic Moment. This is shown plotted against magnetic field strength for RhBi and $RhBi_4$ in **Fig. 23**, p. 51, and **Fig. 24**, p. 51 [1].

Superconductivity. The three superconducting phases and their transition temperatures are: RhBi 2.2 to 2.06 K [3], β-$RhBi_3$ (orthorhombic) 3.2 K, γ-$RhBi_4$ (hexagonal) 2.7 K; α-

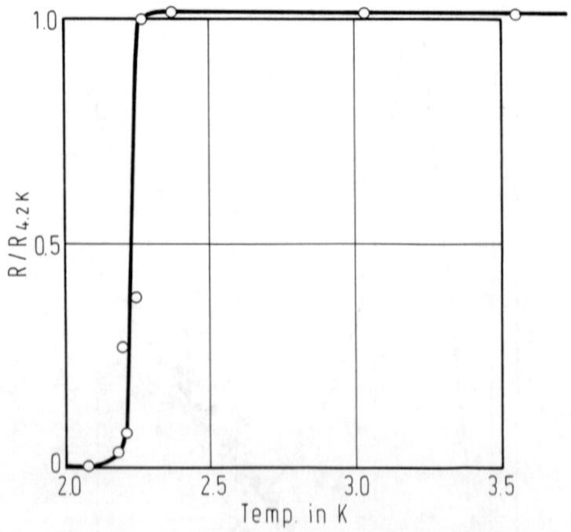

Fig. 21. Resistance of RhBi as a function of temperature.

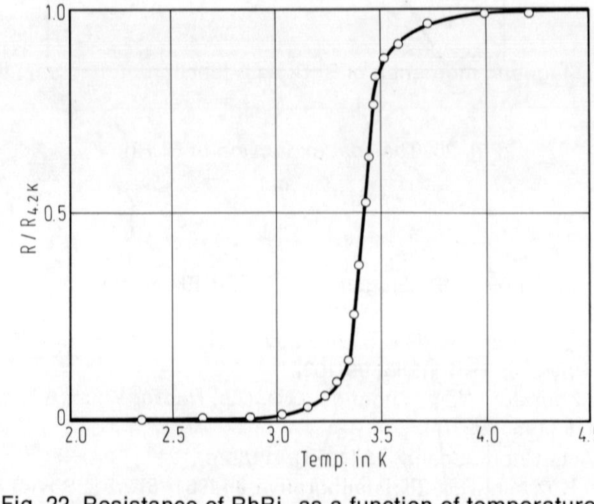

Fig. 22. Resistance of $RhBi_4$ as a function of temperature.

and β-$RhBi_2$ are not superconducting above 1.34 or 1.27 K [4]. α-$RhBi_4$ is not superconducting above 0.1 K [5]. The existence of the $RhBi_4$ modifications referred to would seem to be unlikely from the high-temperature X-ray work carried out by [6].

Magnetic Susceptibility. This was measured for the compounds α- and β-$RhBi_2$ between 90 and 1000 K by the Faraday method (maximum field ≈8000 Oe) using 40 to 130 mg samples, with the following results in emu/g [2]:

α-$RhBi_2$: $\chi = -0.29_5 - 0.00005\,T$; range 80 to 650 K
β-$RhBi_2$: $\chi = -0.17$ range 850 to 1000 K

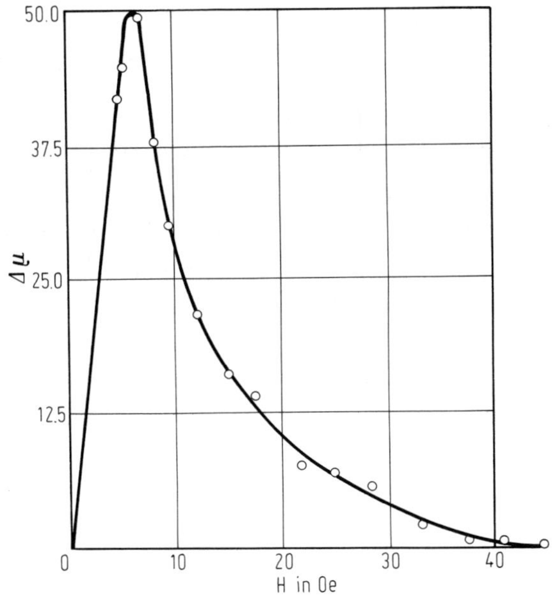

Fig. 23. Magnetic moment μ of RhBi as a function of magnetic field H.

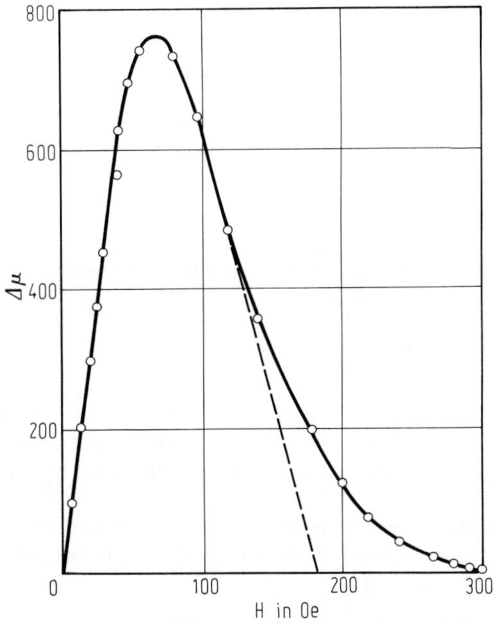

Fig. 24. Magnetic moment μ of $RhBi_4$ as a function of applied magnetic field H.

References:

[1] Alekseevskii, N.E.; Brandt, N.B.; Kostina, T.I. (Izv. Akad. Nauk SSSR Ser. Fiz. **16** [1952] 233/63).
[2] Kjekshus, A. (Acta Chem. Scand. **25** [1971] 411/22).
[3] Glagoleva, V.P.; Zhdanov, G.S. (Zh. Eksperim. Teor. Fiz. **25** [1953] 248/54).
[4] Alekseevskii, N.E.; Zhdanov, G.S.; Zhuralev, N.N. (Zh. Eksperim. Teor. Fiz. **28** [1955] 237/40; Soviet Phys.-JETP **1** [1955] 99/102).
[5] Alekseevskii, N.E.; Gaidukov, Yu.P. (Zh. Eksperim. Teor. Fiz. **25** [1953] 383/4).
[6] Ross, R.G.; Hume-Rothery, W. (J. Less-Common Metals **1** [1959] 304/8).

4 Alloys with Alkali Metals

4.1 Alloys with Lithium

The Rh-Li System

Phase Diagram. A partial phase diagram has been suggested for the system by [1], based on material from [2 to 4]. This is shown in **Fig. 25** [1]. The existence of two intermediate phases has been shown, and their parameters determined by X-ray and neutron diffraction. References [2] and [3] suggest two intermediate phases and a system analogous to that of the iridium-rhodium system; however, only one intermediate phase was found by [4], suggesting a double eutectic system at 8 at% rhodium and 180.5 °C and 76 at% rhodium at 1630 °C; there is negligible mutual solid solubility of lithium and rhodium [5].

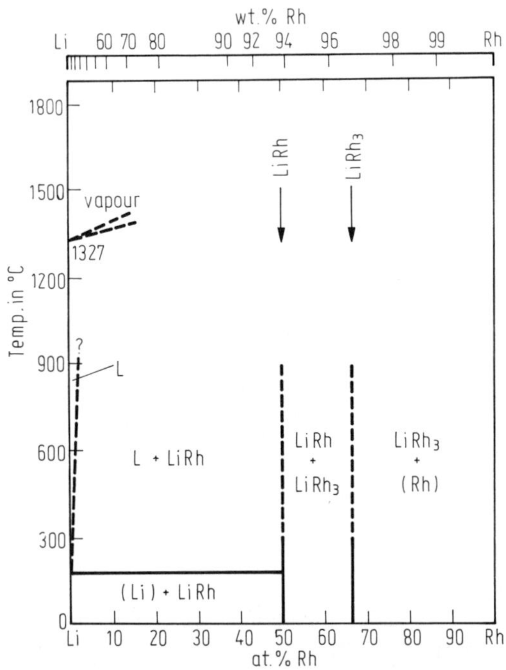

Fig. 25. Tentative phase diagram Li-Rh.

Preparation. The powder specimens of $LiRh_3$ were probably made in the same way as $LiIr_3$ material used in the same research, namely by alloying rhodium powder and pieces of lithium by heating in welded molybdenum capsules for 7 d at 800 °C; the capsules were opened in a glove box under pure argon, and the contents crushed to powder [2]. The samples of LiRh used in the third reference were prepared by melting together 0.002 inch rhodium foil with small pieces of lithium in an iron crucible placed in a lidded Vycor tube heated at 750 to 800 °C for 1 h under dry argon [3]. Some of the difficulties experienced in investigating these and similar alloys are due to the high melting points of the platinum metals and the low boiling points of lithium and other alkali metals. This is described in a review of such alloys by [5]. Rhodium and lithium are said to react at ~300 °C in a two-stage reaction. When the rhodium concentration is ≧50 at%, the solidification of free lithium is not detected on cooling, and it was concluded that complete reaction had occurred [5].

Crystallography. The available data is given below.

compound	type	a	b in Å	c	Ref.
$LiRh_3$	Ir_3Li	2.6586	8.6023	4.6570	[2] a)
LiRh	hexagonal	2.649 ±0.003	—	4.357 ±0.002	[3] b)
LiRh	LiRh	2.649	—	4.359	[4]

a) X-ray diffraction determinations; b) X-ray and neutron diffraction determinations.

References:

[1] Moffatt (Binary Alloy Phase Diagrams, Genium Publ., New York 1976).
[2] Donkersloot, H.C.; Van Vucht, J.H.N. (J. Less-Common Metals **50** [1976] 279/82).
[3] Sidhu, S.S.; Anderson, K.D.; Zauberis, D.D. (Acta Crystallogr. **18** [1965] 906/7).
[4] Wheat, H.G.; Cheng, C-Y.; Bayuzick, R.J.; Sullivan, R.W.; Magee, C.B. (J. Less-Common Metals **58** [1978] P13/P29).
[5] Loebich, O.; Raub, C.J. (Platinum Metals Rev. **25** [1981] 113/20).

4.2 Alloys with Potassium

No diagram has been found on this system. Work carried out on the system using alloys prepared by reaction between rhodium powder and molten potassium in stainless steel or tantalum tubes sealed by flattening the ends, suggests from X-ray diffraction analysis and differential thermal analysis that there was no appreciable solid solubility of rhodium in potassium, and that there are no intermediate phases in the system [1].

Superconductivity. None of the platinum-group metal alloys with the alkali metals is superconducting above 2 K [2].

References:

[1] Loebich, O.; Raub, C.J. (Platinum Metals Rev. **25** [1981] 113/20).
[2] Raub, C.J. (Platinum Metals Rev. **28** [1984] 63/75).

4.3 Alloys with Sodium, Rubidium, and Caesium

There is little information on these systems; the platinum metals, apart from platinum itself, are said to react with sodium above 700 °C, and at higher temperatures the high pressure causes leaks from reaction capsules indicating no reduction in vapour pressure; it was concluded that there was no appreciable solubility of rhodium in sodium. Although caesium and rubidium were not examined, the conclusions reached for sodium and potassium alloys are also probably true for caesium and rubidium.

Loebich, O.; Raub, C.J. (Platinum Metals Rev. **25** [1981] 113/20).

5 Alloys with Alkaline Earth Metals

5.1 Alloys with Beryllium

The Rh-Be System

No diagram has been found for this system. Optical and X-ray examination of a 85.1 wt% alloy showed the existence of the compound Be_2Rh with a powder pattern similar to Be_2Ir [1]. This has been confirmed more recently in an investigation of alloys containing up to 14.3 wt% beryllium, using X-ray, metallography and hardness tests, although the authors give the compound as $RhBe_{\sim 2}$; the compound RhBe was also confirmed, and a eutectic near 13.5 wt% beryllium is suggested. The alloys used were prepared either by vacuum diffusion in a beryllia crucible at 1500 to 1700 K, or by arc-melting under an argon pressure of ~300 Torr. The weight of beryllium in the samples was determined to an accuracy of 1×10^{-4} g by weighing the samples before and after alloying [2]. A melt of 10 wt% rhodium (1 at%) slowly cooled from the melt was found to be two-phased with a eutectic network [3]. Other compounds reported are $RhBe_{6.6}$ and Rh_2Be_{17} [4, 5]. A large number of compounds has been examined for superconductivity, none of which was found to have this property; the following were claimed to be prepared and examined: BeRh, Be_2Rh, $Be_{4.4}Rh$, Be_5Rh, $Be_{6.5}Rh$, $Be_{13}Rh$, $Be_{22}Rh$, and $Be_{49}Rh$ [6].

Crystallography. The available figures are given in the table below.

compound	type	a (in Å)	c (in Å)	Ref.
RhBe	CsCl	2.739(7)	–	[2]
$RhBe_{6.6}$	D2d	4.191 a)	10.886	[4]
$Rh_{22}Be_{17}$	Ru_2Be_{17}	4.190 b)	10.88	[5]

a) accuracy a = ±0.001, c = ±0.003 Å; hexagonal with probable space group P6m2; b) accuracy a = ±0.004, c = ±0.01 Å; space group P6/mm; structural similarities were found between the Ru_2Be_{17}-type structure and the known Th_2Zn_{17} and Th_2Ni_{17} types [5].

Density, Hardness. The compound $RhBe_{\sim 2}$ had a density of 6.8 ± 2.5 g/cm^3 [2].

The optically isotropic compound formed in the diffusion layer had a Vickers VPN hardness of 1400 ± 100 kg/mm^2; this was bounded on one side by rhodium of hardness 270 ± 20 kg/mm^2 and by another isotropic phase of hardness 950 ± 50 kg/mm^2 [2].

References:

[1] Misch, L. (Metallwirtsch. Metallwiss. Metalltech. **15** [1936] 163/6).

[2] Kruglykh, A.A.; Matyushenko, N.N.; Tikhinskii, G.F. (Ukr. Fiz. Zh. [Ukr. Ed.] **13** [1968] 1107/10; Ukr. Phys. J. **13** [1968] 793/5).

[3] Kaufmann, A.R.; Gordon, P.; Lillie, D.W. (Trans. AIME **42** [1950] 801).

[4] Johnson, Q.; Smith, S.G.; Krikorian, O.H.; Sands, D.E. (Acta Crystallogr. B **26** [1970] 109/13).

[5] Verkhorobin, L.F.; Kovtun, G.P.; Kruglykh, A.A.; Matyushenko, N.N.; Pugachev, N.S.; Tikhinskii, G.F. (Izv. Akad. Nauk SSSR Metall. **1971** No. 6, pp. 168/71; Russ. Metal. **1971** No. 6, pp. 121/2).

[6] Alekseevskii, N.E.; Zakosarenko, V.M. (Dokl. Akad. Nauk SSSR **208** [1973] 303/6; Soviet Phys. Dokl. **18** [1973] 45/7).

5.2 Alloys with Magnesium

The Rh-Mg System

As far as is known, no binary phase diagram is in existence; a good analysis of the available information is given in [1]. The intermediate phases RhMg, $Rh_{2-x}Mg_5$ and $RhMg_{\sim5}(Rh_7Mg_{44})$ have been reported by [2 to 5]. Lattice parameter measurements made on an alloy containing 0.24 at% rhodium showed a small solubility of rhodium in magnesium, this reducing the lattice parameter [6]. The available lattice parameter figures are given in the following table, the measurements are made at room temperature.

at% Rh	compound	type	a	c in Å	space group	Ref.
50	RhMg	CsCl (B2)	3.099	—	Pm3m	[2]
13.7	Rh_7Mg_{44} *)	Rh_7Mg_{44}	2.0148	—	F43m	[11]
28.57	$Rh_{2-x}Mg_5$	Al_5Co_2	8.554	8.028	—	[3]
			8.515	8.016	—	[3]

*) $Mg_{\sim5}Rh$ with a cubic structure was reported by [4]. Later papers expressed the compound as $Mg_{44}Rh_7$ [5, 7].

Mg_6Rh was said to have an Na_6Tl-type of structure by [8]. An alternative structure described as the packing of icosahedra, each centred by a rhodium, was proposed by [5]. Another structure suggested describes the compounds as built up of tetrahedra and octahedra [9]. This latter structure is considered the most probable by [10].

References:

[1] Nayeb-Hashemi, A.A.; Clark, J.B. (Bull. Alloy Phase Diagrams **8** [1987] 117/8).
[2] Compton, V.B. (Acta Crystallogr. **11** [1958] 446).
[3] Ferro, R. (Atti Accad. Nazl. Lincei Classe Sci. Fis. Mat. Nat. Rend. [8] **29** [1960] 70/3).
[4] Westin, L.; Edshammar, L.E. (Acta Chem. Scand. **25** [1971] 1480/1).
[5] Westin, L.; Edshammar, L.E. (Acta Chem. Scand. **26** [1972] 3619/26).
[6] Busk, R.S. (Trans. AIME **188** [1950] 1460/4).
[7] Westin, L.; Edshammar, L.E. (Chem. Scr. **3** [1973] 15/22).
[8] Samson, S. (Acta Crystallogr. B **28** [1972] 936/42).
[9] Andersson, S. (Acta Crystallogr. A **34** [1978] 833/5).
[10] Chabot, B.; Cenuzal, K.; Parthe, E. (Acta Crystallogr. B **36** [1980] 2202/5).
[11] Westin, L. (Chem. Scr. **1** [1971] 127/35).

5.3 Alloys with Calcium

No phase diagram exists for the system. A Laves phase $CaRh_2$ with an $MgCu_2$-type structure has been identified, a=7.525 Å [1].

$CaRh_2$ was found to be superconducting with a transition temperature T_c=6.4 K [2].

References:

[1] Wood, E.A.; Compton, V.A. (Acta Crystallogr. **11** [1958] 429/33).
[2] Matthias, B.T.; Corenzwit, E. (Phys. Rev. [2] **107** [1957] 1558).

5.4 Alloys with Strontium

The Rh–Sr System

No phase diagram is available. The existence of an $MgCu_2$-type Laves phase has been established with composition $SrRh_2$; the alloy was prepared by sintering under argon. The lattice parameter was found to be a=7.695 Å [1]. This phase has been confirmed in a later research, but the lattice parameter was found to be a=7.706 Å; in this case the alloy had been prepared by melting in an iron crucible under helium [2]. In the former work, the presence of an Sr_5Rh phase was sought, in analogy with those in the palladium and platinum systems, but this was not found; however, an unknown phase was revealed in which the rhodium : strontium ratio was ~5 : 1, after heating a specimen for 5 h at 1600 °C. This had a face-centred cubic structure with a lattice parameter a=10.0 Å [1].

The compound $SrRh_2$ was found to be superconducting, with a transition temperature T_c=6.2 K [3].

References:

[1] von Heumann, T.; Kniepmeyer, M. (Z. Anorg. Allgem. Chem. **290** [1957] 191/204).
[2] Wood, E.A.; Compton, V.B. (Acta Crystallogr. **11** [1958] 429/33).
[3] Matthias, B.T.; Corenzwit, E. (Phys. Rev. [2] **107** [1957] 1558).

5.5 Alloys with Barium

The Rh–Ba System

No phase diagram is known. An $MgCu_2$-type Laves phase has been identified with lattice parameter a=7.852±0.005 Å; the material was prepared by induction melting in an iron crucible under helium [1].

The compound $BaRh_2$ was found to be superconducting with a transition temperature T_c=6.0 K [2].

References:

[1] Wood, E.A.; Compton, V.B. (Acta Crystallogr. **11** [1958] 429/33).
[2] Matthias, B.T.; Corenzwit, E. (Phys. Rev. [2] **107** [1957] 1558).

6 Alloys with Zinc, Cadmium and Mercury

6.1 Alloys with Zinc and Cadmium

No phase diagram has been determined for either system. It is assumed that compounds of the γ-brass type Rh_5Zn_{21} and Rh_5Cd_{21} exist, based on analogy with the palladium and platinum systems [1 to 3].

The temperature dependence of the quadrupole interaction of ^{100}Rh in a zinc lattice is reported by [4].

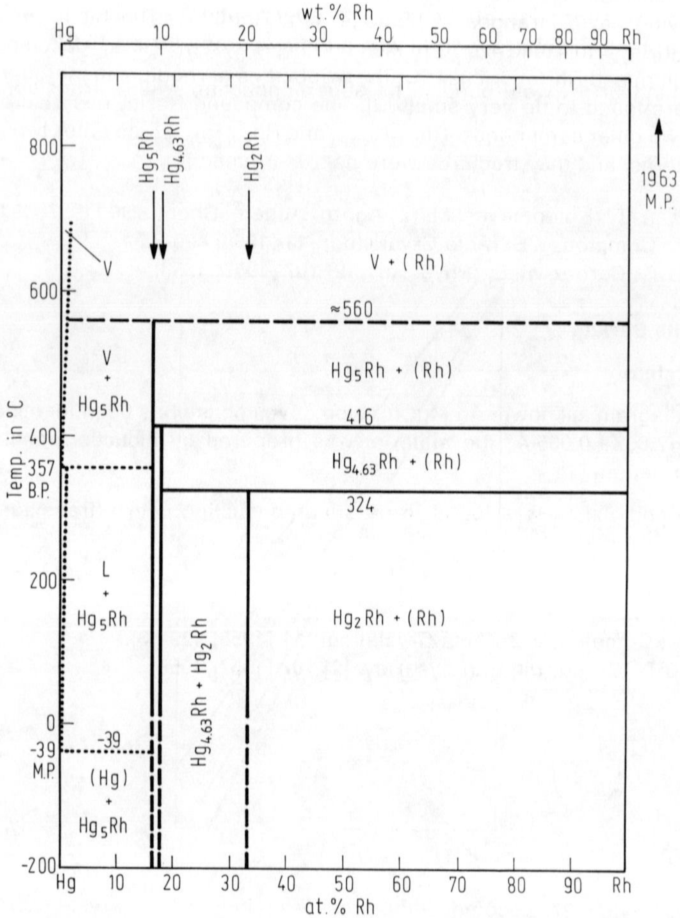

Fig. 26. Phase diagram Rh-Hg. L=liquid, V=vapour, M.P.=melting point, B.P.=boiling point.

References:

[1] Hansen, M.; Anderko, K. (Constitution of Binary Alloys, 2nd Ed., New York-Toronto-London 1958).

[2] Westgren, A.; Ekman, W. (Arkiv Kemi Mineral. Geol. B **10** No. 11 [1931] 1/6).

[3] Savitskii, E.; Polyakova, V.N.; Gorina, N.; Roshan, N.R. (The Physical Metallurgy of the Platinum Metals, MIR Publ., Moscow 1978, p. 183).
[4] Krien, K.; Soares, J.C.; Bibiloni, A.G.; Vianden, R.; Hanser, A. (Z. Physik **266** [1974] 195/9).

6.2 Alloys with Mercury

Phase Diagram. A partial phase diagram is shown in **Fig. 26** due to [1]. It is based on work carried out in the temperature range 0 to 500 °C using vapour pressure and X-ray diffraction measurements. Three intermediate phases were found: $RhHg_5$, $RhHg_{4.63}$ and $RhHg_2$, all having narrow ranges of homogeneity. $RhHg_5$ was found to be stable to 500 to 600 °C and $RhHg_{4.63}$ to 410 to 420 °C in the condensed system. $RhHg_2$ decomposed peritectically into rhodium and $RhHg_{4.3}$ at 324 °C; the solubilities of rhodium in mercury and mercury in rhodium were found to be very small [2]. The compound $RhHg_2$ has also been reported together with two other compounds, $Rh_{0.17}Hg_{0.83}$ and $Rh_{0.18}Hg_{0.82}$; the latter two gave complex diffraction patterns, and the structures were not determined [3].

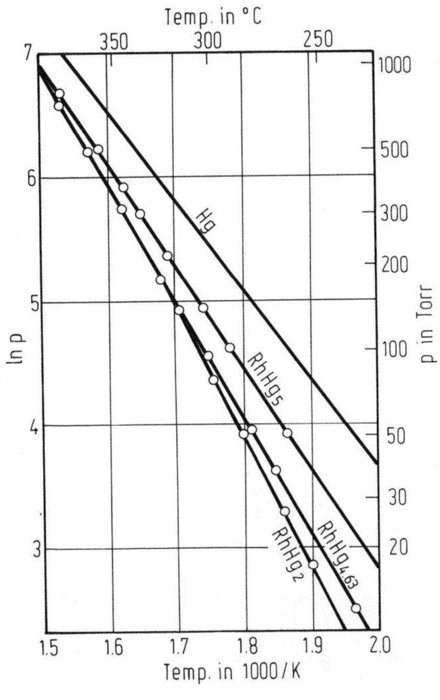

Fig. 27. Decomposition pressure p in the Rh-Hg system.

Crystallography. The data are given below.

compound	type	a	c	Ref.
		in Å		
$RhHg_2$	$PtHg_2$	4.55_1	2.99	[2]
$RhHg_2$	$PtHg_2$	4.55_1	2.99_8	[3]

Thermal Properties. The heat of formation (ΔH in kcal/mol) was determined as follows.

$RhHg_5$: ΔH = −25.8; $RhHg_{4.63}$: ΔH = −25.0; $RhHg_2$: ΔH = −13.6 [2].

The temperature dependence of decomposition vapour pressure for $RhHg_5$, $RhHg_{4.63}$ and $RhHg_2$ is shown in **Fig. 27**, p. 59 [2].

References:

[1] Moffatt (Binary Alloy Phase Diagrams, Genium Publ., New York 1976).
[2] Jangg, G.; Kirchmayr, H.R.; Mathis, H.B. (Z. Metallk. **58** [1967] 724/6).
[3] Ettmayer, P.; Mathis, B. (Monatsh. Chem. **98** [1967] 505/6).

7 Alloys with Aluminium

Phase Diagram, Crystallography

No equilibrium diagram has been determined for this system; an extensive examination of it using metallographic, X-ray powder diffraction photographs and density measurements was carried out revealing interesting features of the system. The intermediate phases Rh_2Al_9, $RhAl_3$, Rh_2Al_5 and RhAl were found during the research using 47 alloys from 0.3 to 83.3 at% Rh. Eutectic reactions at both ends of the system were found, at 0.6 to 0.7 at% rhodium at the aluminium-rich end and at ~75 at% rhodium at the other end; the homogeneity range of the compound RhAl was between 45.2 and 52.2 at% rhodium [1]. The compound RhAl had previously been found to be cubic with a CsCl-type structure and a=2.98 kX (2.99 Å) by [2]. Rh_3Al was said not to be stable above 550 °C, at that temperature, RhAl being in equilibrium with rhodium [3]. Another investigation suggests that the aluminium-rich eutectic occurs at 0.29 at% rhodium and 657.5 °C [4]. A further compound $RhAl_4$ which has a complex structure isomorphous with $PdAl_4$ has been reported by [5]. Specimens of alloys of compositions in the range $RhAl_{1.0-2.5}$ heat-treated at 850 °C were observed to have two phases of cubic CsCl-type with a=2.970±0.002 Å, and a hexagonal phase with a=7.983, c=7.854 Å at the composition $RhAl_{2.5}$; no change in these parameters was observed when comparing powder photographs in the compositional range $RhAl_{2.0-3.0}$. The patterns of arc-melted material of composition $RhAl_{\sim 2.5}$ were rather different, and could be indexed by assuming a cubic cell giving a=15.35 Å in the region $RhAl_{1.5-<2.5}$ but a=15.32 Å at compositions close to Rh_2Al_5 [6].

The relevant information is given below.

at% Rh	compound	a	b in Å	c	β in °	type	Ref.
—	Rh_2Al_9	6.352	6.428	8.721	94.81	Co_2Al_9	[7]
29.1	$\sim Rh_2Al_5$	7.889	—	7.853	—		[1]
—	Rh_2Al_5	7.893	—	7.854	—	Co_2Al_5	[6]
50.0	RhAl	2.980	—	—	—	CsCl	[1]
42.7	~RhAl	2.968	—	—	—	CsCl	[1]
50.0	RhAl	2.99	—	—	—	CsCl	[2]

Melt-Quenched Material. Material containing 2 wt% rhodium rapidly quenched from the liquid state resulted in the formation of a homogeneous solid solution showing a substantial increase in solubility over the equilibrium value. This solid solution was stable up to 640 K, but on decomposition at higher temperatures formed a mixture of aluminium solid solution and another phase of monoclinic structure with a=16.36, b=8.05, c=12.79 Å, β=107.77°, of composition of Rh_4Al_{13}; the equilibrium constitution of the alloy was shown to be a mixture of this compound and the aluminium solid solution [8].

References:

[1] Ferro, R.; Rambaldi, G.; Capelli, R. (Atti Accad. Nazl. Lincei Classe Sci. Fis. Mat. Nat. Rend. [8] **36** [1964] 491/7).

[2] Schubert, K.; Breimer, H.; Burkhardt, W.; Günzel, E.; Haufler, R.; Lukas, H.L.; Vetter, H.; Wegst, J.; Wilkens, M. (Naturwissenschaften **44** [1957] 229/30).

[3] Schubert, K.; Lukas, H.L.; Meissner, H.G.; Bhan, S. (Z. Metallk. **50** [1959] 534/40).

[4] Wright, E.H.; Willey, L.A. (Tech. Paper Alcoa Res. Lab. No. 15 [1960] 1/46).

[5] Magnéli, A.; Edshammar, L.; Dagerhamn, T. (AD-426927 [1963] 46).

[6] Edshammar, L.E. (Acta Chem. Scand. **21** [1967] 647/51).
[7] Edshammar, L.E. (Acta Chem. Scand. **22** [1968] 2822/6).
[8] Chaudhury, Z.A.; Suryanarayana, C. (J. Mater. Sci. **18** [1983] 3011/22).

Miscellaneous Physical Properties

Density. Fig. 28 shows the density in g/cm³ plotted against at% rhodium [1].

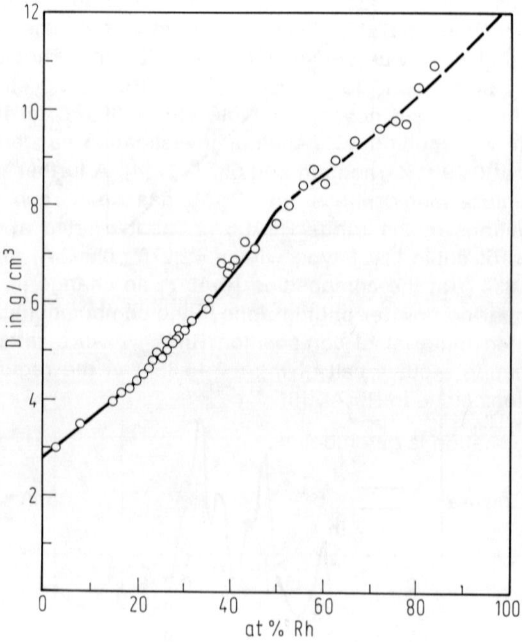

Fig. 28. Density of Rh-Al alloys.

Surface Tension. The surface tension of the whole range of alloys was measured at 2200 °C and an equation was developed in terms of the ideal properties of the surface layer which adequately described the results (see figure in original paper) [2].

Magnetic Susceptibility. By magnetic measurements on the compound RhAl, the atomic susceptibility χ_A was defined by the equation $\chi_A = \chi_0 + C/T$. The constants were determined as $\chi_0 = -4.56 \times 10^{-6}$ cm³/mol, $C = 1750 \times 10^{-6}$ cm³·°C/mol. The measured results are shown in **Fig. 29**; the alloy shows a weak paramagnetism at room temperature, which climbs steeply at lower temperatures. The product $\chi_A \cdot T$ plotted against temperature gave a straight line [3].

Nuclear Magnetic Resonance. Measurements on RhAl of the Al^{27} resonance have been made and the Knight shift, nuclear spin-lattice relaxation rate, linewidth, and intensity determined at temperatures in the range 4.4 to 300 K and at 8 and 12 MHz. The Knight shift and T_1 results implied that the s-character of the conduction electrons is small at Al sites; this is shown to be in accord with the Engel-Brewer model of intermediate phases in alloy systems [4].

Electronic Structure. The mixed states in RhAl were examined by photoemission spectroscopy and band structure calculations. The bonding states revealed a strong mixing

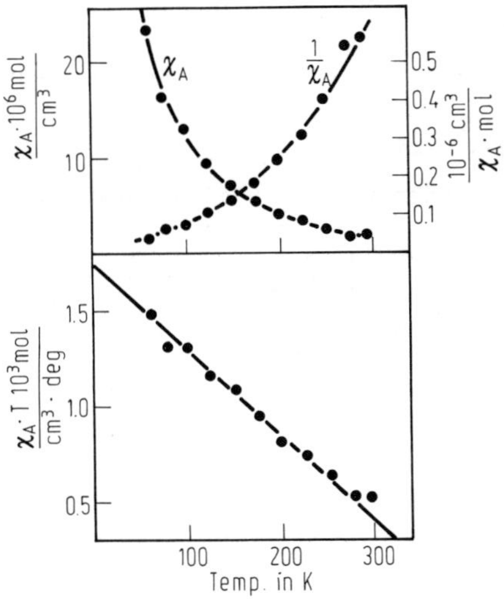

Fig. 29. Magnetic susceptibility of RhAl.

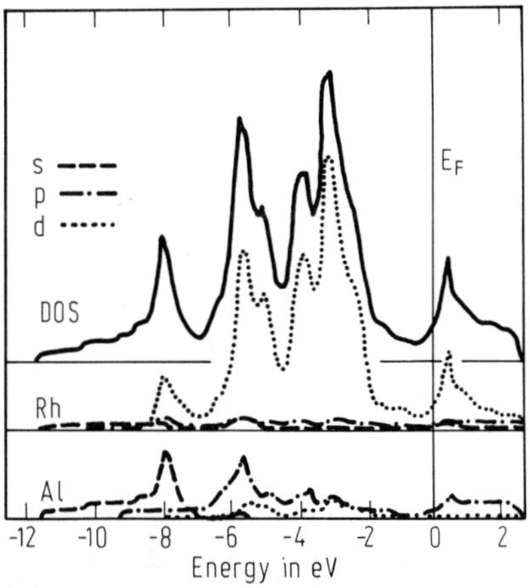

Fig. 30. Energy band structure and density of states for RhAl.

of Rh 4 d and s states of Al. The calculated binding energies agreed well with the experimental results. **Fig. 30** shows the energy band structure and density of states for RhAl [5].

References:

[1] Ferro, R.; Rambaldi, G.; Capelli, R. (Atti Accad. Nazl. Lincei Classe Sci. Fis. Mat. Nat. Rend. [8] **36** [1964] 491/7).

[2] Mit'ko, M.M.; Dubinin, E.L.; Timofeev, A.I.; Chegodaev, A.I. (Soviet Non-Ferrous Metals Res. **4** [1979] 210/4).

[3] Höhl, M. (Ann. Physik [7] **19** [1967] 15/21).

[4] Spokas, J.J.; Sowers, C.H.; van Ostenburg, D.O. (Phys. Rev. [3] B **1** [1970] 2523/31).

[5] Verbeek, B.H.; Rompa, H.W.A.M.; Larsen, P.K.; Methfessel, M.S.; Mueller, F.M. (Phys. Rev. [3] B **28** [1983] 6774/9).

8 Alloys with Gallium and Indium

8.1 Alloys with Gallium

No phase diagram exists as far as is known. The solubility of rhodium in molten gallium has been measured and the following equation was said to represent the variation with temperature: log C = A − B/T, where C is the concentration in at%, T is the absolute temperature and A and B are constants having these values for rhodium: A = 1.04, B = 2250; the compound $RhGa_6$ was said to be in equilibrium with the liquid phase at 673 and 773 K. The heat of fusion ΔH_m and molar entropy of solution ΔS were calculated as 5.1 kcal/g-atom and 2.30 kcal/g-atom/°C, respectively [1]. The compound $RhGa_3$ was first reported to have a C16 $CuAl_2$-type structure with a = 6.475 kX (6.49 Å) and c = 6.55 kX (6.57 Å) by [2]. This compound is in equilibrium with two phases $RhGa_{\sim 2}$ and $RhGa_{\sim 6}$ [3]. RhGa is reported as having a CsCl-type structure with a = 3.01 Å [4]; the same compound has a = 3.20 Å according to [5]. The mixed states in RhGa were studied by photoemission spectroscopy and band structure calculations, and the bonding states revealed a strong mixing of Rh 4d and s states of gallium; **Fig. 31** shows the total and l-projected density of states of RhGa. In the course of this work, the lattice parameter was calculated as a = 3.03 Å and determined as 3.0063_8 Å [6].

The compound $Rh_{10}Ga_{17}$ is said to have a structure which can be derived from the $TiSi_2$-type with a = 5.813, c = 47.46 Å [7].

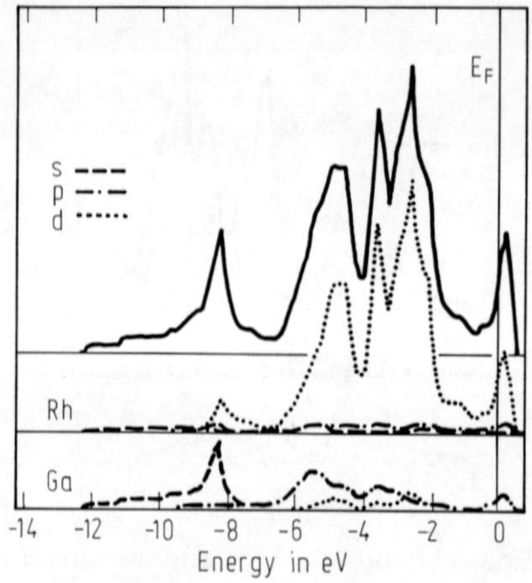

Fig. 31. Total and l-projected density of states for RhGa.

References:

[1] Yatsenko, S.P.; Anikin, Yu.A. (Fiz. Khim. Mekh. Mater. **6** No. 3 [1970] 57/62; Soviet Mater. Sci. **6** [1970] 333/7).

[2] Schubert, K.; Breimer, H.; Gohle, R.; Lukas, H.L.; Meissner, H.G.; Stolz, E. (Naturwissenschaften **45** [1958] 360/1).

[3] Schubert, K.; Lukas, H.L.; Meissner, H.G.; Bhan, S. (Z. Metallk. **50** [1959] 534/40).

[4] Shunk, F.A. (Constitution of Binary Alloys, 2nd Suppl., New York-London-Sydney 1969).

[5] Edshammar, L.E. (Metallovedeniye **1971** 298/305).

[6] Verbeek, B.H.; Rompa, H.W.A.M.; Larsen, P.K.; Methfessel, M.S.; Mueller, F.M. (Phys. Rev. [3] B **28** [1983] 6774/9).

[7] Völlenkle, H.; Wittmann, A.; Nowotny, H. (Monatsh. Chem. **98** [1967] 176/83).

8.2 Alloys with Indium

No phase diagram has been determined; a recent review of the system has been carried out by [1]. The solubility of rhodium in liquid indium at a temperature of 1250 K has been determined under vacuum conditions; the solubility was described by the equation, $\log C = 2.93 - 2780/T$, where C is the concentration in at%, and T the temperature in K [2]. Immiscibility in the molten state was predicted by [3]. Two intermediate phases have been identified: RhIn and $RhIn_3$, the former having a B2 CsCl-type structure with a = 3.19 kX (3.2 Å) and the latter a C16 $CuAl_2$-type structure with a = 7.01 kX (7.01 Å), c = 7.13 kX (7.15 Å) [4, 5]. These two compounds are said to be in equilibrium at temperatures between 200 and 500 °C [6].

$RhIn_3$ is said not to be superconducting at 1.02 K [7]. Mixed states in polycrystalline RhIn have been studied by photoemission spectroscopy using synchrotron radiation and band-structure calculations. The bonding states show a strong mixing of Rh 4d and s states with In. **Fig. 32** shows the total and l-projected density of states [8].

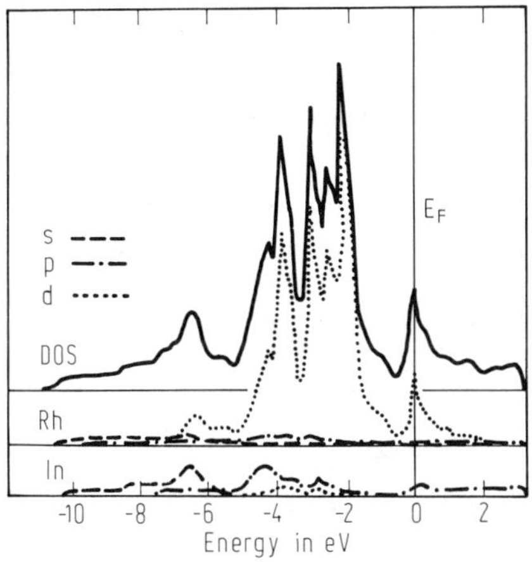

Fig. 32. Density of states for RhIn.

References:

[1] Okamoto, H. (Bull. Alloy Phase Diagrams **9** [1988] 703).
[2] Yatsenko, S.P.; Dieva, E.N. (Zh. Fiz. Khim. **47** [1973] 2948; Russ. J. Phys. Chem. **47** [1973] 1658).
[3] Dasarathy, C. (Trans. AIME **245** [1969] 2015/9).
[4] Schubert, K.; Breimer, H.; Burkhardt, W.; Günzel, E.; Haufler, H.; Lukas, H.L.; Vetter, H.; Wegst, J.; Wilkens, M. (Naturwissenschaften **44** [1957] 229/30).
[5] Schubert, K.; Breimer, H.; Gohle, R.; Lukas, H.L.; Meissner, H.G.; Stolz, E. (Naturwissenschaften **45** [1958] 360/1).
[6] Schubert, K.; Lukas, H.L.; Meissner, H.G.; Bhan, S. (Z. Metallk. **50** [1959] 534/40).
[7] Matthias, B.T.; Geballe, T.H.; Compton, V.B. (Rev. Mod. Phys. **35** [1963] 1/22).
[8] Verbeek, B.H.; Rompa, H.W.A.M.; Larsen, P.K.; Methfessel, M.S.; Mueller, F.M. (Phys. Rev. [3] B **28** [1983] 6774/9).

9 Alloys with Scandium and Yttrium

9.1 Alloys with Scandium

Crystallography. No equilibrium diagram has been found; there is however some data on the intermediate phases in the system, and this is summarized in the table below.

compound	RhSc	RhSc	RhSc	RhSc	$RhSc_3$	Rh_3Sc	Rh_3Sc	$RhSc_4$	$Rh_{13}Sc_{57}$ a)	$Rh_{13}Sc_{57}$ b)
type . .	CsCl	B2	CsCl	CsCl	—	Cu_3Au	Cu_3Au	—	—	—
a in Å	3.206	3.204	3.206	3.255 c)	—	3.900	3.898	—	14.4051	14.414
Ref. . .	[1]	[3]	[6]	[6]	[4]	[2]	[4]	[4]	[5]	[5]

a) Ordered sample; annealed at 1213 to 1163 K having been annealed for 8 d at 1023 K, later heated to 1213 K for 1 h and allowed to cool slowly (10 K/h) to 1163 K. This sample had space group Pm3; b) disordered sample as cast; this had space group Im3; c) computed figure [6] converted from au.

The structure of $RhSc_{\sim 4}$ is represented as a slightly deformed form of a hypothetical bcc structure built up of icosahedra centred by the minority atoms [5]. The lattice constant for RhSc was computed as 6.15 au, compared with a value of 6.058 au [6].

Superconductivity. The compound Sc_3Rh becomes superconducting at 0.88 K; however, only about a quarter of the compound was superconducting. Rh_3Sc with an $L1_2$-type structure was not superconducting at 0.32 K [4].

Electronic Structure. The energy bands and density of states for RhSc have been calculated by [6, 7]. The heat of formation ΔH, bulk modulus K and the bulk moduli due to s and p electrons, together with the electron density at the Wigner-Seitz radius were calculated by [6]. The results are shown below.

ΔH in Ryd/atom	K in Mbar	K_{s+p}	electron density *)
−0.085	1.69	1.84	2.5

*) units (10^{-2} au^{-3}); 1 au = 0.5292 Å

References:

[1] Compton, V.B. (Acta Crystallogr. **11** [1958] 446).
[2] Dwight, A.E.; Downey, J.W.; Conner, R.A. (Acta Crystallogr. **14** [1961] 75/6).
[3] Dwight, A.E.; Conner, R.A.; Downey, J.W. (Acta Crystallogr. **18** [1965] 835/9).
[4] Geballe, T.H.; Matthias, B.T.; Compton, V.B.; Corenzwit, E.; Hull, G.W.; Longinotti, E. (Phys. Rev. [2] **137** [1965] A119/A127).
[5] Cenuzal, K.; Chabot, B.; Parthé, E. (Acta Crystallogr. C **41** [1985] 313/9).
[6] Kübler, J. (J. Phys. F **8** [1978] 2301/11).
[7] Nemoshkalenko, V.V.; Plotnikov, V.N.; Antonov, V.N. (Dopov. Akad. Nauk Ukr. RSR A **1983** No. 2, pp. 56/8).

9.2 Alloys with Yttrium

Crystallography. Superconductivity. No phase diagram is available for the system. Several intermediate phases are known and have been examined, particularly for superconducting and magnetic properties; Rh_2Y was found to have an $MgCu_2$-type structure (C15) which was non-superconducting [1]. The available data on crystallography and superconductivity is given in the following table.

compound	type	a	b in Å	c	T_C in K	Ref.
YRh	B2	3.410	—	—	—	[2]
YRh	—	3.409	—	—	—	[3]
YRh[a)]	B2	3.407[b)]	—	—	—	[4]
YRh	CsCl	3.410	—	—	—	[5]
YRh_2	C15	7.489	—	—	—	[2]
YRh_2	$MgCu_2$	7.459[b)]	—	—	—	[1]
YRh_2	$MgCu_2$	7.500 to 7.464		—	—	[5]
YRh_2	—	—	—	—	<0.35	[8]
YRh_3	C15	7.424	—	—	1.07	[2]
YRh_3	$CeNi_3$	5.230	—	17.38	—	[3, 6]
YRh_3	$CeNi_3$	5.222	—	17.36	—	[5]
YRh_5	—	5.141	—	4.294	—	[3, 6]
YRh_5	—	—	—	—	0.56	[2]
Y_2Rh	—	—	—	—	—	[2]
Y_3Rh	Fe_3C	7.138	9.438	6.319	—	[7]
Y_3Rh	—	—	—	—	0.65	[2]
Y_3Rh_2	—	—	—	—	1.48	[2]
Y_3Rh_2	—	—	—	—	—	[3]
Y_5Rh_3	—	—	—	—	—	[3]
Y_7Rh_3	D10/2	9.793	—	6.196	—	[2]
Y_7Rh_3	Th_7Fe_3	9.775	—	6.190	—	[7]

a) superlattice lines not detected; b) accuracy: [4] = ±0.002 Å, [1] = ±0.001 Å.

Magnetic Properties. Measurements of magnetic susceptibility of the whole compositional range at room temperature show Pauli paramagnetic behaviour (free electron paramagnetism); the results indicate the spread of readings at compositions corresponding to the compounds Y_3Rh, Y_7Rh_3, Y_5Rh_3, Y_3Rh_2, YRh, YRh_2, YRh_3, and YRh_5.

Samples were annealed in high vacuum before testing for 100 to 200 h, at 800 °C for low-rhodium alloys, and for 8 to 200 h at 1000 to 1300 °C in the case of rhodium-rich material. Individual figures for the susceptibility for some compounds are given below [9].

compound	Y_7Rh_3	YRh	YRh_2	YRh_3	YRh_5
χ in $cm^3 \cdot g^{-1} \times 10^{-6}$.	0.7	0.15	0.47	0.72	1.25

References:

[1] Compton, V.B.; Matthias, B.T. (Acta Crystallogr. **12** [1959] 651/4).

[2] Geballe, T.H.; Matthias, B.T.; Compton, V.B.; Corenzwit, E.; Hull, G.W.; Longinotti, L.D. (Phys. Rev. [2] **137** [1965] A119/A127).

[3] Ghassem, H.; Raman, A. (Z. Metallk. **64** [1973] 197/9).

[4] Dwight, A.E.; Conner, R.A.; Downey, J.W. (Acta Crystallogr. **18** [1965] 835/9).

[5] Loebich, O.; Raub, E. (J. Less-Common Metals **46** [1976] 1/6).

[6] Raman, A.; Ghassem, H. (J. Less-Common Metals **30** [1973] 185/97).

[7] Raman, A. (J. Less-Common Metals **26** [1972] 199/206).

[8] Suhl, N.; Matthias, B.T.; Corenzwit, E. (J. Phys. Chem. Solids **11** [1959] 346).

[9] Loebich, O.; Raub, E. (Mater. Res. Bull. **10** [1975] 1017/22).

10 Alloys with Lanthanum

10.1 The Rh-La System

Phase Diagram. The system has been investigated by X-ray powder diffraction, metallographic, and differential thermal analysis methods. The crystallographic data of nine of the eleven intermediate phases was established; the resultant diagram is shown in **Fig. 33**. The compounds $LaRh_2$, $LaRh_3$, and La_2Rh_7 show polymorphic changes. La_3Rh and La_2Rh_7 melt congruently at ~950 and 1610 °C, respectively. La_2Rh_7 transforms by eutectoid reaction into $LaRh_3$ and Rh at 1205 °C. La_5Rh_3 is formed by peritectoid reaction between La_3Rh and La_4Rh_3 at 750 °C. There are three eutectics in the system formed by $Rh+La_2Rh_7$, La_4Rh+La_3Rh and La_3Rh+La, at 1500, 820 and 695 °C at compositions of 85, 32, and 11.5 at% rhodium [1].

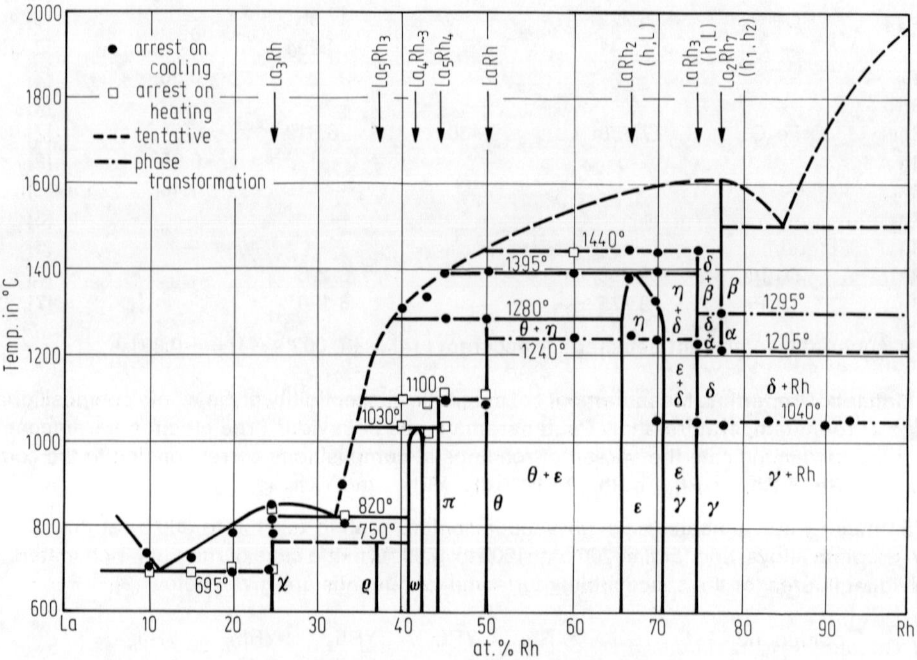

Fig. 33. Tentative phase diagram La-Rh.

Crystallography. The data from the above paper together with that from some earlier and later papers is given below.

phase	at% Rh	type	a	b in Å	c	Ref.
La_3Rh	25	—	15.01	10.50	13.34	[1]
La_3Rh	—	Th_7Fe_3	10.200	—	6.500	[6]
La_7Rh_3	—	D10/2	10.145	—	6.434	[4]
La_5Rh_3	37.5		not determined			[1]
$La_4Rh_{\sim 3}$	41.5	anti-				
	42.5	Th_3P_4	8.922 a)	—	—	[1, 5]

phase	at% Rh	type	a	b in Å	c	Ref.
La_5Rh_4	44.5	not determined			—	[1]
LaRh	—	CrB	3.986 b)	11.144 b)	4.245 b)	[3]
LaRh	50	CrB	3.986	11.14	4.253	[1]
$LaRh_2$	—	C15	7.646 b)	—	—	[2]
$LaRh_2$ l	66 to 71	$MgCu_2$	7.646	—	—	[1]
$LaRh_2$ h	66 to 71	undetermined		—	—	[1]
$LaRh_3$ l	75	$CeNi_3$	5.305	—	17.59	[1]
$LaRh_3$ h	75	$PuNi_3$	5.326	—	26.46	[1]
La_2Rh_7 c)	78	Ce_2Ni_7	5.296	—	26.13	[1]
La_2Rh_7 d)	78	Gd_2Co_7	5.333	—	39.46	[1]

a) ±3 [5]; b) accuracy ±0.001; c) h_1; d) h_2, these represent two different modifications of the phase stable at different high temperature ranges.

The accuracy of figures from [1] is ±0.005 for values ≦10 Å, and ±0.01 for values >10 Å; l and h represent low- and high-temperature forms, respectively.

Superconductivity. The Laves compound $LaRh_2$ was not superconducting [2, 4]. The following compounds were found to be superconducting and to have the following transition temperatures (in K): Rh_3La_7 2.58; Rh_3La 2.60; Rh_5La 1.62 [2].

Small additions of rhodium to lanthanum have been found to render the alloy superconducting, as little as 0.1 at% producing a distinct result. The effect is correlated with the microstructure, the most striking results being obtained in specimens showing grains surrounded by a film of the compound $LaRh_5$. A figure in the original paper shows effects of temperature and composition for alloys with 0.1 to 1 at% La [7].

References:

[1] Singh, P.P.; Raman, A. (Trans. AIME **245** [1969] 1561/8).
[2] Compton, V.B.; Matthias, B.T. (Acta Crystallogr. **12** [1959] 651/4).
[3] Dwight, A.E.; Conner, R.A.; Downey, J.W. (Acta Crystallogr. **18** [1965] 837/9).
[4] Geballe, T.H.; Matthias, B.T.; Compton, V.B.; Corenzwit, E.; Hull, G.W.; Longinotti, L.D. (Phys. Rev. [2] **137** [1965] A 119/A 127).
[5] Virkar, A.V.; Singh, P.P.; Raman, A. (Inorg. Chem. **9** [1970] 353/5).
[6] Raman, A. (J. Less-Common Metals **26** [1972] 199/206).
[7] Arrhenius, G.; Fitzgerald, R.; Hamilton, D.C.; Holm, B.A.; Matthias, B.T.; Corenzwit, E.; Geballe, T.H.; Hull, G.W. (J. Appl. Phys. **35** [1964] 3487/90).

10.2 Ternary Alloys

Rh-La-As. Alloys of the type Rh_2LaAs_2 crystallize in the tetragonal $CaBe_2$-type structure with a = 4.3150, c = 9.8620 Å; the structure is discussed in detail [1].

Rh-La-In. The compound RhLaIn crystallizes in the Fe_2P-type structure with a = 7.610, c = 4.129 Å. The electric resistivity of the compound plotted against temperature is shown in **Fig. 34**, p. 70 [2].

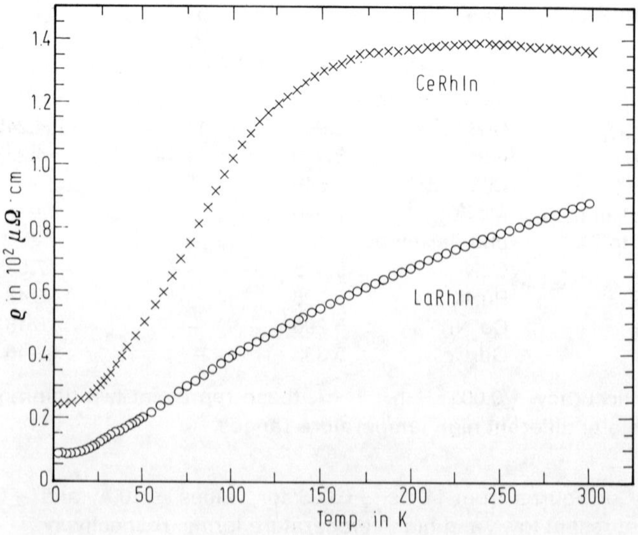

Fig. 34. Resistivity vs. temperature for LaRhIn and CeRhIn.

References:

[1] Madar, R.; Chaudouet, P.; Senateur, J.P. (J. Less-Common Metals **133** [1987] 303/11).

[2] Adroja, D.T.; Malik, S.K.; Padalia, B.D.; Vijayaraghavan, R. (Phys. Rev. [3] **39** B [1989] 4831/3).

11 Alloys with Cerium

11.1 Binary Alloys

Crystallography. No diagram exists for this system, but a large number of intermediate phases have been identified and crystallographic data established; considerable interest has been shown in the valency states existing in these materials. The following table gives the crystallographic data.

compound	type	a	b in Å	c	Ref.
$CeRh_2$	$MgCu_2$	7.538	—	—	[1]
$CeRh_2$ 1)	$MgCu_2$	7.547	—	—	[2]
$CeRh_2$ 2)	$MgCu_2$	7.534	—	—	[2]
$CeRh_2$	—	7.546	—	—	[6]
$CeRh_3$	Cu_3Au	4.012	—	—	[3]
$CeRh_3$	Cu_3Au	4.023 3)	—	—	[4]
$CeRh_3$	Cu_3Au	4.020	—	—	[2]
Ce_3Rh_7	Th_7Fe_3	10.005	—	6.356	[3]
Ce_5Rh_3	Er_5Ir_3	15.850	—	—	[2]
Ce_5Rh_3	La_5Rh_3	—	—	—	[2]
CeRh	CrB	3.852 4)	10.986	4.152	[5]
CeRh	CrB	3.855 5)	10.966	4.153	[7]
Ce_4Rh_3	Nd_4Rh_3	—	—	—	[2]

1) [2] = Ce-rich alloy; 2) [2] = Rh-rich alloy; 3) = conversion from 4.0148 ± 0.0002 kX; 4) = accuracies a ± 0.001, b ± 0.001, c ± 0.001 Å; 5) = accuracies a ± 0.005, b ± 0.01, c ± 0.004 Å.

[2] in a later paper, Ce_4Rh_3 is shown to be Ce_5Rh_3 of Gd_5Si_4-type with a = 7.434, b = 14.86, c = 7.604 Å [8]. Another paper confirms that Ce_4Rh_3 does not exist and that a proposed Ce_3Rh is also non-existent [9]. Ce_3Rh was shown to be Ce_7Rh_3 with a Th_7Fe_3-type structure with a = 10.023, c = 6.376 Å [10]. The X-ray photoemission spectrum of $CeRh_3$ is given by [11].

References:

[1] Compton, V.B.; Matthias, B.T. (Acta Crystallogr. **12** [1959] 651/4).
[2] Ghassem, H.; Raman, A. (Z. Metallk. **64** [1973] 197/9).
[3] Raman, A. (J. Less-Common Metals **26** [1972] 199/206).
[4] Harris, I.R.; Norman, M. (J. Less-Common Metals **13** [1967] 629/31).
[5] Dwight, A.E.; Conner, R.A.; Downey, J.W. (Acta Crystallogr. **18** [1965] 835/9).
[6] Olcese, G.L. (Boll. Sci. Fac. Chim. Ind. Bologna **24** [1966] 165/73).
[7] Canepa, F.; Minguzzi, M.; Olcese, G.L. (J. Magn. Magn. Mater. **63/64** [1987] 591/3).
[8] Raman, A. (J. Less-Common Metals **48** [1976] 111/7).
[9] Kappler, J.P.; Lehmann, P.; Schmerber, G.; Nieva, G.; Sereni, J.G. (J. Phys. Colloq. [Paris] **49** [1988] C8-721/C8-722).
[10] Olcese, G.L. (J. Less-Common Metals **33** [1973] 71).
[11] Raaen, S.; Parks, R.D. (Phys. Rev. [3] B **32** [1985] 4241/4).

Thermodynamic Data

The molecule of CeRh was identified by means of Knudsen cell mass spectrometry and its dissociation energy determined as D = 130.4 ± 6.0 kcal/mol [1].

Gingerich, K.A. (J. Chem. Soc. Faraday Trans. II **1974** 471/6).

Magnetic and Electrical Properties

Magnetic Properties. The magnetic change in some cerium intermetallics has been investigated by measurements of the magnetic susceptibility of molten alloys; in some alloys of cerium with La, Pr, Cu and Sn, cerium retains its localised magnetic moment over the whole range of composition, whilst in the case of alloys with rhodium a continuous transition into the non-magnetic state is found, see figure in original paper [1].

The magnetic susceptibilities of several intermetallic compounds in the system have been examined, often in relation to their temperature dependence as an adjunct to investigations of the valency states in them; the available data is given below.

The value of χ for $CeRh_2$ at temperatures between 77 and 473 K was measured by [2]. Similar measurements between ~1 and 300 K were made by [3]. The results for $CeRh_3$, $CeRh_2$ and CeRh are shown in **Fig. 35**; CeRh shows, in contrast to the other two compounds, considerable variation with temperature. Low-temperature measurements between <1 and 10 K for Ce_3Rh_2, Ce_5Rh_3, Ce_5Rh_4, and Ce_7Rh_3 are shown in **Fig. 36** [3]. The table below gives figures obtained at higher temperatures on $CeRh_3$ [4].

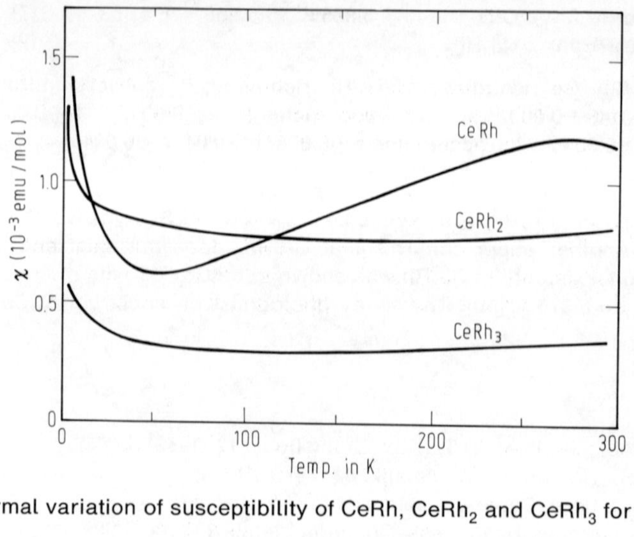

Fig. 35. Thermal variation of susceptibility of CeRh, $CeRh_2$ and $CeRh_3$ for $4 \leq T \leq 300$ K.

t in °C	15.4	50.5	71.3	99.7	150.0	171.9	194.0
$\chi \cdot 10^6$ in emu/g .	0.751	0.752	0.754	0.760	0.770	0.774	0.778
t in °C	246.7	289.7	350.0	440.0	495.0	543.7	585.3
$\chi \cdot 10^6$ in emu/g .	0.788	0.800	0.817	0.842	0.856	0.871	0.883

The small paramagnetic susceptibility increasing steadily with temperature is essentially that of a Pauli paramagnetic material and suggests that the cerium atoms in this compound

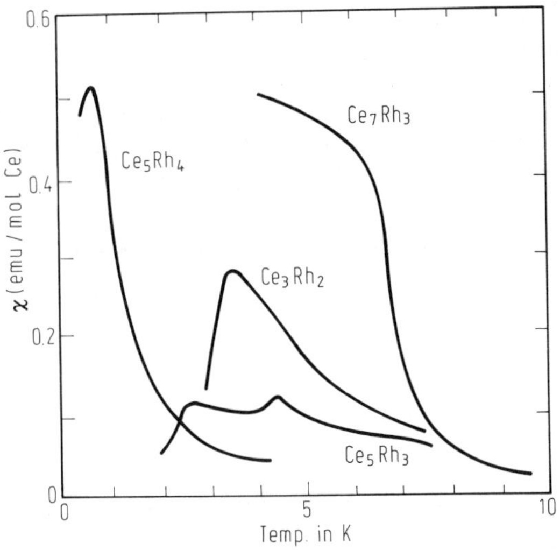

Fig. 36. Magnetic behaviour of Ce_7Rh_3, Ce_3Rh_2, Ce_5Rh_3 and Ce_5Rh_4 for $T \leqq 10$ K.

have no localized moment with them due to reduction of electrons in the 4f-shell [4]. Measurements between room temperature and 800 K are shown in a figure of the original paper; the results indicate that CeRh is a valence fluctuation compound [5].

Superconductivity. No superconducting phases were found in this system containing 67 at% or more rhodium [6].

Electronic Structures. Photoemission studies on $CeRh_3$ gave strong evidence for a 4f to conduction d-band transfer [1]. M-edge X-ray absorption spectroscopy was used to study $CeRh_3$, and interpretation of the spectra with the recent Anderson impurity theory of Gunnarsson and Schönhammer leads to 4f occupancies greater than 0.7 [2]. Valence band photoemission at the giant 4d-4f resonance was measured on $CeRh_2$; double-peaked structures in the 4f-derived photoemission spectrum were observed [3]. The results of self-consistent APW band calculations on $CeRh_2$ suggests 4d-4f hybridization contributes to the bonding of the compound [4, 5]. The Ce 4d X-ray absorption spectrum of $CeRh_3$ is examined and discussed on the basis of a simplified Anderson model combined with interactions described by the Slater integrals and spin-orbit interactions [6].

References:

[1] Schlapbach, L. (Phys. Condens. Matter **18** [1974] 189/215).
[2] Olcese, G.L. (Boll. Sci. Fac. Chim. Ind. Bologna **24** [1966] 165/73).
[3] Kappler, J.P.; Lehmann, P.; Schmerber, G.; Nieva, G.; Sereni, J.G. (J. Phys. Colloq. [Paris] **49** [1988] C8-721/C8-722).
[4] Harris, I.R.; Norman, M. (J. Less-Common Metals **13** [1967] 629/31).
[5] Canepa, F.; Minguzzi, M.; Olcese, G.L. (J. Magn. Magn. Mater. **63/64** [1987] 591/3).
[6] Geballe, T.H.; Matthias, B.T.; Compton, V.B.; Corenzwit, E.; Hull, G.W.; Longinotti, L.D. (Phys. Rev. [2] **137** [1965] A119/A127).

11.2 Ternary Alloys

Rh-Ce-As. The compound $CeRh_2As_2$ was found to crystallize in a tetragonal $CaBe_2Ge_2$-type structure with a=4.283(2), c=9.850(2) Å; the three-dimensional aspects of the crystal structure were analysed. The compound is not superconducting above 1.5 K [1].

Rh-Ce-In. The compound RhCeIn prepared by argon-arc melting on a water-cooled copper hearth crystallizes in the hexagonal Fe_2P-type form with a=7.547, c=4.05 Å. The magnetic susceptibility shows a broad maximum at ~150 K, and the electric resistivity due to magnetic scattering also shows a broad maximum at ~140 K; on the basis of these measurements and of the lattice volume, the cerium ions are considered to be in the mixed valence state. **Fig. 37** and **Fig. 38** show χ, ρ and the resistivity due to magnetic scattering in this compound [2].

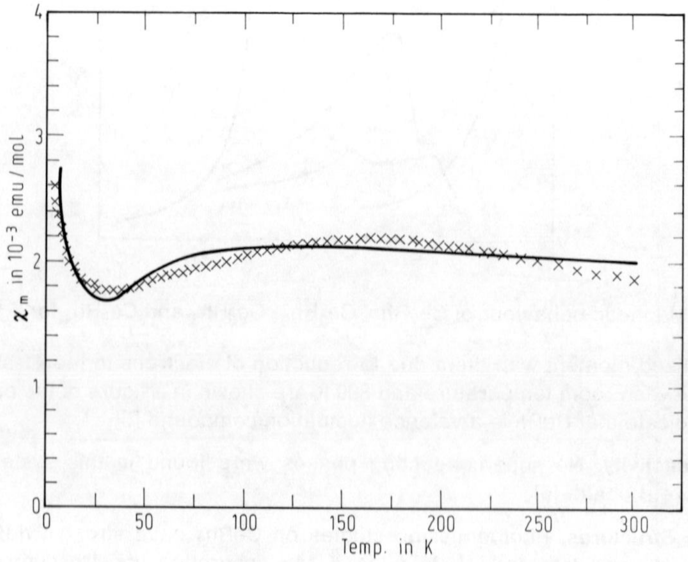

Fig. 37. Magnetic susceptibility of RhCeIn as a function of temperature.

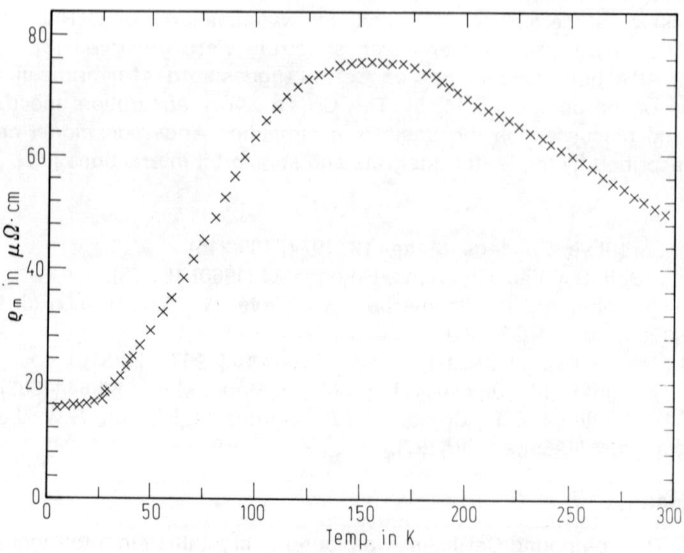

Fig. 38. Magnetic contribution to the resistivity of RhCeIn after subtraction of the resistivity of RhLaIn.

Rh-Ce-La. $Rh_2La_{(1-x)}Ce_x$ mixed valence compounds have been examined by electric resistivity and magnetic susceptibility measurements, and it is deduced therefrom that the valence of the cerium ion decreases with increasing substitution of lanthanum. Samples with $0.5<x<1.0$ display essentially the same resistivities (~80 μΩ·cm), even though their residual resistivities varied between 90 and 5 μΩ·cm; samples with $x<0.5$ do not show this effect. **Fig. 39** and **Fig. 40** show the total resistivity minus phonon resistivity against temperature and the susceptibility against temperature, respectively [3].

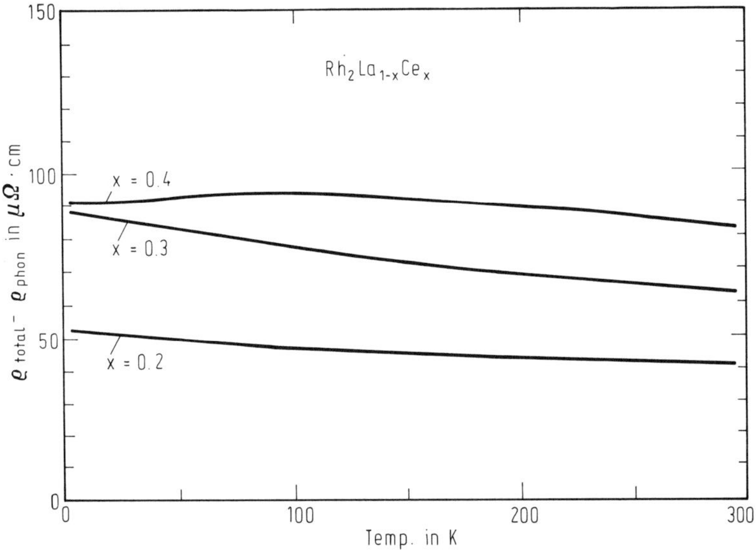

Fig. 39. Total minus phonon resistivity for $RhLa_{1-x}Ce_x$ with x=0.4, 0.3 and 0.2.

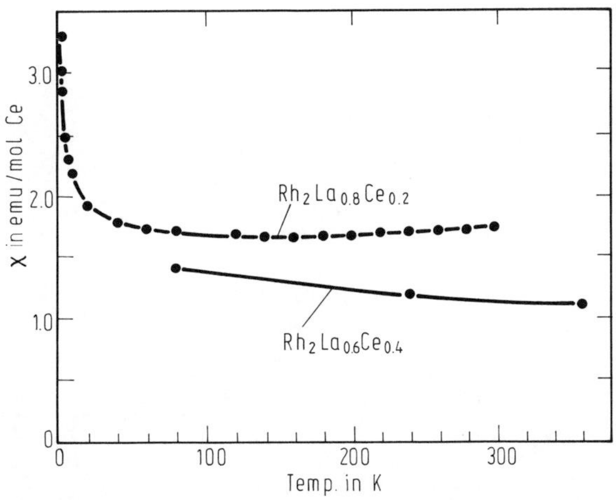

Fig. 40. Magnetic susceptibility vs. temperature for $Rh_2La_{1-x}Ce_x$ with x=0.2 and 0.4.

References:

[1] Madar, R.; Chaudouet, P.; Senateur, J.P. (J. Less-Common Metals **133** [1987] 303/11).
[2] Adroja, D.T.; Malik, S.K.; Padalia, B.D.; Vijayaraghavan, R. (Phys. Rev. [3] B**39** [1989] 4831/3).
[3] Harrus, A.; Mihalisin, T.; Batlogg, B. (J. Appl. Phys. **55** [1984] 1993/5).

12 Alloys with Praseodymium, Neodymium, Samarium, and Europium

12.1 Alloys with Praseodymium

No phase diagram is available for this system. The existing data is given below.

Crystallography.

compound	type	a (in Å)	b (in Å)	c (in Å)	Ref.
$PrRh_2$	$MgCu_2$	7.575 1)	—	—	[1]
PrRh 2)	CrB	3.905 1)	10.910 1)	4.210	[2]
$PrRh_3$ 3)	—	—	—	—	[3]
Pr_7Rh_3	Th_7Fe_3	—	—	—	[4]

1) = accuracy ±0.001; 2) = contained 5% of a second phase; 3) metallographic examination of the $PrRh_3$ composition showed the alloy not to be single-phased in the as-cast condition; annealing at 900 °C for one week resulted in the appearance of a complex X-ray diffraction pattern [3].

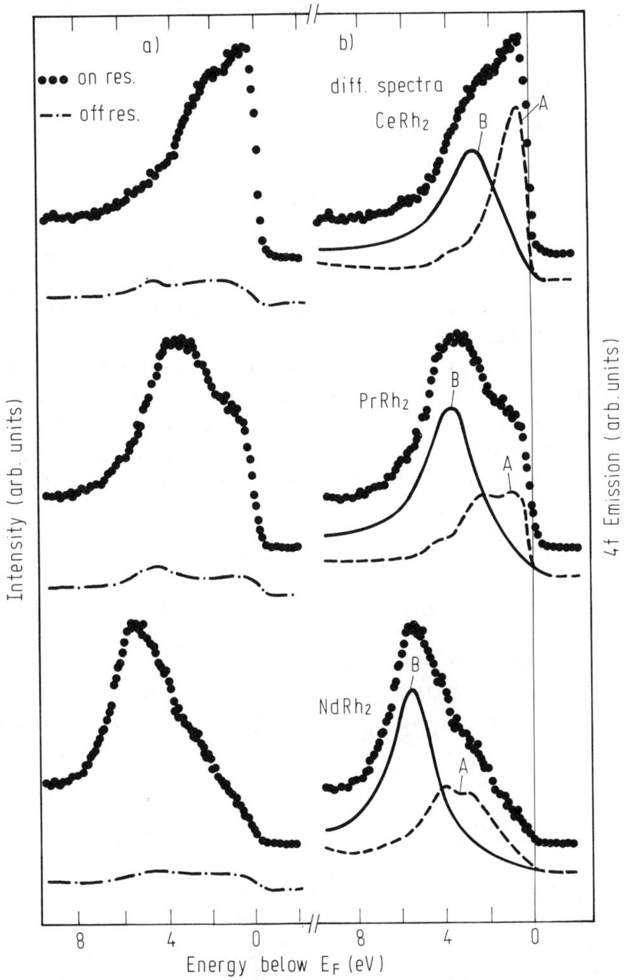

Fig. 41. a) Photoemission spectra of $RERh_2$ compounds (RE = Ce, Pr, Nd), "on" and "off" the 4d→4f resonance. b) Difference spectra and their decomposition into subspectra A and B.

Valence-band Photoemission. The results of measurements at the 4d-4f resonance are reported for the intermetallic compound $PrRh_2$; the similar cerium and neodymium compounds were also investigated. Double-peaked structures in the 4f derived spectra were observed in all cases, with approximately constant separation between the peaks. It is suggested that the peak closest to the Fermi level results from the occupation of a 4f/5d hybridized screening orbital, with decreasing 4f weight from the Ce to the Nd systems. The spectrum for $CeRh_2$, $PrRh_2$ and $NdRh_2$ showing the separation into the two peaks is reproduced in **Fig. 41**, p. 77 [5].

References:

[1] Compton, V.B.; Matthias, B.T. (Acta Crystallogr. **12** [1959] 651/4).
[2] Dwight, A.E.; Conner, R.A.; Downey, J.W. (Acta Crystallogr. **18** [1965] 835/9).
[3] Harris, I.R.; Norman, M. (J. Less-Common Metals **13** [1967] 629/31).
[4] Loebich, O.; Raub, E. (J. Less-Common Metals **46** [1976] 1/6).
[5] Kalkowski, G.; Sampathkumaran, E.V.; Laubschat, C.; Domke, M.; Kaindl, G. (Solid State Commun. **55** [1985] 977/79).

12.2 Alloys with Neodymium.

12.2.1 The Rh-Nd System

Phase Diagram. Two compounds occurring in the system, $NdRh_2$ and NdRh, were identified prior to the determination of a phase diagram; $NdRh_2$ had a cubic $MgCu_2$-type structure with $a = 7.564 \pm 0.001$ Å, NdRh was found to have a CrB-type structure with $a = 3.890 \pm 0.002$ Å, $b = 10.839 \pm 0.007$ Å, $c = 4.247 \pm 0.004$ Å [1, 2]. A tentative diagram of the system determined by differential thermal analysis, metallography and X-ray diffraction showed eight intermediate phases, the compound Nd_5Rh_3 having high- and low-temperature forms. Nd_5Rh_3 and NdRh formed by peritectic reactions. The five phases Nd_4Rh, Nd_3Rh, Nd_4Rh_3, $NdRh_2$ and $NdRh_3$ all melt congruently. The alloys were prepared by argon arc-melting in the form of small buttons which were remelted three times to ensure homogeneity; the specimens were wrapped in molybdenum foil and annealed in vacuum for 1 week at temperatures below 900 °C, for 4 d at 1100 °C and for 12 h at 1250 °C. Stress-relieving anneals on the crushed powders for X-ray analysis were carried out in molybdenum crucibles at 600 °C. The diagram is shown in **Fig. 42** [3].

Crystallography. The relevant data is given below.

phase	at% Rh	type	a	b in Å	c	Ref.
Nd_4Rh	19.5	Fe_3C	7.258	9.84	6.431	[3]
Nd_3Rh	25.0	La_3Rh	14.66	10.38	13.13	[3]
Nd_5Rh_3 l	37.5	—	—	—	—	[3]
Nd_5Rh_3 h	—	—	—	—	—	[3]
Nd_4Rh_3	43.0	—	—	—	—	[3]
NdRh	47.0	CrB	3.890	10.84	4.247	[2]
NdRh	—	—	3.876	10.83	4.234	[3]
$NdRh_2$	66 to 68.5	$MgCu_2$	7.564	—	—	[1]
$NdRh_2$	Nd-rich	—	7.585	—	—	[3]
$NdRh_2$	Rh-rich	—	7.559	—	—	[3]
$NdRh_3$	75.0	$CeNi_3$	5.282	—	17.52	[3]

l and h indicate low- and high-temperature forms.

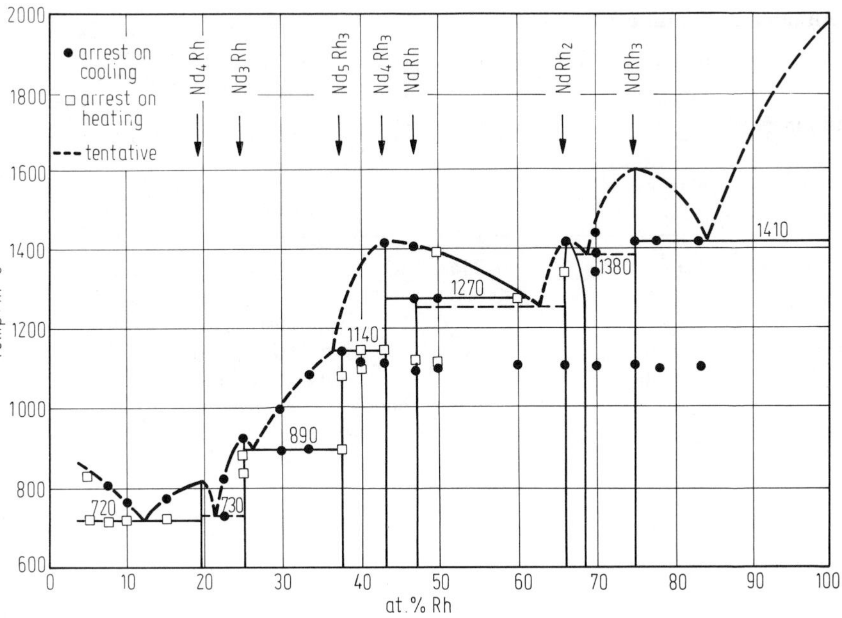

Fig. 42. Tentative equilibrium diagram Nd-Rh.

Accuracies for [3] are ±0.005 Å for values <10 Å and ±0.01 Å for values >10 Å. The accuracies for [1] are ±0.001 Å and for [2] a ±0.002, b ±0.007 and c ±0.004 Å.

Photoemission Studies. Valence-band photoemission measurements on the 4d-4f resonance in $NdRh_2$ suggest a 4f/5d hybridized screening orbital [4].

References:

[1] Compton, V.B.; Matthias, B.T. (Acta Crystallogr. **12** [1959] 651/4).
[2] Dwight, A.E.; Conner, R.A.; Downey, J.W. (Acta Crystallogr. **18** [1965] 835/9).
[3] Singh, P.P.; Raman, A. (Metall. Trans. **1** [1970] 233/7).
[4] Kalkowski, G.; Sampathkumaran, E.V.; Laubschat, C.; Domke, M.; Kaindl, G. (Solid State Commun. **55** [1985] 977/9).

12.2.2 Ternary Alloys

Rh-Nd-As. The compound Rh_2NdAs was found to crystallize in the tetragonal form with a $CaBe_2Ge_2$-type structure; the lattice parameters were a=4.240(6), c=9.859(1) Å. The positional structure is discussed [1].

Rh-Nd-In. An alloy of composition RhNdIn was examined by metallographic, X-ray diffraction, and density determinations; it was found to have a hexagonal Fe_2P-type structure with a=7.534, c=4.028 Å and a density of 9.1 g/cm^3. These results are compared with those of other rare-earth compounds [2].

References:

[1] Madar, R.; Chaudouet, P.; Sénateur, J.P. (J. Less-Common Metals **133** [1987] 303/11).
[2] Ferro, R.; Marazza, R.; Rambaldi, G. (Z. Anorg. Chem. **410** [1974] 219/24).

12.3 Alloys with Samarium

No phase diagram exists for this system.

Crystallography

compound	type	a	b (in Å)	c	Ref.
Sm_4Rh	Fe_3C	7.245	9.675	6.368	[1]
Sm_7Rh_3	Th_7Fe_3	9.903	—	6.248	[1]
Sm_5Rh_3 1)	La_5Rh_3		not determined		[1]
Sm_5Rh_3 2)	Er_5Rh_3 2)		not determined		[1]
Sm_4Rh_3	Nd_4Rh_3		not determined		[1]
SmRh	CsCl	3.464±0.002	—	—	[3]
SmRh	CsCl	3.466	—	—	[1]
SmRh	CsCl	3.466	—	—	[5]
$SmRh_2$ 3)	$MgCu_2$	7.540	—	—	[1]
$SmRh_2$ 4)	$MgCu_2$	7.492	—	—	[1]
$SmRh_2$	$MgCu_2$	7.539	—	—	[4]
$SmRh_2$	$MgCu_2$	7.505 to 7.542	—	—	[5]
$SmRh_3$ 2)	$CeNi_3$	5.255	—	17.46	[1, 2]
$SmRh_3$	$CeNi_3$	5.251	—	17.43	[5]

1) = high-temperature form; 2) = low-temperature form; 3) = Sm-rich sample; 4) = Rh-rich sample

The compound Sm_4Rh_3 is to be Sm_5Rh_4 of Gd_5Si_4-type with a = 7.380, b = 14.63, c = 7.560 Å [6].

References:

[1] Ghassem, H.; Raman, A. (Z. Metallk. **64** [1973] 197/9).
[2] Raman, A.; Ghassem, H. (J. Less-Common Metals **30** [1973] 185/97).
[3] Dwight, A.E.; Conner, R.A.; Downey, J.W. (Acta Crystallogr. **18** [1965] 835/9).
[4] Dwight, A.E.; Downey, J.W.; Conner, R.A. (Trans. Metall. Soc. AIME **236** [1966] 1509/10).
[5] Loebich, O.; Raub, E. (J. Less-Common Metals **46** [1976] 1/6).
[6] Raman, A. (J. Less-Common Metals **48** [1976] 111/7).

12.4 Alloys with Europium

No phase diagram is known for the system; the available data are given under "Crystallography".

Crystallography

compound	type	a	c (in Å)	Ref.
$EuRh_2$	$MgCu_2$	7.444	—	[1]
EuRh	CsCl	3.364	—	[1]
Eu_7Rh_3	Th_7Fe_3	9.643	6.070	[1]

Electrical and Magnetic Properties

Amorphous alloys of composition Eu_xRh_{1-x} were prepared by a co-evaporation technique in a vacuum of $\sim 1 \times 10^{-18}$ Torr, the vapour condensing onto a substrate cooled by liquid nitrogen. The samples so prepared within the compositional range of $0.28 \leqq x \leqq 0.70$ were checked for their amorphous state by electron microscopy at room temperature; this was

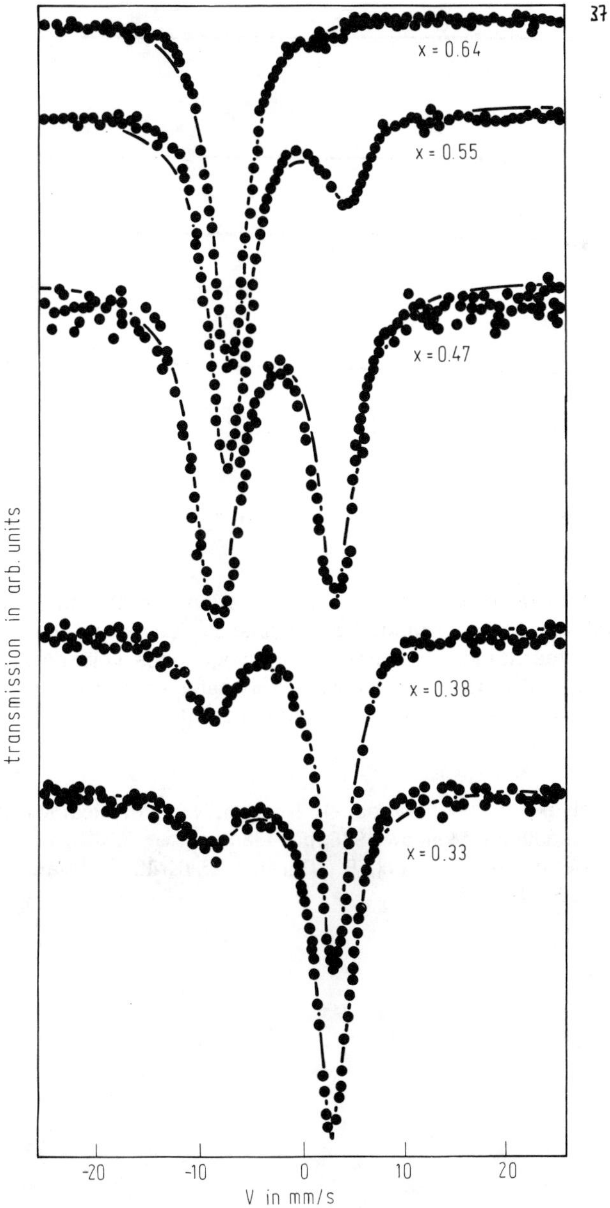

Fig. 43. Mössbauer spectra at 90 K for some amorphous $Rh_{1-x}Eu_x$ alloys.

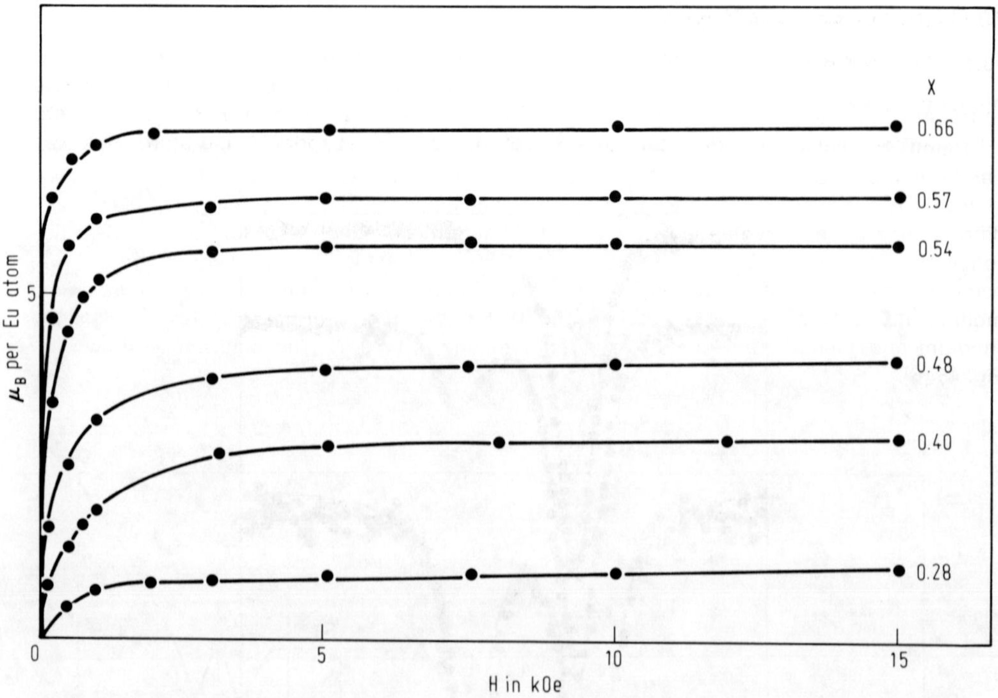

Fig. 44. Low-temperature magnetization per Eu atom in amorphous $Rh_{1-x}Eu_x$ alloys (T = 2 K).

confirmed by the lack of an extended X-ray absorption fine structure signal after the L_{III} edge. These specimens were examined by magnetic, Mössbauer and L_{III} edge measurements; the results clearly indicated that an inhomogeneous valence admixture occurs in the disordered phase. **Fig. 43**, p. 81, shows the Mössbauer spectra taken at 90 K, **Fig. 44** the low-temperature magnetization curves made at 2 K [2].

References:

[1] Savitskii, E.; Polyakova, V.N.; Gorina, N.; Roshan, N.R. (Physical Metallurgy of Platinum Metals, MIR Publications, Moscow 1978, pp. 183/5).

[2] Malterre, D.; Menny, A.; Delcroix, P.; Durand, J.; Krill, G.; Marchal, G. (Phys. Rev. [3] B **39** [1989] 10665/8).

13 Alloys with Gadolinium

13.1 The Rh-Gd System

Phase Diagram

Before the establishment of a complete diagram in 1976, several intermediate phases had been found, $GdRh_2$, $GdRh_3$, $GdRh_5$, Gd_3Rh, Gd_7Rh_3, Gd_5Rh_3, Gd_3Rh_2, Gd_4Rh_3 and GdRh [1 to 5]. The diagram was determined using thermal analysis, metallography and X-ray diffraction; nine intermediate phases were identified, five of which were formed by peritectic reactions. The compounds GdRh, $GdRh_2$ and $GdRh_5$ were found to melt congruently. Two compounds were not stable at room temperature. Eutectics were found at ~18 at% rhodium melting at 920 °C, at ~55 at% rhodium and 1395 °C between GdRh and $GdRh_2$, at 80 at% rhodium and 1460 °C, and at ~87 at% rhodium and ~1470 °C. The diagram is shown in **Fig. 45** [6].

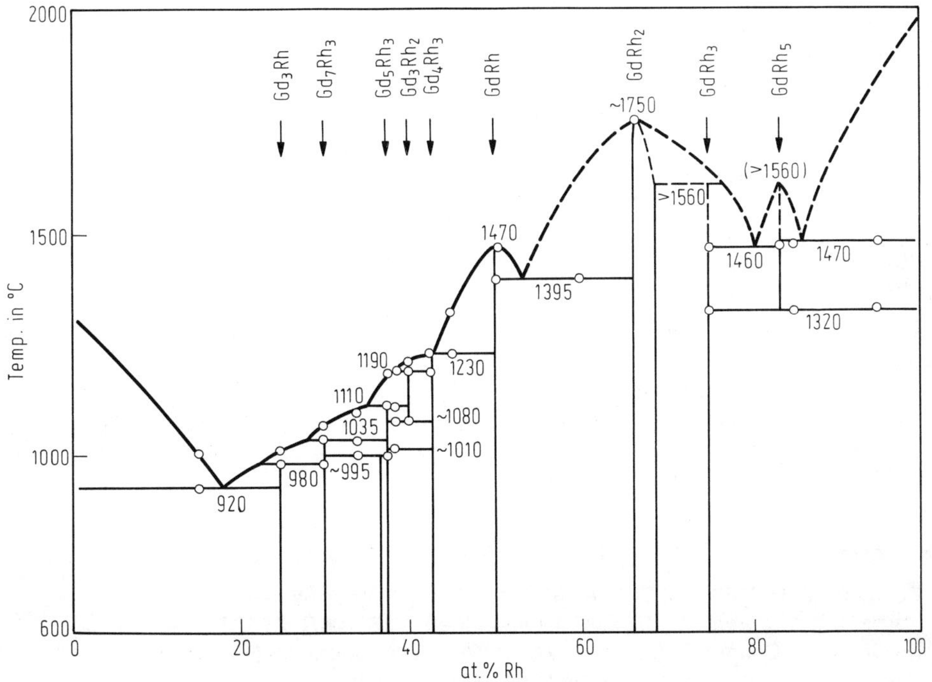

Fig. 45. Phase diagram Gd-Rh.

References:

[1] Compton, V.B.; Matthias, B.T. (Acta Cryst. **12** [1959] 651/4).
[2] Harris, I.R.; Norman, M. (J. Less-Common Metals **13** [1967] 629/31).
[3] Dwight, A.E.; Conner, R.A.; Downey, J.W. (Acta Cryst. **18** [1965] 835/9).
[4] Raman, A. (J. Less-Common Metals **26** [1972] 199/206).
[5] Ghassem, H.; Raman, A. (Z. Metallk. **64** [1973] 197/9).
[6] Loebich, O.; Raub, E. (J. Less-Common Metals **46** [1976] 1/6).

Crystallography

The available data is given below.

compound	type	a	b	c	Ref.
			in Å		
GdRh	CsCl	3.442	–	–	[6]
GdRh	CsCl	3.443	–	–	[7]
GdRh	B2	3.440 1)	–	–	[3]
GdRh	CsCl	3.435	–	–	[5]
$GdRh_2$	$MgCu_2$	3.514 2)	–	–	[1]
$GdRh_2$+Gd	$MgCu_2$	7.514	–	–	[5]
$GdRh_2$+Rh	$MgCu_2$	7.470	–	–	[5]
$GdRh_2$	$MgCu_2$	7.525	–	–	[6]
$GdRh_2$	$MgCu_2$	7.488	–	–	[6]
$GdRh_3$ 5)	–	not single-phased			[2]
$GdRh_3$ 5)	$CeNi_3$	5.242	–	17.38	[6]
$GdRh_3$ 5) 3)	$CeNi_3$	5.235	–	17.40	[5]
$GdRh_5$ 5) 4)	$CaZn_5$	5.168	–	4.306	[5]
Gd_7Rh_3	Th_7Fe_3	9.840	–	6.210	[4]
Gd_3Rh	Fe_3C	7.195	9.540	6.328	[4]
Gd_5Rh_3 4)	La_5Rh_3	–	–	–	[5]
Gd_5Rh_3	Er_5Rh_3 4)	–	–	–	[5]
Gd_5Rh_3 3)	Mn_5Si_3	8.244	–	6.455	[5]
Gd_3Rh_2	Er_3Rh_2	–	–	–	[5]
Gd_4Rh_3	Nd_4Rh_3	–	–	–	[5]

1) = accuracy ±0.002 Å; 2) = accuracy ±0.001 Å; 3) = high-temperature form; 4) = low-temperature form

5) Investigators of the magnetic properties of Rh-Gd alloys were unable to find any trace of the $GdRh_3$ compound; also, whilst retaining some $GdRh_5$ of $CaCu_5$-type by quenching, this was found to contain another unidentified phase. Following annealing, the $GdRh_5$ disappeared and rhodium and an unidentified phase appeared [8].

References:

[1] Compton, V.B.; Matthias, B.T. (Acta Crystallogr. **12** [1959] 651/4).
[2] Harris, I.R.; Norman, M. (J. Less-Common Metals **13** [1967] 629/31).
[3] Dwight, A.E.; Conner, R.A.; Downey, J.W. (Acta Crystallogr. **18** [1965] 835/9).
[4] Raman, A. (J. Less-Common Metals **26** [1972] 199/206).
[5] Ghassem, H.; Raman, A. (Z. Metallk. **64**, [1973] 197/9).
[6] Loebich, O.; Raub, E. (J. Less-Common Metals **46** [1976] 1/6).
[7] Gschneider, K.A. (Acta Crystallogr. **18** [1965] 1082).
[8] Smith, J.L.; Fisk, Z.; Roof, R.B. (J. Appl. Phys. **52** [1981] 1684/5).

Thermal Properties

Material arc-melted under argon was used for direct-current heat pulse measurements to give specific heat data for the polycrystalline compound GdRh; **Fig. 46** shows the temperature dependence of the specific heat C and its lattice contribution, **Fig. 47** shows the magnetic contribution to the specific heat. The magnetic entropy was calculated from the figures shown in Fig. 46, giving 17.4±0.5 $J \cdot mol^{-1} \cdot K^{-1}$ [1]. It is in good agreement with previously obtained 17.3 $J \cdot mol^{-1} \cdot K^{-1}$ by [2]. A Debye temperature of 198±1 K was found [1].

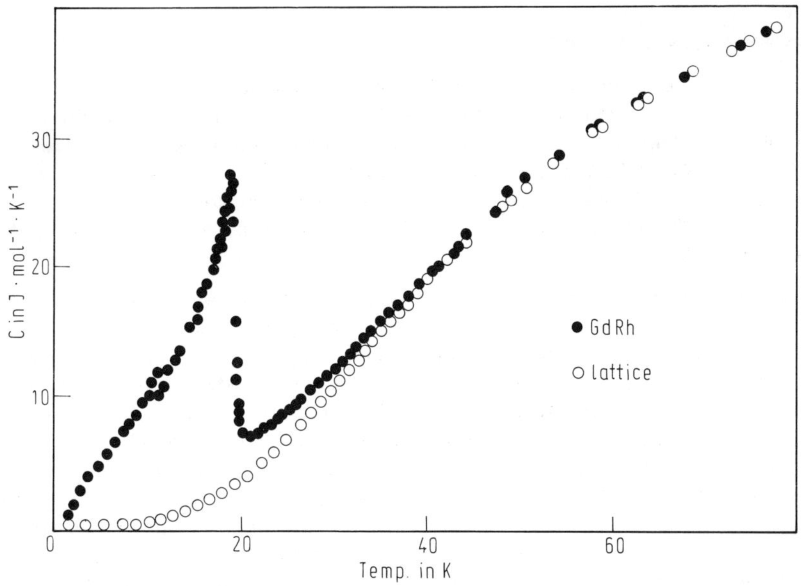

Fig. 46. Specific heat of polycrystalline GdRh and its lattice contribution.

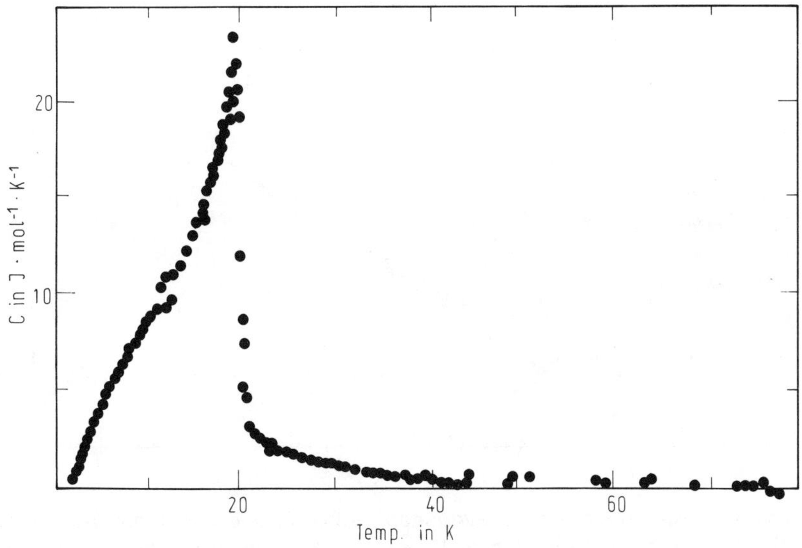

Fig. 47. Magnetic contribution to the specific heat of GdRh.

References:

[1] Azhar, A.A.; Mitescu, C.D.; Johanson, W.R.; Zimm, C.B.; Barclay, J.A. (J. Appl. Phys. **57** [1985] 3235/7).

[2] Olijhoek, J.F.; Nauts, H.C.A. (Proc. 14th Intern. Conf. Low Temp. Phys., Otaniemi, Finland, 1975, Vol. 3, p. 196).

Electrical and Magnetic Properties

The interaction between conduction electrons and paramagnetic ions in metals has been investigated by the observation of the paramagnetic resonance of gadolinium in alloys and intermetallic compounds; a 3 at% gadolinium-rhodium alloy was amongst those tested. Spectra of powdered materials were observed at temperatures between 1.4 and 500 K and frequencies between 10 and 80 kMc/s. Values for g and spectrum half-width at half-power are given for the 3% Gd-Rh alloy as $g = 1.989 \pm 0.010$ (gyromagnetic ratio) and $DH = 760 \pm 80$ G at 47 K and 48.4 kMc/s [1].

Gd_7Rh_3. The magnetic properties of this compound have been studied in some detail using samples with a weak fibrous texture of c-axis alignment within $\pm 30°$; samples were arc-melted and heat-treated under vacuum; some material was crushed to powder for comparative measurements against the orientated material. Measurements were made by the Faraday method at temperatures betwen 100 and 800 K; the fibrous specimens showed paramagnetic behaviour, with an increasing field-independent magnetism below 200 K depending on specimen orientation. Measurements indicated an antiferromagnetic ordering, and the c-axis or a direction near it was the preferred easy direction of magnetization. On fixing this orientation in a bulk sample at 90° to the field, a marked antiferromagnetic ordering was observed at 142 ± 1 K, independent of field strength; at this orientation, the field-independent part of the magnetization remained very small after the antiferromagnetic ordering. The powdered sample showed only a small faint peak at 142 K. Above 170 K, no anisotropic effects were found. Above 470 K all samples followed the Curie-Weiss law

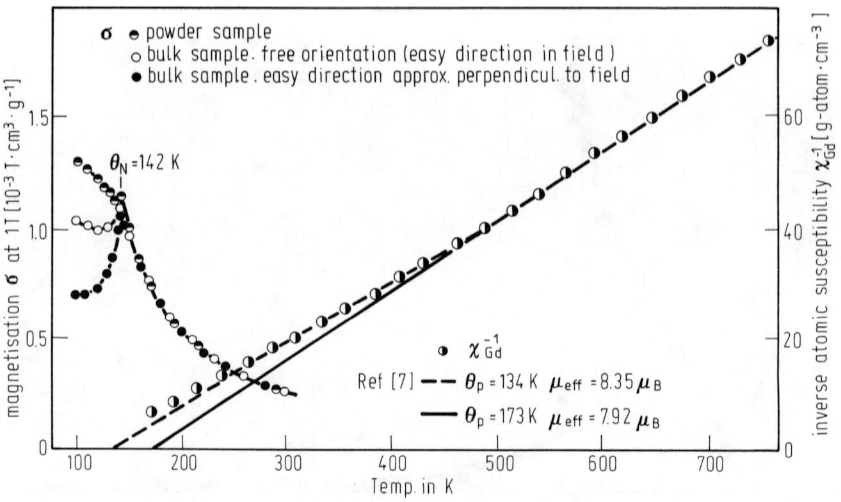

Fig. 48. Temperature variation of the inverse susceptibility and low-temperature magnetization at 1 Tesla of different samples of Gd_7Rh_3 at different orientations of the preferred direction.

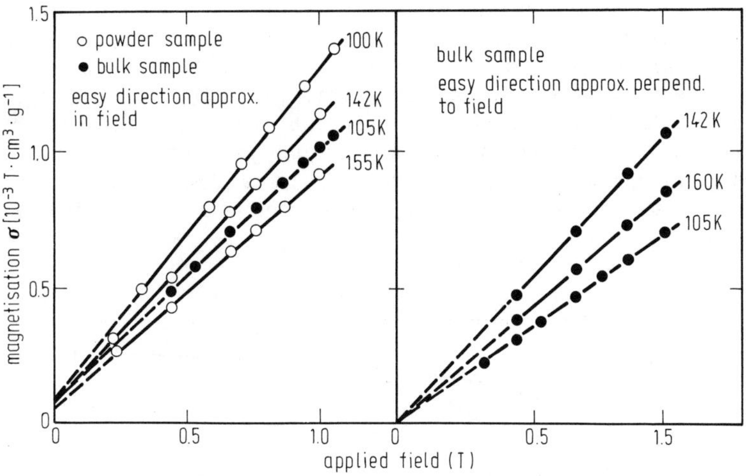

Fig. 49. Field dependence of magnetization of powder and bulk samples of Gd_7Rh_3 at different orientations.

with $\mu_{eff}=7.9\pm0.05\ \mu_B$ and $\theta_P=175\pm7$ K (paramagnetic transition temperature). Some of the results are shown in **Fig. 48** and **Fig. 49** [2].

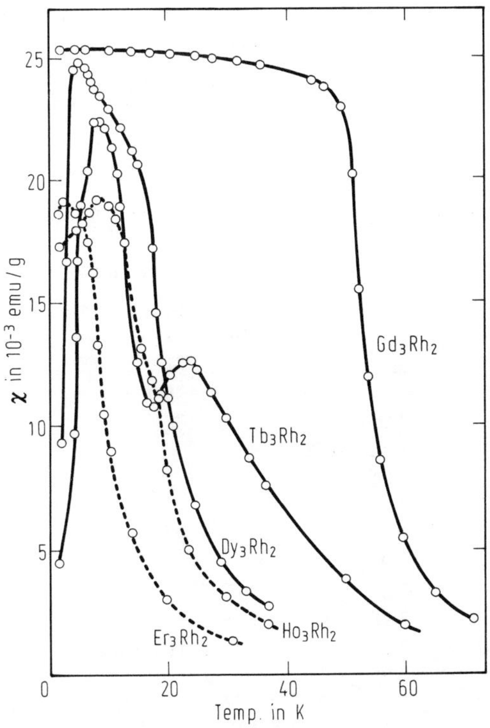

Fig. 50. Thermal dependence of the initial susceptibility when temperature is increased.

Gd_3Rh_2. The bulk compound was examined magnetically with other R_3Rh_2 compounds where R = Tb, Dy, Ho and Er; all these compounds were found to crystallize in the tetragonal Y_3Rh_2-type structure. Magnetic measurements were made between 1.5 and 300 K, in fields up to 60 kOe. The thermal dependence of the magnetic susceptibility of polycrystalline samples measured in low fields is shown in **Fig. 50**, p. 87, and the first magnetization curves measured at 1.5 K in **Fig. 51**. The compound is ferromagnetic below its Curie temperature of 51 K. The paramagnetic transition temperature (θ_P) was 56 K, the effective magnetization 8.35 μ_B, the calculated figure being 7.94 μ_B; in the strong field of 60 kPe at 1.5 K, the figure was 6.85 μ_B [3].

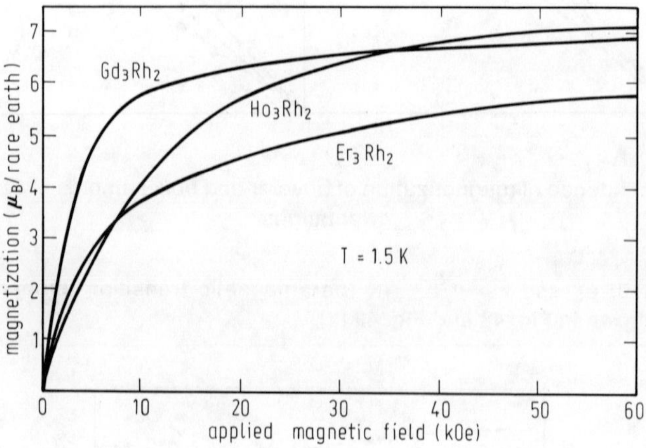

Fig. 51. First magnetization curves of Gd_3Rh_2, Ho_3Rh_2 and Er_3Rh_2 measured at 1.5 K.

Some further results on rhodium-gadolinium compounds are given below.

phase	paramagnetic state		ordered state		Ref.
	θ_P in K	μ_{eff} in μ_B	θ_f in K	θ_N in K	
$GdRh_2$	–	–	>77	–	[4]
$GdRh_2$	–	–	~75	–	[6]
$GdRh_2$	–	7.95±0.23	<80	–	[11]
$GdRh_2$	73±2	7.65±0.25	72±2	–	[12]
$GdRh_5$*)	38	7.95	–	–	[5]
$GdRh_3$	53	7.85	–	–	[5]
GdRh	25	7.95	–	–	[5]
Gd_4Rh_3	33	7.92	–	–	[5]
Gd_3Rh_2*)	73	7.9	–	–	[5]
Gd_5Rh_3	98	7.98	–	–	[5]
Gd_7Rh_3	173	7.92	anisotropic	142	[3]
Gd_7Rh_3	134	8.35	–	–	[7]
Gd_3Rh	150	8.05	–	115	[5]

*) high-temperature phase measured in unannealed state

Fig. 52 shows the paramagnetic Curie temperature θ_P, Néel temperature θ_N and ferromagnetic Curie temperature θ_f plotted against composition [5]. Magnetic studies on dilute gadolinium alloys (~1 at% Gd) with rhodium showed that at low temperatures they behave ferromagnetically like the similar iridium alloys examined in the same work; also, like the iridium alloys they showed two Curie temperatures attributed to the dilute solid solution and a disordered eutectic phase, respectively. The following figures were obtained:

Saturation magnetic moment 6.94 μ_B/Gd atom; 1st Curie temperature (T_c) 35 K; 2nd Curie temperature 19.7 K [8].

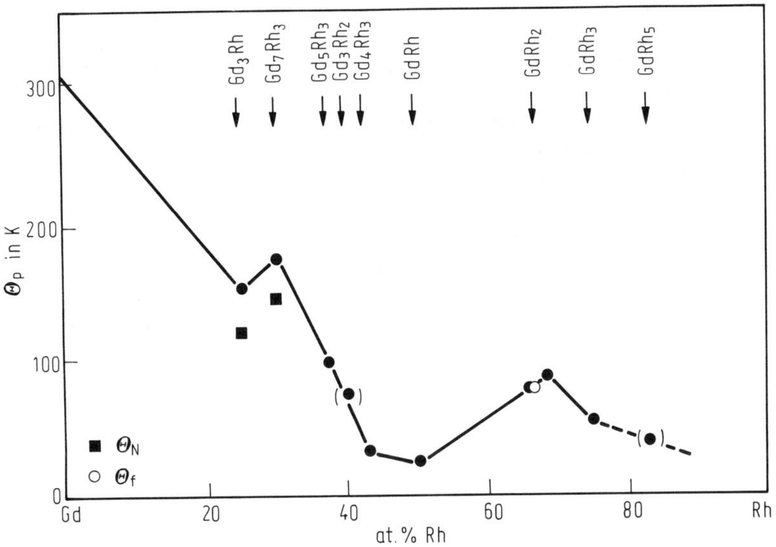

Fig. 52. Paramagnetic (θ_P) and ferromagnetic (θ_f) Curie temperature, Néel temperature θ_N of Gd-Rh compounds.

Somewhat similar work covering a wider alloy range showed three transition temperatures for the 1 at% gadolinium alloy; the results are shown below. As reported earlier, the authors were unable to find the compound $GdRh_3$, and there were some anomalies with regard to $GdRh_5$; however, an attempt was made to identify the phases associated with the different transition temperatures as follows [9]:

T_C in K	attributed to T
76	$GdRh_2$
56	$GdRh_2$
49	$GdRh_2$ or unidentified phase in eutectic layers
33	dilute Gd in Rh in eutectic layers
20	$GdRh_5$
~10	ordering of dilute Gd in rhodium grains

For **GdRh**, a Curie temperature of 19.93±0.01 K and a Debye temperature of 198±1 K was found by [18].

Hyperfine Magnetic Field. This was studied in ferromagnetic GdRh using the zero-field nuclear resonance of ^{155}Gd, ^{157}Gd, ^{103}Rh and ^{139}La, the latter in the compound diluted with La. Prior magnetic measurements showed the compound to have a Curie temperature of 24 K and a magnetization corresponding to 6.6 μ_B/atom Gd at 4.2 K. The NMR results on the dilute La-containing alloy allowed the hyperfine field at the gadolinium site to be separated into its different contributions. The results do not support an earlier investigation suggesting considerable d-spin polarization at the rhodium site, antiparallel to the Gd spin [10].

Nuclear Magnetic Resonance. Electron Spin Resonance. These techniques have been used to study the compound $GdRh_2$ and other Laves-phase compounds. Nuclear magnetic resonance (NMR) using ^{165}Ho introduced as a dilute impurity revealed a strong crystalline electric field in $GdRh_2$; the electron-spin resonance (ESR) results suggest that the gadolinium ESR is bottlenecked in all the compounds examined. The reciprocal mass susceptibility against temperature, the temperature dependence of the g-value and linewidth obtained from the ESR measurements, and the hyperfine parameters from the NMR measurements are shown in figures in the original paper [11].

A detailed nuclear magnetic resonance (NMR) of the quadrupolar splitting was reported for $^{155,157}Gd$ in cubic, ferromagnetically ordered $GdRh_2$; the zero-field ^{157}Gd NMR spin-echo spectrum of a $GdRh_2$ powder sample at T=4.2 K (Curie temperature 75 K) is shown in **Fig. 53**. The table below gives all the pertinent data of the NMR spectra for the compound [13].

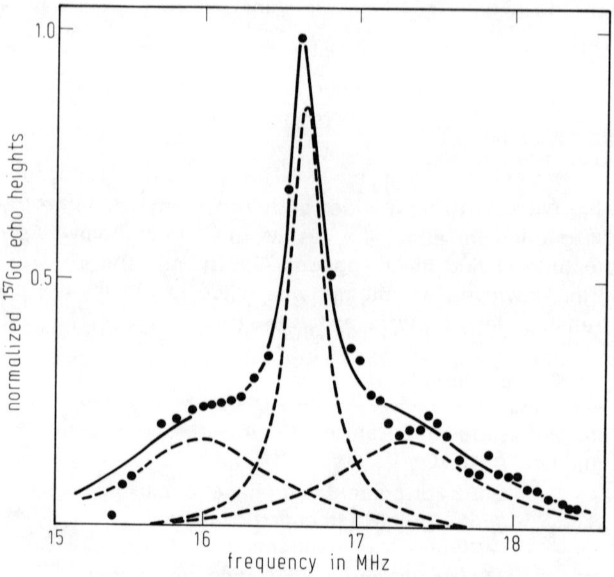

Fig. 53. Zero-field ^{157}Gd NMR spin-echo spectrum of a $GdRh_2$ powder sample at T=4.2 K (T_C=75 K). The best fit with a symmetrically split three-line pattern is also shown. (The solid line gives the sum of the three broken line curves.)

Observed magnetic dipole and electric quadrupole interaction for ^{155}Gd and ^{157}Gd in $GdRh_2$ (powder sample) at T = 4.2 K: central line Zeeman frequencies ν_z, width of $+{}^1/_2 \leftrightarrow -{}^1/_2$ transition $\delta\nu_z$, average (av) and maximum (max) quadrupole splitting (satellite to satellite) $2\Delta\nu_Q$ and width of quadrupole satellite $\delta(\Delta\nu_Q)$.

	^{155}Gd	^{157}Gd	$^{157}X/^{155}X$
ν_z(MHz)	12.98	16.93	1.30
$\delta\nu_z$(MHz)	0.11	0.11	0.96
$(2\Delta\nu_Q)_{av}$(MHz)	1.12	1.17	1.04
$\delta(\Delta\nu_Q)$ (MHz)	1.05	1.06	0.99
$\delta(\Delta\nu_Q)/\delta\nu_z$	9.6	10.1	—
$(2\Delta\nu_Q)_{max}$(MHz)	1.8	2.2	1.2

Rapidly Quenched Amorphous Alloys. The alloy $Gd_{82}Rh_{18}$ has been the subject of magnetic and electrical property studies connected with the applicability of the RKKY model to describe the coupling between the 4f moments in amorphous alloys containing rare earth elements; the alloy was prepared by melt-spinning in purified argon. The alloy was found to behave ferrogmanetically.

Samples of the amorphous alloy were prepared by melt-spinning under a cover of purified argon to produce a tape. X-ray diagrams were taken with CuKα radiation in combination with an X-ray monochromator; the scattering vectors $Q = 4\pi \sin\theta/\lambda$ corresponding to the first diffuse halo are given. The Curie temperature and magnetic moment at 4.2 K in a field of 18 kOe were determined. The following figures are given: $Q = 2.18\ Å^{-1}$, d_x (closest approach between Gd atoms) = 3.54 Å, Curie temperature (T_c) = 111 K, M (magnetic moment) = 5.8 μ_B/Gd atom.

The results are interpreted as indicating that a direct 5d-5d electron coupling mechanism is responsible for the ferromagnetic behaviour rather than coupling between 4f moments in terms of a RKKY model [14, 15].

Further work on the same alloy using FMR (ferromagnetic resonance) and saturation magnetic measurements, DSC (differential scanning calorimetric) measurements of the crystallization temperature, and determinations of the Curie temperature further supported the 5d-5d coupling mechanism. Some of the results are given below.

Crystallization temperature 563 K; the uniaxial anisotropy constant K_u was determined from the FMR spectrum with the external field parallel to the plane of the ribbon; the significance of a positive value is that it aligns the Gd atom spins perpendicular to the plane of the ribbon.

$K_u = 8.9 \times 10^6$ erg/cm³, Curie temperature = 111 K, saturation magnetic moment $\mu = 5.6\ \mu_B$/ Gd atom at 4.2 K in a field of 18 kOe [16, 17].

References:

[1] Peter, M.; Shaltiel, D.; Wernick, J.H.; Williams, H.J.; Mock, J.B.; Sherwood, R.C. (Phys. Rev. [2] **126** [1962] 1395/402).

[2] Loebich, O.; Raman, A. (J. Less-Common Metals **43** [1975] 89/92).

[3] Gignoux, D.; Gomez-Sal, J.C.; Paccard, D.; Aramburu-Zabala, J.A. (Solid State Commun. **50** [1984] 43/5).

[4] Compton, V.B.; Matthias, B.T. (Acta Crystallogr. **12** [1959] 651/4).
[5] Loebich, O.; Raub, E. (Mater. Res. Bull. **10** [1975] 1017/22).
[6] Crangle, J.; Ross, J.W. (Proc. Intern. Conf. Nottingham. Inst. Phys., Phys. Soc., London 1964, p. 240).
[7] Olcese, G.L. (J. Less-Common Metals **33** [1973] 71).
[8] Smith, J.L.; Matthias, B.T. (J. Magn. Magn. Mater. **21** [1980] 2203/8).
[9] Smith, J.L.; Fisk, Z.; Roof, R.B. (J. Appl. Phys. **52** [1981] 1684/5).
[10] Dormann, E.; Buschow, K.H.J. (J. Appl. Phys. **47** [1976] 1662/7).

[11] Tari, A.; Larica, C.; Popplewell, J. (J. Less-Common Metals **78** [1981] P7/P19).
[12] Hrubeck, J.; Steinen, W. (J. Phys. Colloq. [Paris] **40** [1979] C5-198).
[13] Dressel, U.; Meister, U.; Dormann, E.; Buschow, K.H.J. (J. Magn. Magn. Mater. **74** [1988] 91/100).
[14] Buschow, K.H.J.; Beekmans, N.M. (Rapidly Quenched Metals Proc. 3rd Intern. Conf., Brighton, Engl., 1978, Vol. 2, pp. 133/6).
[15] Buschow, K.H.J. (Solid State Commun. **27** [1978] 275/8).
[16] Algra, H.A.; Buschow, K.H.J.; Henskens, R.A. (J. Magn. Magn. Mater. **15/18** [1980] 1395/6).
[17] Buschow, K.H.J.; Algra, H.A.; Henskens, R.A. (J. Appl. Phys. **51** [1980] 561/6).
[18] Azhar, A.A.; Mitescu, C.D.; Johanson, W.R.; Zimm, C.B.; Barclay, J.A. (J. Appl. Phys. **57** [1985] 3235/7).

14 Alloys with Terbium

No phase diagram is established for this system.

Crystallography. The available crystallographic data are given below.

compound	type	a	b	c	Ref.
			in Å		
Tb_3Rh	Fe_3C	7.156	9.505	6.308	[1]
Tb_7Rh_3	Th_7Fe_3	9.788	—	6.175	[1]
Tb_5Rh_3 1)	Er_5Rh_3 1)	—	—	—	[1]
Tb_5Rh_3 2)	Mn_5Si_3	8.176	—	6.385	[1]
Tb_3Rh_2	Er_3Rh_2	—	—	—	[1]
Tb_3Rh_2	Y_3Rh_2	tetragonal	not determined		[2]
TbRh	CsCl	3.417	—	—	[1]
$TbRh_2$	$MgCu_2$	—	—	—	[1]
$TbRh_2$ 3)	$MgCu_2$	7.492	—	—	[1]
$TbRh_4$ 4)	$MgCu_2$	7.452	—	—	[1]
$TbRh_5$ 2)	$CaZn_5$	5.134	—	4.290	[1]

1) = low-temperature form; 2) = high-temperature form; 3) = terbium-rich; 4) = rhodium-rich

High-temperature Tb_5Rh_3 was found to be stable only above 900 °C. The low-temperature compound had a narrow range of homogeneity. The $TbRh_5$ compound with $CaZn_5$-type structure was only found in as-cast material; after annealing at 900 °C, a mixture of $TbRh_2$ + rhodium was obtained; from this it was concluded that the $CaZn_5$-type phase was only stable at high temperatures and underwent eutectoid reactions on cooling [1].

Electrical and Magnetic Properties. Neutron crystal field spectroscopy has been used to establish the crystal field parameters for the intermetallic compound TbRh using polycrystalline material prepared by induction levitation melting. The results obtained on this material were less precise than those obtained on the similar compounds HoRh and ErRh, due to a larger exchange interaction than in the other compounds, some parameters were deduced. The results gave $x = -0.72 \pm 0.02$, $W = 0.0742$ meV which gave $A_4^0\langle r^4\rangle = -7.27$ meV and $A_6^0\langle r^6\rangle = -2.45$ meV [3].

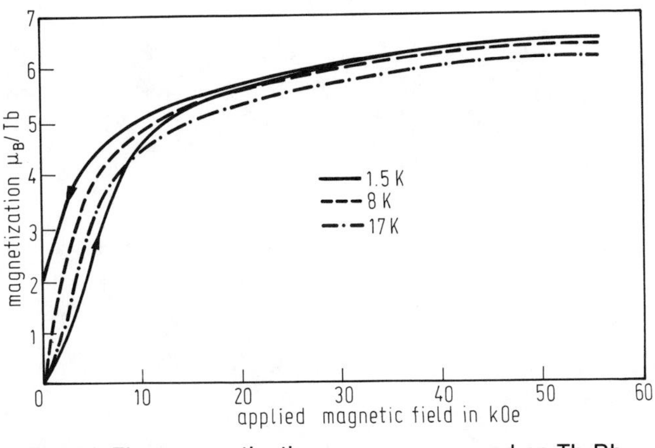

Fig. 54. First magnetization curves measured on Tb_3Rh_2.

Bulk magnetic measurements have been made on polycrystalline samples of the tetragonal compound Tb_3Rh_2; antiferromagnetic behaviour was found between 14 and 24 K, in contrast with the similar type Gd, Dy, Ho and Er compounds examined. **Fig. 54**, p. 93, shows the first magnetization curves measured at 1.5, 8, and 17 K, the temperature dependence of the initial susceptibility is shown in Fig. 50, p. 87. The magnetic properties were: T_N (Néel temperature) = 24 K, θ_P (paramagnetic transition temperature) = 34 K, experimental effective magnetic moment = 10.01 μ_B, calculated magnetic moment = 9.72 μ_B.

The implications of the differences in the experimental and calculated magnetic moments are associated with the polarization of the Rh 4d band by the spin of the rare earth [2].

References:

[1] Ghassem, H.; Raman, A. (Z. Metallk. **64** [1973] 197/9).

[2] Gignoux, D.; Gomez-Sal, J.C.; Paccard, D.; Aramburu-Zabala, J.A. (Solid State Commun. **50** [1984] 43/5).

[3] Rossat-Mignod, J.; Chamard-Bois, R.; Knorr, K.; Drexel, W. (Proc. 11th Rare Earth Res. Conf., Traverse City, Mich., 1974, Vol. 1, pp. 317/26).

15 Alloys with Dysprosium and Holmium

15.1 The Rh-Dy System

Crystallography

As far as is known, no diagram exists for this system. There is some data on the intermediate phases and this is given under crystallography, below.

compound	type	a	b	c	Ref.
			in Å		
Dy_3Rh	Fe_3C	7.138	9.397	6.276	[1]
Dy_7Rh_3	Th_7Fe_3	9.740	—	6.135	[1]
Dy_5Rh_3 1)	Er_5Rh_3 1)		not determined		[1]
Dy_5Rh_3 2)	Mn_5Si_3	8.152	—	6.288	[1, 2]
Dy_5Rh_3 1)	Er_5In_3	15.30	—	—	[1, 2]
Dy_5Rh_3	Mn_5Si_3	8.153(5)	—	6.365(5)	[7]
Dy_3Rh_2	Er_3Rh_2		not determined		[1]
DyRh	CsCl	3.403 ± 0.02	—	—	[4]
DyRh	CsCl	3.403	—	—	[6]
DyRh	CsCl	3.401	—	—	[1]
DyRh	CsCl	3.407	—	—	[3]
$DyRh_2$	$MgCu_2$	7.483	—	—	[5]
$DyRh_2$ 3)	$MgCu_2$	7.488	—	—	[1]
$DyRh_2$ 4)	$MgCu_2$	7.428	—	—	[1]
$DyRh_2$ 3)	$MgCu_2$	7.490 to 7.447	—	—	[6]
$DyRh_5$ 2)	$CaZn_5$	5.144	—	4.294	[1, 2]

1) = low-temperature form; 2) = high-temperature form; 3) = dysprosium-rich alloy; 4) = rhodium-rich alloy

References:

[1] Ghassem, H.; Raman, A. (Z. Metallk. **64** [1973] 197/9).
[2] Raman, A.; Ghassem, H. (J. Less-Common Metals **30** [1973] 185/97).
[3] Belakhovsky, M.; Pierre, J.; Ray, D.K. (J. Phys. F **5** [1975] 2274/82).
[4] Dwight, A.E.; Conner, R.A.; Downey, J.W. (Acta Crystallogr. **18** [1965] 837/9).
[5] Dwight, A.E. (Trans. Am. Soc. Metals **53** [1961] 479/500).
[6] Loebich, O.; Raub, E. (J. Less-Common Metals **46** [1976] 1/6).
[7] Le Roy, J.; Moreau, J.M.; Paccard, D. (J. Less-Common Metals **86** [1982] 63/7).

Electrical and Magnetic Properties

Electronic Structure. Two papers report the results of studies of the electronic structure of several cubic intermetallic compounds of dysprosium, including DyRh, using self-consistent augmented-plane-wave calculations (APW). The energy bands, density of states and the nature of the electrons in the conduction band are discussed [1, 2]. Two theoretical papers deal with the contribution of the conduction electrons to the crystalline electric field in five equiatomic intermetallic compounds including DyRh. The first paper examines the direct coulombic part using electronic symmetrized APW wave functions; the second paper deals with the exchange coulombic part [3, 4]. **Fig. 55**, p. 96, shows the total and partial densities of states for DyRh [2].

Magnetic Properties. Mössbauer effect on ^{161}Dy has been used on several equiatomic compounds, including DyRh to establish their hyperfine parameters including the hyperfine

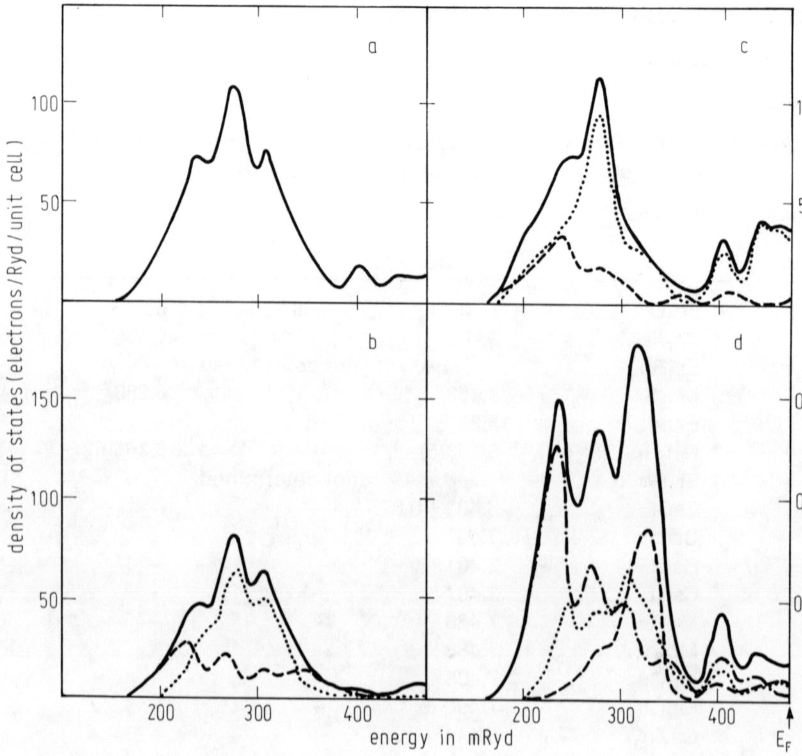

Fig. 55. Total and partial densities of states for DyRh compound. Full curves denote: a) total; b) rhodium d electrons; c) dysprosium d electrons; d) dysprosium 5f electrons (Fermi energy $E_f = 474$ mRyd). In b) and c) broken curve denotes e_g electrons and dotted curve denotes t_{2g} electrons. In d) broken curve denotes a_{2u} electrons, dotted curve denotes t_{1u} electrons and chain curve denotes t_{2u} electrons.

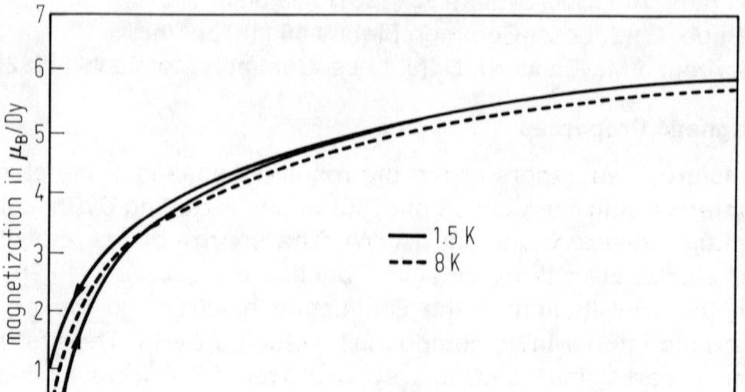

Fig. 56. First magnetization curves of DyRh.

field, quadrupole interaction and isomer shift. The saturation magnetization was found to be 6.5 μ_B, and the paramagnetic transition temperature (θ_P) 2 K. The isomer shift was 2.52 mm/s [5]. Bulk magnetic measurements on polycrystalline samples of a number of R_3Rh_2 tetragonal compounds where R = Gd, Tb, Dy, Ho and Er have been determined. Dy_3Rh_2 was found to be ferromagnetic below 18 K, and θ_P is given as 15 K; the effective magnetic moment determined experimentally was 10.77 μ_B compared with a calculated figure of 10.54 μ_B. The significance of the degree of agreement between these figures is discussed in terms of the polarization of Rh-4d band electrons by the rare earth spin. **Fig. 56** shows the first magnetization curves at 1.5 and 8 K, for thermal dependence of the initial susceptibility see Fig. 50, p. 87 [6].

References:

[1] Belakhovsky, M.; Ray, D.K. (Phys. Rev. [3] B**12** [1975] 3956/63).
[2] Belakhovsky, M.; Pierre, J.; Ray, D.K. (J. Phys. F**5** [1975] 2274/82).
[3] Schmitt, D. (J. Phys. F**9** [1979] 1745/58).
[4] Schmitt, D. (J. Phys. F**9** [1979] 1759/70).
[5] Belakhovsky, M.; Pierre, J. (Solid State Commun. **9** [1971] 1409/15).
[6] Gignoux, D.; Gomez-Sal, J.C.; Paccard, D.; Aramburu-Zabala, J.A. (Solid State Commun. **50** [1984] 43/5).

15.2 The Rh-Ho-System

Crystallography. No phase diagram has been found for the system. Some intermetallic compounds have been examined, and these are detailed below.

compound	type	a (in Å)	c (in Å)	Ref.
$HoRh_2$	$MgCu_2$	7.426	—	[1]
HoRh	CsCl	3.387	—	[2]
HoRh	B2	3.388 ± 0.002	—	[3]
HoRh	—	3.388	—	[4]
$HoRh_2$	$MgCu_2$	7.475 to 7.425	—	[4]
Ho_3Rh_2	Y_3Rh_2	tetragonal	not determined	[6]
Ho_5Rh_3	Mn_5Si_3	8.100	6.337	[8]

Neutron crystal-field spectroscopy has been carried out on several rhodium-rare earth compounds, including HoRh; the crystal-field energy-level scheme for Ho^{3+} was obtained from the observed crystal-field transitions at temperatures in the range 6.7 to 300 K. The paper should be consulted for further details. **Fig. 57**, p. 98, and **Fig. 58**, p. 98, show the time-of-flight spectra and crystal-field level scheme for Ho^{3+} in HoRh [7].

Magnetic Properties. Ho_3Rh_2 should show ferromagnetism rather than superconductivity according to [5]. The low-temperature ferromagnetism of Ho_3Rh_2 has been shown in magnetic measurements of bulk polycrystalline material; measurements were made at temperatures between 1.5 and 300 K in fields up to 60 kOe. Fig. 50, p. 87, shows the thermal dependence of the initial susceptibility when the temperature was increased. Fig. 51, p. 88, shows the first magnetization curve measured at 1.5 K plotted against the applied magnetic field. The following properties were found for the compound: Curie temperature 19 K, paramagnetic transition temperature (θ_P) 11 K, effective magnetic moment 10.63 μ_B which agreed well with the calculated value of 10.60 μ_B, magnetization at 1.5 K in a field of 60 kOe 7.07 μ_B/Ho atom. The discrepancies or agreement between the experimental values of the magnetic moments of several rare earth-rhodium compounds of the R_3Rh_2 type suggest that

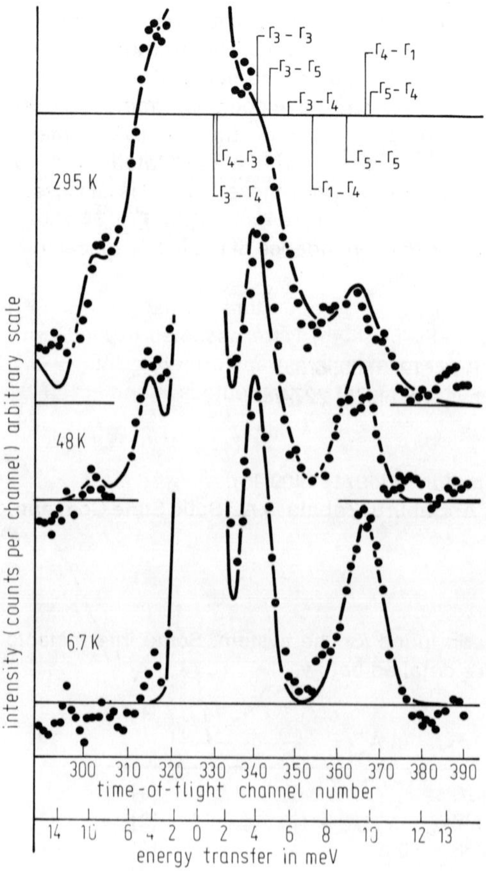

Fig. 57. Neutron time-of-flight spectra of HoRh.

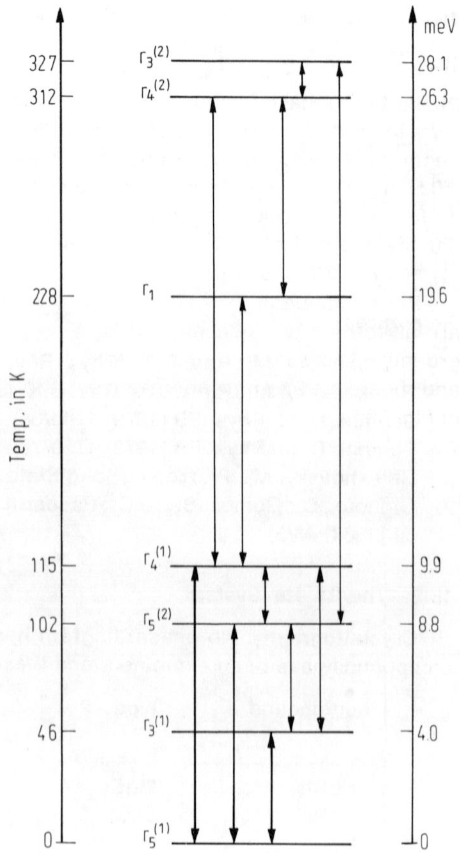

Fig. 58. Crystal field level scheme for Ho^{3+} in HoRh for x = 0.433, W = 0.0498 meV.

this could be due to the polarization of the rhodium 4d band by the spin of the added rare earth element [6].

With ^{165}Ho introduced as a dilute impurity in Gd_3Rh_2, the mass susceptibility and the temperature dependence of the g-value and the linewidth have been determined, see Fig. 50, p. 87 and Fig. 51, p. 88.

References:

[1] Dwight, A.E. (Trans. Am. Soc. Metals **53** [1961] 479/500).

[2] Dwight, A.E. (ANL-6868 [1963] 303/5).

[3] Dwight, A.E.; Conner, R.A.; Downey, J.W. (Acta Crystallogr. **18** [1965] 835/8).

[4] Loebich, O.; Raub, E. (J. Less-Common Metals **46** [1976] 1/6).

[5] Raub, C.J. (Platinum Metals Rev. **28** [1984] 63/75).

[6] Gignoux, D.; Gomez-Sal, J.C.; Paccard, D.; Aramburu-Zabala, J.A. (Solid State Commun. **50** [1984] 43/5).

[7] Rossat-Mignod, J.; Chamard-Bois, R.; Knorr, K.; Drexel, W. (Proc. 11th Rare Earth Res. Conf., Traverse City, Mich., 1974, Vol. 1, pp. 317/26).

[8] Le Roy, J.; Moreau, J.M.; Paccard, D. (J. Less-Common Metals **86** [1982] 63/7).

16 Alloys with Erbium

16.1 The Rh-Er System

Phase Diagram

The phase diagram has been investigated by differential thermal analysis, metallography and X-ray diffraction techniques; as in other rhodium-rare earth systems, this one shows a large number of intermediate phases, eight in this case [1]. Four of these had been identified before this work was completed, namely ErRh, $ErRh_2$, Er_3Rh, and Er_7Rh_3 [2, 3, 4].

The compounds Er_3Rh, Er_7Rh_3, Er_5Rh_3 (high- and low-temperature forms), and Er_3Rh_2 are all formed peritectically, whilst ErRh, $ErRh_2$, and $ErRh_6$(h) were assumed to melt congruently. This latter compound was found to undergo eutectoid transformation into $ErRh_2$ and rhodium containing some dissolved erbium. The system has four eutectics and is shown in **Fig. 59** [1].

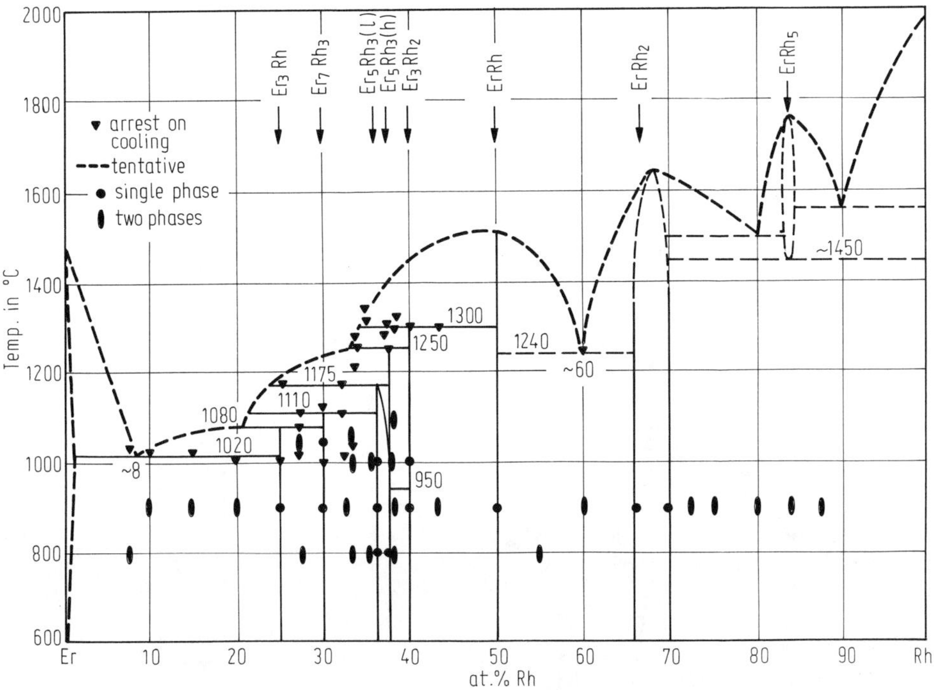

Fig. 59. Proposed equilibrium diagram Er-Rh.

Preparation. Alloys were prepared by arc-melting under argon and annealing was carried out with the samples wrapped in molybdenum foil in evacuated quartz capsules; annealing times varied between 4 days and 1 week, the longer times being given to anneals below 900 °C. The differential thermal analysis was carried out under vacuum and the estimated accuracy of the arrest temperatures in the diagram is ± 10 °C [1].

Crystallography

The table shows the available data; lattice parameters are in Å (explanations see p. 101).

phase	type	a	b	c	Ref.
Er[α]	Mg	3.559	–	5.587	[1]
Er[Rh]	Mg	3.567	–	5.608	[1]
Er_3Rh	Fe_3C	7.075	9.235	6.218	[4]
Er_7Rh_3	Th_7Fe_3	9.643	–	6.070	[4]
Er_5Rh_3 h	Mn_5Si_3 h	8.084	–	6.306	[1, 6]
Er_5Rh_3 l		not determined			[1]
Er_5Rh_3 [1)] h		8.122	–	6.259	[6]
Er_3Rh_2		tetragonal	not determined		[1]
Er_3Rh_2					[9]
ErRh	CsCl	3.372	–	–	[2]
ErRh	–	3.377 to 3.367	–	–	[1]
$ErRh_2$	$MgCu_2$	7.444	–	–	[3]
$ErRh_2$	Er-rich	7.465	–	–	[1]
$ErRh_2$	Rh-rich	7.414	–	–	[1]
$ErRh_5$ h	$CaZn_5$	5.118	–	4.292	[1]
Rh[Er]	Cu	3.806	–	–	[1]
Rh	Cu	3.804	–	–	[5]

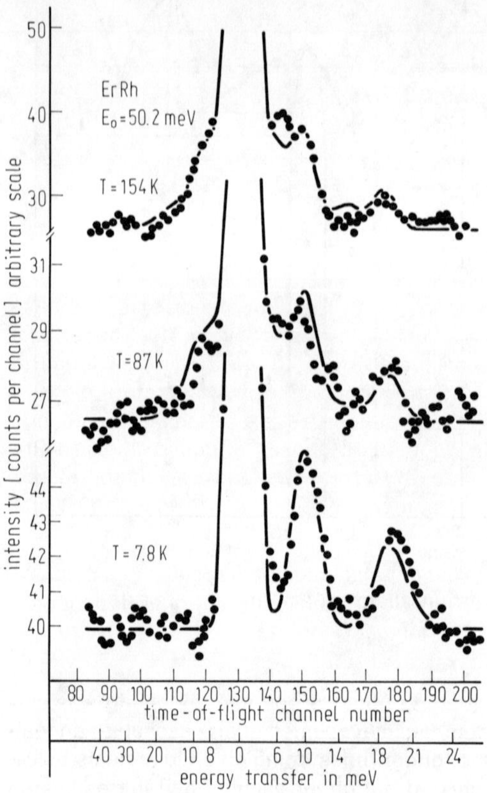

Fig. 60. Neutron time-of-flight spectra of ErRh.

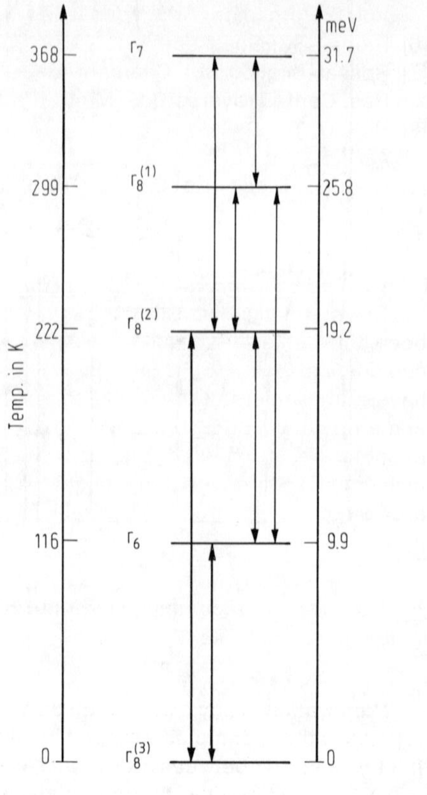

Fig. 61. Level scheme for Er^{3+} in ErRh.

h and l indicate high- and low-temperature forms, respectively; the square brackets used in the phase column indicate a solubility of Er or Rh; 1) = sample containing some oxygen and nitrogen.

Neutron Crystal Field Spectroscopy

This has been carried out on the compound ErRh and the crystal field parameters given as $A_4^0\langle r^4\rangle$ (meV) = -11.7 ± 1.6 and $A_6^6\langle r^6\rangle$ (meV) = -1.64 ± 0.25; the neutron time-of-flight spectra and level scheme for Er^{3+} in ErRh are shown in **Fig. 60** and **Fig. 61** [7].

Electron spin resonance measurements on rhodium alloys containing 0.183 to 0.3 at% erbium have been made; they showed a change in slope of the erbium resonance thermal broadening above ~7 K. This behaviour was interpreted as being caused by off-diagonal matrix elements of the exchange coupling between the Γ_7 ground state and the Γ_8 first excited state [8].

References:

[1] Ghassem, H.; Raman, A. (Metall. Trans. **4** [1973] 745/8).

[2] Dwight, A.E.; Conner, R.A.; Downey, J.W. (Acta Crystallogr. **18** [1965] 835/9).

[3] Dwight, A.E. (Trans. Am. Soc. Metals **53** [1961] 479/500).

[4] Raman, A. (J. Less-Common Metals **30** [1972] 199/206).

[5] Pearson, W.B. (A Handbook of Lattice Spacings and Structures of Metals and Alloys, Vol. 2, Pergamon, New York 1967).

[6] Raman, A.; Ghassem, H. (J. Less-Common Metals **30** [1973] 185/97).

[7] Rossat-Mignod, J.; Chamard-Bois, R.; Knorr, K.; Drexel, W. (Proc. 11th Rare Earth Res. Conf., Traverse City, Minn., 1974, Vol. 1, pp. 317/26).

[8] Rettori, C.; Davidov, D.; Kim, H.M. (AD-774090-5GA [1973] 16 pp.; C.A. **81** [1974] No. 43703).

[9] Gignoux, D.; Gomez-Sal, J.C.; Paccard, D.; Aramburu-Zabala, J.A. (Solid State Commun. **50** [1984] 43/5).

Magnetic Properties

The magnetic properties of lanthanide compounds with rhodium are diverse and have been the subject of much investigation and a good review is found in [1]. The erbium compounds $ErRh_2$, ErRh and Er_7Rh_3 have been studied by [2, 3, 4]. $ErRh_2$ was found to have a θ_c temperature of ~7 K [2]. ErRh had a paramagnetic change temperature of −4 K and a magnetic moment μ_{eff} of 9.4 μ_B [3]. Er_7Rh_3 was found to have a paramagnetic change temperature of $\theta_P = 26$ K with $\mu_{eff} = 9.22\ \mu_B$ [4]. Lanthanide phases of composition Ln_3Rh_2 including Er_3Rh_2 have been examined for their magnetic properties, and some of the results are given below. μ is the effective magnetic moment. Er_3Rh_2: $\theta_c = 8$ K, $\mu_{eff} = 9.59\ \mu_B$, $\mu_{calc} = 9.58\ \mu_B$.

There is good agreement between the experimental and calculated results. Fig. 51, p. 88, shows the first magnetization curve for Er_3Rh_2 and Fig. 50, p. 87, the thermal dependence of the initial susceptibility when the temperature is increased [5].

References:

[1] Loebich, O.; Raub, E. (Mater. Res. Bull. **10** [1975] 1017/22).

[2] Crangle, J.; Ross, J.W. (Proc. Intern. Conf. Nottingham. Inst. Phys., Phys. Soc., London 1964, p. 240).

[3] Chamard-Bois, R.; Nguyen Van Nhung; Pierre, J. (Phys. Status Solidi B**49** [1972] 161/6).

[4] Olcese, G.L. (J. Less-Common Metals **33** [1973] 71/81).
[5] Gignoux, D.; Gomez-Sal, J.C.; Paccard, D.; Aramburu-Zabala, J.A. (Solid State Commun. **50** [1984] 43/5).

16.2 Ternary Alloy RhErIn

The compound RhErIn was found to have an Fe_3P-type structure with a=7.461, c=3.837 Å; the alloys were prepared by induction melting under argon, the crucible material not being specified. The formation of the alloy was strongly exothermic, and the molten material very brittle; annealing was carried out for 150 to 200 h at 500 °C.

Ferro, R; Marazza, R; Rambaldi, G. (Z. Anorg. Allgem. Chem. **410** [1974] 219/24).

17 Alloys with Thulium, Ytterbium and Lutetium

17.1 The Rh-Tm System

Crystallography. No phase diagram has been produced for the system; knowledge is confined to some intermetallic compounds, and this data is given below:

compound	type	a (in Å)	c (in Å)	Ref.
$TmRh_2$	$MgCu_2$	7.416	—	[1]
TmRh[1)]	CsCl (B2)	3.358±0.002	—	[2]
Tm_5Rh_3	Mn_5Si_3	8.058(5)	6.252(5)	[3]

[1)] = sample contained 2% second phase

Mössbauer Spectroscopy. The cubic Laves phase $TmRh_2$ was examined using ^{169}Tm in the temperature range 2 to 300 K, and well resolved quadrupolar splitting found; this was taken to indicate the presence of static or dynamic distortion. The compound did not show magnetic ordering at temperatures ≧2 K, nor could it be characterized as a van Vleck paramagnet at low temperatures. **Fig. 62** shows the ^{169}Tm Mössbauer spectrum; **Fig. 63**, p. 104, shows the temperature dependence of the quadrupolar splitting in $TmRh_2$ [4].

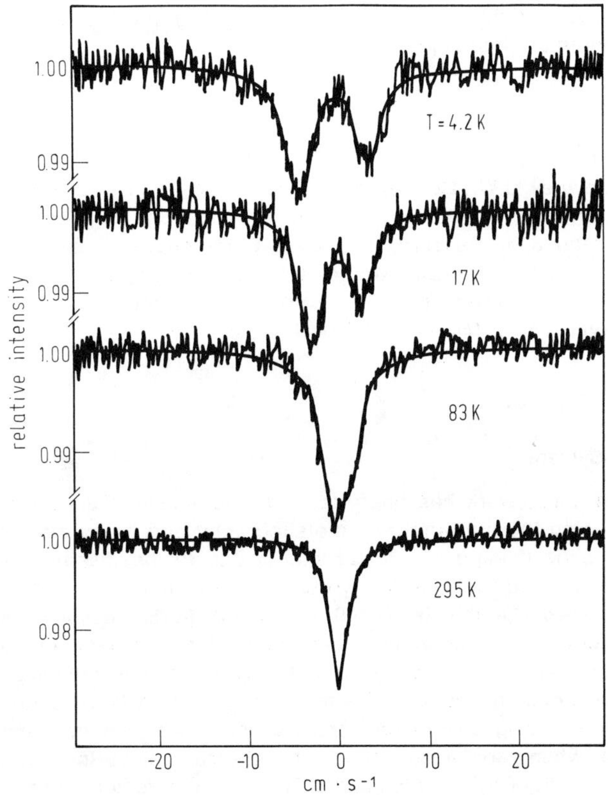

Fig. 62. ^{169}Tm Mössbauer spectra of $TmRh_2$.

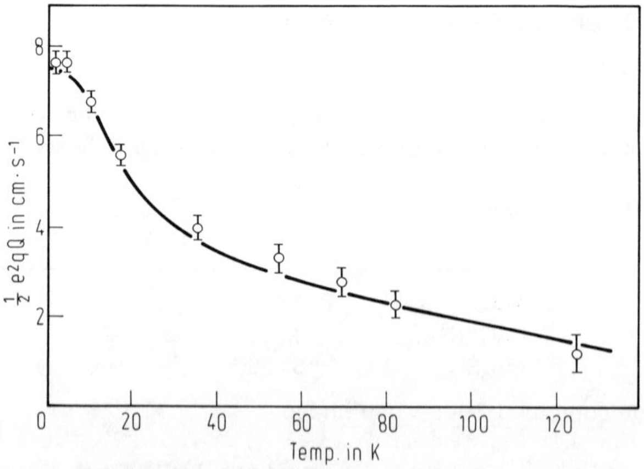

Fig. 63. Temperature dependence of the quadrupolar splitting of $TmRh_2$.

Neutron Crystal Field Spectroscopy. This has been used to examine a number of RRh compounds where R=Tb, Ho, Er, and Tm, and the crystal field parameters have been determined. The polycrystalline samples used were melted in a high-frequency levitation furnace. For the time of flight spectrum (with that of YRh) and the level scheme for Tm^{3+} in TmRh see figures in original paper [5].

References:

[1] Dwight, A.E.; Downey, J.W.; Conner, R.A. (Trans. Metall. Soc. AIME **236** [1966] 1509/10).

[2] Dwight, A.E.; Conner, R.A.; Downey, J.W. (Acta Crystallogr. **18** [1965] 835/8).

[3] Le Roy, J.; Moreau, J.M.; Paccard, D. (J. Less-Common Metals **86** [1982] 63/7.

[4] Gubbens, P.C.M.; van der Kraan, A.M.; Buschow, K.H.J. (J. Phys. F**14** [1984] 2195/201).

[5] Rossat-Mignod, J.; Chamard-Bois, R.; Knorr, K.; Drexel, W. (Proc. 11th Rare Earth Res. Conf., Traverse City, Mich., 1974, Vol. 1, pp. 317/26).

17.2 The Rh-Yb-System

Phase Diagram. Little work has been done on the system, but a partial diagram has been determined using thermal analysis, metallography and X-ray diffraction techniques; some of the very difficult experimental problems due to the disparity in melting points of the constituents and the high vapour pressure of ytterbium are outlined in the paper. Ytterbium melts at ~815 °C and boils at 1700 °C at atmospheric pressure. At higher ytterbium contents, the problem was overcome by encapsulating the crucible in tantalum or molybdenum sheet, and thermal analysis was possible. However, these difficulties prevented thermal analysis of the alloys high in rhodium. The following compounds were found: YbRh, $YbRh_2$ and $YbRh_5$, the latter being a high-temperature phase which decomposed into rhodium and $YbRh_2$, except when rapidly quenched. The structure of $YbRh_5$ was not determined; it is not isomorphous with Rh_5Er, which has a $CaCu_5$-type structure. The diagram is shown in **Fig. 64** [1].

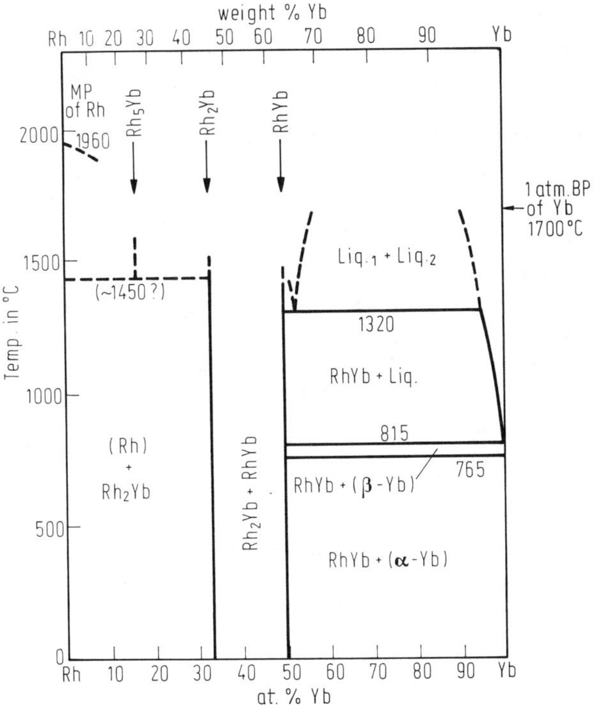

Fig. 64. Phase diagram Rh-Yb.

Crystallography. The available data is given below:

compound	type	a in Å	Ref.
YbRh	CsCl	3.347 [1)]	[1]
YbRh	CsCl	3.328 [2)]	[1]
$YbRh_2$	$MgCu_2$	7.432 [1)]	[1]
$YbRh_2$	$MgCu_2$	7.432	[2]
$YbRh_2$	$MgCu_2$	7.379 [2)]	[1]

1) indicates stoichiometric composition; 2) indicates rhodium-rich composition

Rh_2Yb is of $MgCu_2$ type, a = 7.432 Å (stoichiometric) or a = 7.379 Å (Rh-rich); RhYb is of CsCl type, a = 3.347 Å (stoichiometric) or a = 3.328 Å (Rh-rich). Rh_5Yb is stable only at high temperature; the structure is unknown, except that it is not isomorphous with Rh_5Er ($CaCu_5$ type) which is similarly a high-temperature phase. Decomposition temperature of Rh_5Yb not stated, but shown in the diagram of [2] at about 1450 °C (temperature for Rh_5Er) [1].

References:

[1] Iandelli, A.; Palenzona, A. (Rev. Chim. Miner. **13** [1976] 55/61).
[2] Elliott, R.P. (IITRI-578P19-13 [1964] 35 pp.; C.A. **62** [1965] 11488).

17.3 The Rh–Lu System

No phase diagram has been found. Some intermetallic compounds have been examined, mainly because of interest in the possibility of their being superconducting.

Crystallography. The available data is given below:

compound	type	a in Å	Ref.
LuRh	CsCl	3.525	[1]
LuRh	B2	3.334	[2]
$LuRh_2$	$MgCu_2$	7.412	[1]
$LuRh_2$	C 15	7.404	[2]
$LuRh_{\sim 2.6}$ [1)]	C 15	7.355	[2]
$LuRh_5$	not determined		[2]

1) non-stoichiometric Laves phase of composition $Lu_{0.275}Rh_{0.725}$

Superconductivity. There are two superconducting compounds at the rhodium-rich end of the system, a non-stoichiometric Laves phase and $LuRh_5$. The C 15-type compound $LuRh_2$ is not superconducting. $Lu_{0.275}Rh_{0.725}$ has a superconducting transition temperature of 1.27 K, whilst that of $LuRh_5$ is 0.49 K. The transition temperature plotted against composition is shown in **Fig. 65** [2].

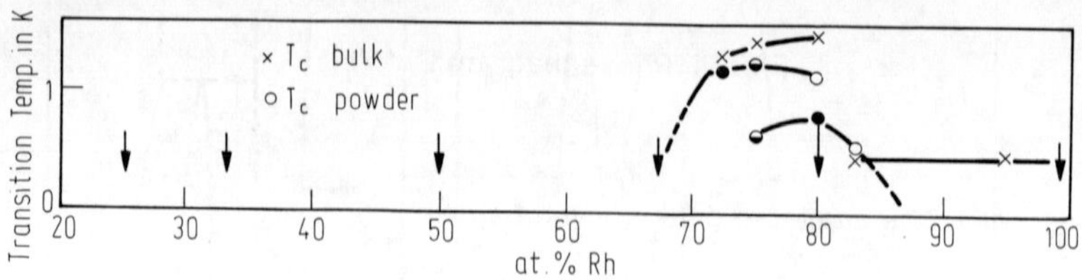

Fig. 65. Transition temperature in Lu–Rh alloys.

References:

[1] Dwight, A.E. (ANL-6516 [1961] 259/60).

[2] Geballe, T.H.; Matthias, B.T.; Compton, V.B.; Corenzwit, E.; Hull, G.W.; Longinotti, L.D. (Phys. Rev. [2] **137** [1965] A 119/A 127).

18 Alloys with Titanium

18.1 The Rh-Ti System

Diffusion. Studies of interdiffusion in thin films of Ti-Rh and some ternary alloys have been carried out using Rutherford backscattering. The binary system showed very low levels of interdiffusion at temperatures of 400 °C or less [10].

Phase Diagram. The rhodium-titanium equilibrium has not yet been completely established, and further work needs to be done. Two important practical investigations have been made on this system, and it is from these that the diagrams in **Fig. 66** [1] and **Fig. 67**, p. 108 [2] have been taken.

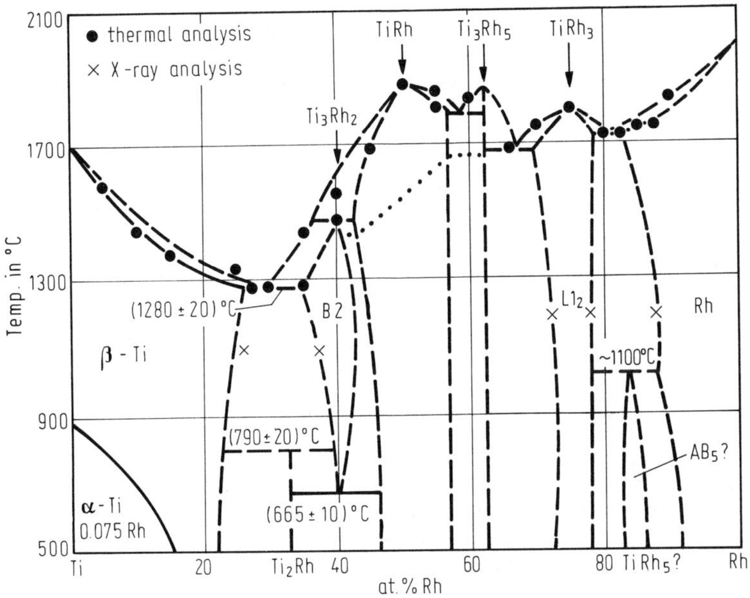

Fig. 66. Proposed phase diagram Rh-Ti.

The available evidence has been assessed, modified diagrams derived from the above papers were drawn and compared with a diagram calculated from thermodynamic data by [3]. Another thermodynamically calculated diagram has been drawn by [4].

The principal areas of disagreement in the two main practical investigations which were carried out in both cases using thermal analysis, metallography and X-ray diffraction methods, are the extent of the single phase β-titanium field, the method of formation of the compound Ti_2Rh, the existence of the compounds Ti_3Rh_5 and $TiRh_5$ and also whether $TiRh_3$ forms congruently from the melt or peritectically. As well, the method of decomposition of TiRh and its resultant crystallography demanded more investigation, and this has recently been carried out by X-ray diffraction studies between room temperature and 1000 °C; the high-temperature cubic form was found to transform to a tetragonal structure at 845 ± 20 °C and to a monoclinic one at 83 ± 5 °C. The transition appeared to occur continuously [5]. It is recommended that the individual papers are consulted.

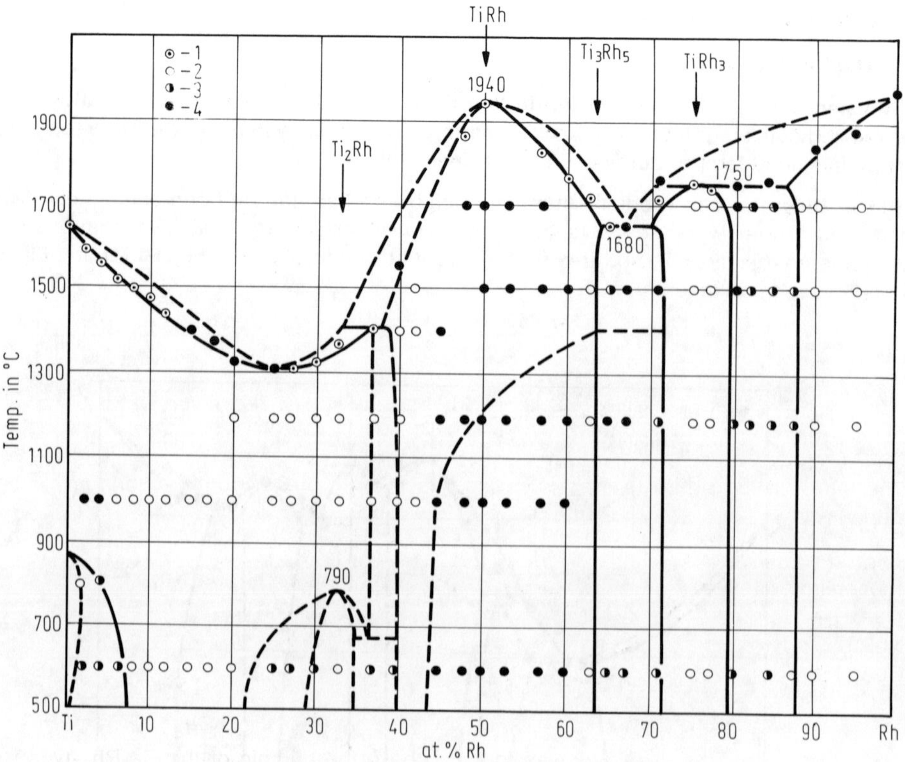

Fig. 67. Tentative phase diagram Rh-Ti.

Crystallography. The original X-ray work on the system showed seven intermediate phases [6]. All these and an additional one, $TiRh_5$, were found by [1]; not all were found by [2]. The structure of Ti_3Rh_5 has been very carefully examined by [7]. The following table gives a summary of the available data.

at% Rh	phase	a	b in Å	c	annealing t in °C	Ref.
5	β-Ti	3.264	—	—	—	[8]
10	β-Ti	3.242	—	—	—	[8]
12.5	β-Ti	3.240	—	—	—	[8]
15	β-Ti	3.228	—	—	—	[8]
12	β-Ti	3.217	—	—	700	[1]
15	β-Ti	3.214	—	—	600 to 1000	[1]
20	β-Ti	3.178	—	—	600 to 650	[1]
25	β-Ti	3.164	—	—	800 to 1100	[1]
27.5	β-Ti	3.157	—	—	1100	[1]
33.3	Ti_2Rh	3.078	—	9.882	—	[14, 16]
62.5	Ti_3Rh_5	5.36	10.42	4.08	—	[7]
75	$TiRh_3$	3.845	—	—	600 to 900	[12]
75	$TiRh_3$	3.821	—	—	—	[1]

	at% Rh	phase	a	b in Å	c	annealing t in °C	Ref.
	77.5 to 80	$TiRh_3$	3.815	—	—	600 to 1630	[1]
	73 to 80	$TiRh_3$	3.823	—	—	1500	[14, 16]
a	50	TiRh	2.735	—	3.679	—	[13]
b	50	TiRh	2.96	2.86	3.41	—	[16]
c	50	TiRh	4.173	—	3.54	—	[14]
d	55	—	4.15	4.111	3.40	—	[15]
e	57	—	4.11	—	—	—	[15]
f	35	—	3.11	—	—	—	[6]
g	45	—	4.27	—	3.33	1190 to 1300	[1]
h	55	—	3.12	—	—	640 to 1630	[1]
i	37.5	—	3.120	—	—	1200	[1]
j	40	—	3.121	—	—	800 to 1200	[1]
k	42.5	—	3.100	—	—	1200	[1]

The letters a to k indicate the structures observed in these roughly equiatomic alloys as follows:

a = tetragonal distorted CsCl structure
b = monoclinic, $\alpha = 90°37'$
c = CuAu I
d = NbRu
e = CuAu I
f = CsCl
g = CuAu I
h = CuAu I
i = CsCl
j = CsCl
k = CsCl

The crystal structure of Ti_3Rh_5 was found to be orthorhombic of the Ge_3Rh_5-type; the structure is shown in (001) projection in **Fig. 68** [7]. The X-ray diffraction results obtained on RhTi between room temperature and 1000 °C are given below [5].

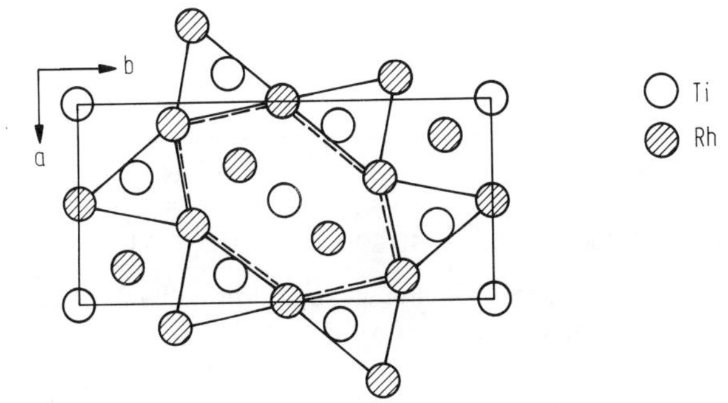

Fig. 68. Crystal structure of Ti_3Rh_5 shown in (001) projection.

t in °C	structure	a	b	c	γ
		lattice parameters in Å			
1000	cubic CsCl-type	3.126	—	—	—
83	tetragonal CuAu	4.226	—	3.350	—
25	monoclinic	4.178	4.185	3.371	90.89°

Recrystallization Temperature. Small additions of up to 0.1 at% rhodium were found to raise the recrystallization temperature of titanium [11].

References:

[1] Raub, E.; Röschel, E. (Z. Metallk. **57** [1966] 546/51).
[2] Jeremenko, W.; Stepa, T. (Colloq. Intern. Centre Natl. Rech. Sci. [Paris] No. 205 [1972] 403/13).
[3] Murray, J.L. (Bull. Alloy Phase Diagrams **3** [1982] 335/42).
[4] de Fontaine, D. (Mater. Sci. Forum **37** [1989] 25/38).
[5] Yi, S.S.; Chen, B.H.; Franzen, H.F. (J. Less-Common Metals **143** [1988] 243/9).
[6] Raman, A.; Schubert, K. (Z. Metallk. **55** [1964] 704/10).
[7] Giessen, B.C.; Wang, R.; Grant, N.J. (Trans. Metall. Soc. AIME **245** [1969] 1207/10).
[8] Buckel, W.; Dummer, G.; Gey, W. (Z. Angew. Physik **14** [1962] 703/6).
[9] Raub, C.J. (Z. Physik **178** [1964] 216/20).
[10] De Bonte, W.J.; Poate, J.M.; Melliar-Smith, C.M.; Levesque, R.A. (J. Appl. Phys. **46** [1975] 4284/90).

[11] Abrahamson, E.P. (Trans. Metall. Soc. AIME **224** [1962] 265/7).
[12] Dwight, A.E.; Beck, P.A. (Trans. Metall. Soc. AIME **215** [1959] 976/9).
[13] Dwight, A.E. (ANL-6677 [1962] 258/9).
[14] Schubert, K.; Meissner, H.G.; Raman, A.; Rossteutscher, W. (Naturwissenschaften **51** [1964] 506/7).
[15] Schubert, K.; Meissner, H.G.; Raman, A.; Rossteutscher, W. (Naturwissenschaften **51** [1964] 287/8).
[16] Eremenko, V.N.; Shtepa, T.D.; Sirotenko, V.G. (Poroshk. Metall. **6** [1966] 68/72; Soviet Powder Metall. Metal Ceram. **37** [1966] 487/90).

Mechanical and Thermal Properties

Density. The measured and calculated densities for Ti_2Rh, TiRh, and $TiRh_3$ are given in the following table [1].

compound	density in g/cm^3 calculated	measured
Ti_2Rh	6.88	6.76
TiRh	8.66[1)]	8.31
$TiRh_3$	10.60	10.51

1) Calculated on the assumption of two atoms per elementary cell

Hardness. The microhardness of Ti_2Rh is given as 900, TiRh as 500, and $TiRh_3$ as 550 daN/mm^2 by [1]. The table below gives the hardness H_v 5 of some alloys containing up to 8 at% rhodium after quenching and tempering [2] (Table see p. 111).

The hardness of Rh_3Ti fell from about 220 to 140 VPN (10 mPa) between 300 and 1200 K [3].

Specific Heat. This has been measured between 0.9 and 8 K for alloys containing 0 to 2 at% rhodium for hexagonal and 3 to 10 at% rhodium for cubic material; it was found to increase with rhodium concentration [5]. Measurements between 0.4 and 4 K for rhodium

at% Rh	H_v 5 in kg/mm^2 after quenching from 950 °C	after tempering (t in °C) 200	400	600
0.2	363	291	218	257
0.5	362	412	232	232
1.0	442	362	271	331
2.0	442	412	332	319
4.5	413	401	386	378
6.0	442	362	441	362
8.0	362	363	441	383

concentrations between 0 and 3 at% were made by [7]. Results from the former paper are given in **Fig. 69** [5]. The specific heats for the compounds Ti_2Rh, TiRh, and $TiRh_3$ are given as 1.58, 2.46, and 2.85 $mJ \cdot g-at^{-1} \cdot K^{-2}$ [6].

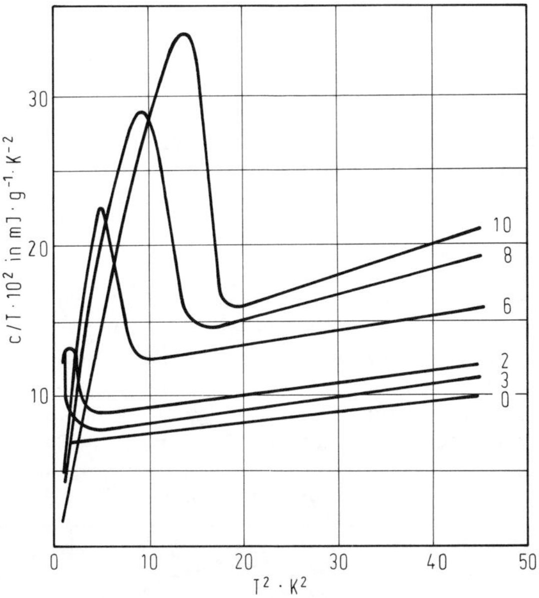

Fig. 69. Specific heat of Rh-Ti alloys. The figures at the curves denote Rh concentration in at%.

Thermodynamic Data. The dissociation energy of the molecule TiRh has been determined as 88.9±5 kcal/mol; this value supports the Brewer-Engel metallic theory [8]. The thermodynamic parameters used in the calculation of the equilibrium diagram by [9] were obtained from [10]. The standard enthalpy of formation of RhTi has been determined by high-temperature mixing calibrimetry at 1400 K; $\Delta H = -143.0 \pm 9.8$ kJ/mol [4].

Debye Temperature. The following table gives some of the figures.

at% Rh	θ in K	Ref.
1	385	[5]
2	378	[5]
3	379	[5]
6	315	[5]
8	290	[5]
10	271	[5]
33.3	478	[6]
50	383	[6]
75	526	[6]

References:

[1] Eremenko, V.N.; Shtepa, T.D.; Sirotenko, V.G. (Poroshk. Metall. **6** [1966] 68/72; Soviet Powder Metall. Metal Ceram. **37** [1966] 487/90).
[2] Raub, E.; Röschel, E. (Z. Metallk. **57** [1966] 546/51).
[3] Wee, D.M.; Suzuki, T. (Trans. Japan Inst. Metals **20** [1979] 634/46).
[4] Topor, L.; Kleppa, O.J. (J. Less-Common Metals **135** [1987] 67/75).
[5] Dummer, G.; Oftedal, E. (Z. Physik **208** [1968] 238/248).
[6] Kuentzler, R.; Waterstrat, R.M. (Solid State Commun. **68** [1988] 85/91).
[7] Danner, S.; Dummer, G. (Z. Physik **222** [1969] 243/52).
[8] Gingerich, K.A.; Cocke, D.L. (J. Chem. Soc. Chem. Commun. **9** [1972] 536).
[9] Murray, J.L. (Bull. Alloy Phase Diagrams **3** [1982] 335/42).
[10] Kaufman, L.; Bernstein, H. (Computer Calculation of Phase Diagrams, Academic, New York 1970).

Electrical and Magnetic Properties

Electronic Structure. The electronic properties of TiRh, Ti_2Rh and $TiRh_3$ have been studied by low-temperature specific heat, magnetic susceptibility and resistance measurements [1]. The cubic → tetragonal phase transition of the compound RhTi has been studied using band structure calculations; it was found that the Jahn-Teller effect arises from the 25% filled Ti 3d e_g-like band. The lowering of the energy of the compound by the tetragonal distortion was calculated to be 0.04 eV/unit cell. The density for RhTi is shown in a figure of the original paper [2].

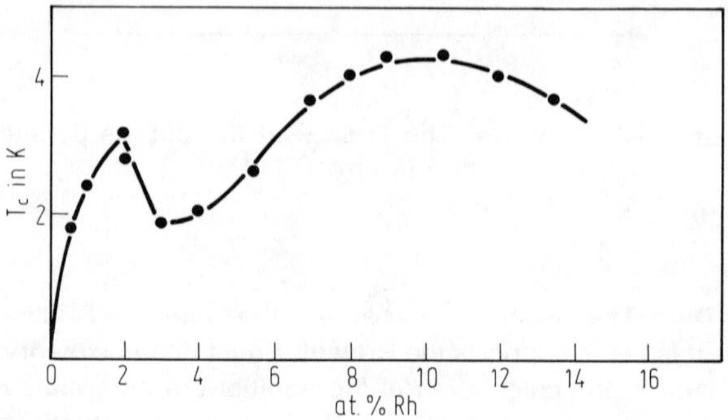

Fig. 70. Transition temperatures T_c of Rh-Ti alloys as a function of Rh contents.

Magnetic Susceptibility. The three compounds Ti_2Rh, TiRh, and $TiRh_3$ are all paramagnetic with low susceptibilities independent of the field; the values found were 66.1, 75.8, and 66.1×10^{-6} emu/g-atom [1].

Superconductivity. An investigation of the effect of rhodium on α- and β-titanium with respect to the transition temperature for superconductivity showed that small amounts dissolved in α-titanium increased T_c by a factor of 2 to 4; the results are shown in **Fig. 70** [3]. The following table shows some other results.

at% Rh	T_c in K	Ref.
0.5	0.79	[4]
1	1.16	[4]
1.5	1.6	[4]
2	1.73	[4]
2.5	1.79	[4]
3	1.34	[4]
1	0.9	[5]
2	1.7	[5]
3	1.0	[5]
6	2.6	[5]
8	3.5	[5]
10	4.0	[5]
33.3	<1.4	[1]
50	<1.4	[1]
75	<1.4	[1]

Fig. 71 summarizes the results obtained by [6] and [7]. The results of the former paper are at variance with those of [5] and [4] at low concentrations of rhodium.

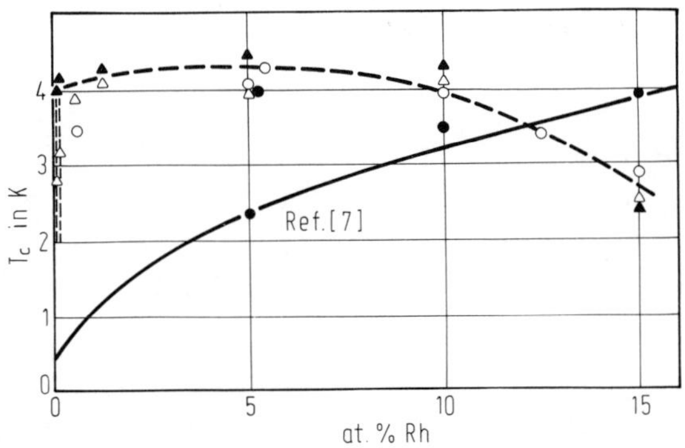

Fig. 71. Transition temperatures of Rh-Ti alloys according to [6] and [7]; empty circles = massive samples, not annealed; full circles = massive samples, annealed; empty triangles = chips, not annealed; full triangles = chips, annealed. The vertical dotted lines near the values for alloys with very low Rh contents indicate that these samples did not become completely superconducting down to ~2 K [6].

References:

[1] Kuentzler, R.; Waterstrat, R.M. (Solid State Commun. **68** [1988] 85/91).
[2] Folkerts, W.; Haas, C. (J. Less-Common Metals **147** [1989] 181/4).
[3] Raub, C.J.; Hull, G.W. (Phys. Rev. [2] **133** [1964] A932/A934).
[4] Danner, S.; Dummer, G. (Z. Physik **222** [1969] 243/52).
[5] Dummer, G.; Oftedal, E. (Z. Physik **208** [1968] 238/48).
[6] Buckel, W.; Dummer, G.; Gey, W. (Z. Angew. Physik **14** [1962] 703/6).
[7] Matthias, B.T. (IBM J. Res. Develop. **6** [1962] 250/5).

18.2 Ternary Alloys

Rh-Ti-As. The 3d and 4d transition metal ordering in RhTiAs was studied using Mössbauer spectroscopy, neutron and X-ray diffraction and magnetic measurements; the order could be correlated with the non-metal surrounding of the transition metal site. The effects of size factor, d-electron number, crystalline field and s-d electron transfer are considered [3]. The compound RhTiAs was found to have an orthorhombic structure of type Co_2P with a = 6.334, b = 3.816, c = 7.388 Å [4].

Rh-Ti-Sb. The structure of the compound RhTiSb is of type MgAgAs with a = 6.088 Å [1].

Rh-Ti-Al. A new G-type phase in this system was found with lattice parameter a = 12.135 Å [2].

Rh-Ti-Ga. The structure of RhTiGa was found to be of type MgAgAs with a = 6.11 Å [1].

References:

[1] Dwight, A.E. (J. Less-Common Metals **34** [1974] 279/84).
[2] Ganglberger, E.; Nowotny, H.; Benesovsky, F. (Monatsh. Chem. **97** [1966] 829/32).
[3] Sénateur, P.J.P.; Fruchart, D.; Boursier, D.; Rouault, A. (J. Phys. Colloq. [Paris] **38** [1977] C7-61/C7-66).
[4] Deyris, B.; Roy-Montreuil, J.; Rouault, A.; Krumbügel-Nylund, A.; Sénateur, J.-P.; Fruchart, R.; Michel, A. (Compt. Rend. C **278** [1974] 237/9).

19 Alloys with Zirconium

19.1 The Rh-Zr System

Phase Diagram

Superconductivity in rhodium-zirconium alloys was discovered by [1]. This has encouraged a number of investigations on the equilibrium diagram [2 to 5]. These resulted in the diagram shown in [6]. The diagram shown in **Fig. 72** and **Fig. 73**, p. 116, is the result of the most complete investigation so far carried out using differential thermal analysis, levitation thermal analysis, X-ray analysis, electron microprobe analysis, metallography and superconductivity measurements [7]. The diagram is a complex one with some phases transforming martensitically and consequently the equiatomic region is shown enlarged in the second figure; reactivity of the constituents with gases and refractories poses many experimental problems in such work, and the role of oxygen in stabilizing a Ti_2Ni phase found by some investigators has been demonstrated by [8, 9]. More recently, the effect of hydrogen on the superconducting properties has been shown [10].

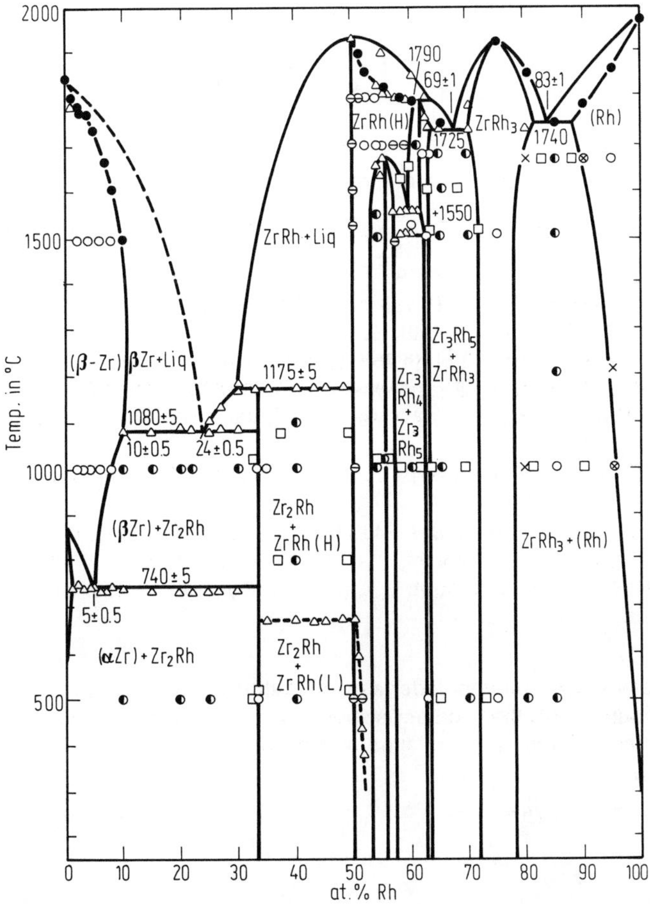

Fig. 72. The Zr-Rh phase diagram: ● levitation; △ DTA; × phase limits from X-rays; □ phase limits from microprobe; ◐ two-phase sample; ○ single-phase sample; ⊖ samples showing a martensitic-like pattern.

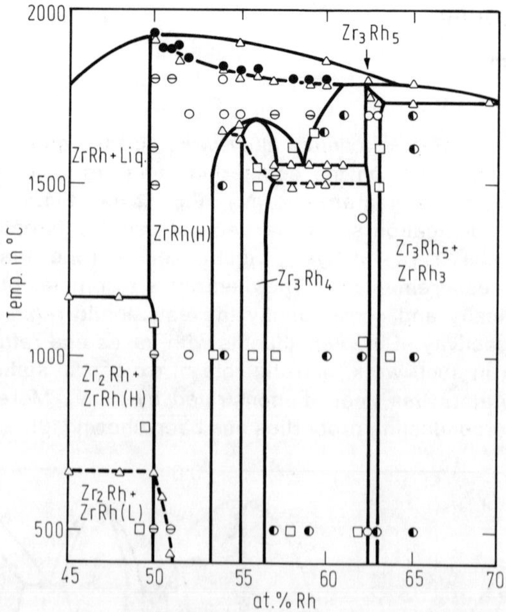

Fig. 73. Equiatomic region of the Rh-Zr system.

References:

[1] Matthias, B.T. (Phys. Rev. [2] **97** [1955] 74/6).

[2] Raub, C.J.; Andersen, C.A. (Z. Physik **175** [1963] 105/14).

[3] Raman, A.; Schubert, K. (Z. Metallk. **55** [1964] 704/10).

[4] Zegler, S.T. (J. Phys. Chem. Solids **26** [1965] 1347/9).

[5] Eremenko, V.N.; Semenova, E.L.; Shmena, T.D. (Metallofizika **52** [1974] 112/6).

[6] Moffat, W.G. (Handbook of Binary Phase Diagrams, Vol. 4, Gen. Elect. C., Schenectady, N.Y., 1978).

[7] Jorda, J.L.; Graf, T.; Schellenberg, L.; Muller, J. (J. Less-Common Metals **136** [1987/88] 313/28).

[8] Nevitt, M.V.; Schwartz, L.H. (Trans. Metall. Soc. AIME **212** [1958] 700/3).

[9] Nevitt, M.V.; Downey, J.W.; Morris, R.A. (Trans. Metall. Soc. AIME **218** [1960] 1019/23).

[10] Narang, P.P.; Taylor, K.N.R.; Paul, G.L. (J. Less-Common Metals **70** [1980] P25/P37).

Preparation

The materials were arc-melted under an argon atmosphere from 99.9% rhodium powder and 99.9% zirconium ingot; the rhodium powder was premelted to avoid spitting. Annealing was carried out under high vacuum, and a diamond saw was used for cutting. No gas analyses are given.

Jorda, J.L.; Graf, T.; Schellenberg, L.; Muller, J. (J. Less-Common Metals **136** [1987/88] 313/28).

Crystallography

The diagram presented shows six intermediate single-phase areas in addition to the terminal solid solutions and rhodium. The existence of the compound Zr_3Rh_4 was confirmed, but the structure not determined; the compound Zr_3Rh with the Ti_2Ni structure was not found, and its presence in some earlier work was attributed to either oxygen contamination or crystallization from amorphous material; the latter phenomenon has been reported where the experimental conditions would seem to exclude oxygen contamination [11, 12]. The lattice parameters obtained experimentally by various investigators are given in the table below.

Temperatures are in °C, in the heat treatment column, Q = quenched, R = reacted in calorimeter, A = annealed. With regard to figures from [7], except where preface R has been used, the at% rhodium is that found by microprobe analysis.

at% Rh	heat treatment	phases	lattice parameter in Å a	b	c	Ref.
0.1 to 1.0	Q1000	α'		martensitic		[1]
3	Q1000	ω	5.055 ± 0.009		3.103 ± 0.005	[1]
3.5	—	β-Zr	3.568	—	—	[2]
4	—	β-Zr	3.568	—	—	[2]
5	—	β-Zr	3.565	—	—	[2]
5	Q1500	β-Zr	3.510	—	—	[3]
7	Q1000	β-Zr	3.549	—	—	[1]
7.5	—	β-Zr	3.538	—	—	[2]
8.5	Q1065	β-Zr	3.546	—	—	[1]
9	—	β-Zr	3.542	—	—	[2]
10	Q1000	β-Zr	3.500	—	—	[3]
12	—	Zr_3Rh	3.540	—	—	[2]
—	—	Zr_2Rh	6.4937 ± 0.0005		5.6058 ± 0.0006	[1]
25	—	Zr_3Rh	$E9_3$-type			[2]
32	Q1000	Zr_2Rh	6.493		5.604	[3]
33	R1130	Zr_2Rh	6.490		5.605	[3]
37	Q800	Zr_2Rh	6.493		5.604	[3]
37	Q1000	Zr_2Rh	6.492		5.607	[3]
50	—	ZrRh "1−"	ZrIr-type		—	[4]
50	—	ZrRh "1+"	CsCl-type		—	[4]
55	A1200?		distorted	CsCl-type	—	[4]
49	Q800	ZrRh	3.318a		—	[3]
49	Q1800	ZrRh	3.264a		—	[3]
50	Q1000	ZrRh	3.318a		—	[3]
50	R1130	ZrRh	3.305a		—	[3]
57	cast?	Zr_3Rh_4	12.46	Fe_3W_3C-type		[4]
57	Q1500	Zr_3Rh_4	—		—	[3]
58	Q1600	Zr_3Rh_4	—		—	[3]
62	Q1500	Zr_3Rh_5	8.663	6.986	8.622	[3]
62.5	—	Zr_3Rh_5	4.40	4.33	3.42	[4]
63	Q1500	Zr_3Rh_5	—		—	[3]
64	Q1000	Zr_3Rh_5	—		—	[3]
70	Q1000	$ZrRh_3$	3.923		—	[3]
72	Q1500	$ZrRh_3$	3.922		—	[3]
74.85	7D/1000	$ZrRh_3$	3.927	Cu_3Au	—	[5]
75	—	$ZrRh_3$	3.927	Cu_3Au-type		[6]

at% Rh	heat treatment	phases	lattice parameter in Å a	b	c	Ref.
75	–	$ZrRh_3$	3.926	Cu_3Au-type		[4]
75	–	$ZrRh_3$	3.921(1)	Cu_3Au-type		[9]
75	R 1130	$ZrRh_3$	3.900	–		[3]
81	Q 1000	$ZrRh_3$	3.891	–		[3]
83	Q 1600	$ZrRh_3$	3.887	–		[3]
90	Q 1000	Rh	3.835	–		[3]
90	Q 1600	Rh	3.840	–		[3]

References:

[1] Zegler, S.T. (J. Phys. Chem. Solids **26** [1965] 1347/9).
[2] Raub, C.J.; Andersen, C.A. (Z. Physik **175** [1963] 105/14).
[3] Jorda, J.L.; Graf, T.; Schellenberg, L.; Muller, J. (J. Less-Common Metals **136** [1987/88] 313/28).
[4] Raman, A.; Schubert, K. (Z. Metallk. **55** [1964] 704/10).
[5] Wee, D.M.; Suzuki, T. (Trans. Japan Inst. Metals **20** [1979] 634/46).
[6] Dwight, A.E.; Beck, P.A. (Trans. Metall. Soc. AIME **215** [1959] 976/9).
[7] Wang, F.E. (J. Appl. Phys. **38** [1967] 822/4).
[8] Wang, F.E.; Ernst, D.W. (J. Appl. Phys. **39** [1968] 2192/5).
[9] Gachon, J.C.; Charles, J.; Hertz, J.; Jorda, J.L. (J. Calorim. Anal. Therm. Thermodyn. Chim. **17** [1986] 125/9).
[10] Giessen, B.C.; Wang, R.; Grant, N.J. (Trans. Metall. Soc. AIME **245** [1969] 1207/10).
[11] Cemuzal, K.; Chabot, B.; Parthé, E. (Acta Crystallogr. C **41** [1985] 313/9).
[12] Cantrell, J.S.; Wagner, J.E.; Bowman, R.C. (J. Appl. Phys. **57** [1985] 545/53).

Crystal Structure of the Compounds

Zr_2Rh. This was found to be a peritectic compound melting at 1175±5 °C, with a $CuAl_2$-type structure and lattice parameters a=6.493(2), c=5.604(1) Å [2]. These results agreed well with those of [1].

ZrRh. This compound was found to melt congruently by both [2, 3]. The melting temperature is 1935 °C according to the former reference and 1890 °C according to the latter. It is now generally agreed that the high-temperature form has a CsCl-type structure and a=3.255(3) Å [2]. The structure of the low-temperature form was determined as ZrIr-type by [4].

The martensitic transformation in alloys of equiatomic composition has been investigated by single-crystal X-ray diffraction methods [5, 6]. A martensitic transformation was found at 500 °C by [7].

Zr_3Rh_4. Confirmation of the existence of this compound formed by solid state reaction at 1660±20 °C is given by [2]. It was first found by [4]. The latter workers assigned it an Fe_3W_3C-type structure with a=12.46 Å. The structure has not been satisfactorily determined.

Zr_3Rh_5. The crystal structure was examined with those of other A_3B_5 compounds and the authors noted that the structure was not of the Ge_3Rh_5-type found in the corresponding titanium and hafnium compounds [9]. This compound was found to be isotypic with Pu_3Pd_5 with space group Cmcm, which can be derived from a CsCl-type structure. The lattice parameters were found to be a = 8.66(3), b = 6.986(1), c = 8.622(1) Å [8].

$ZrRh_3$. This compound was first reported by [10] and confirmed by [4]. It was found to melt congruently at 1920 °C, the structure was of the Cu_3Au-type with lattice parameter a = 3.887(1) Å at the rhodium-rich end and a = 3.928(1) Å at the zirconium-rich end [2].

References:

[1] Zegler, S.T. (J. Phys. Chem. Solids **26** [1965] 1347/9).
[2] Jorda, J.L.; Graf, T.; Schellenberg, L.; Muller, J. (J. Less-Common Metals **136** [1988] 313/28).
[3] Eremenko, V.N.; Semenova, E.L.; Shtepa, T.D. (Diagrammy Sostoyaniya Tugoplav. Syst. **1980** 119/32).
[4] Raman, A.; Schubert, K. (Z. Metallk. **55** [1964] 704/10).
[5] Wang, F.E. (J. Appl. Phys. **38** [1967] 822/4).
[6] Wang, F.E.; Ernst, D.W. (J. Appl. Phys. **39** [1968] 2192/5).
[7] Gachon, J.C.; Charles, J.; Hertz, J.; Jorda, J.L. (J. Calorim. Anal. Therm. Thermodyn. Chim. **17** [1986] 125/9).
[8] Cemuzal, K.; Chabot, B.; Parthé, E. (Acta Crystallogr. C **41** [1985] 313/9).
[9] Giessen, B.C.; Wang, R.; Grant, N.J. (Trans. Metall. Soc. AIME **245** [1969] 1207/10).
[10] Dwight, A.E.; Beck, P.A. (Trans. Metall. Soc. AIME **215** [1959] 976/9).

Mechanical Properties

Hardness. Available Vickers figures are given in the following table; Q = quenched, A = annealed, h = hours, SC = slow-cooled, R.T. = room temperature.

at% Rh	heat treatment t in °C	H_v in kg/mm^2	Ref.
3	cast	536	[1]
3	A 100 h 700	176	[1]
3	A 100 h 800	490	[1]
12	Q 1700	400	[2]
14	cast	355	[1]
14	A 100 h 700	270	[1]
14	A 100 h 800	366	[1]
33	cast	542	[1]
33.3	cast	440	[2]
62.5	A 15 h 1500 SC	870 ± 10	[2]
75	—	650	[2]
74.85	A 7 h 1000	(R.T.) 225	[3]
74.85	A 7 h 1000	(770 °C) 150	[3]

References:

[1] Togano, K.; Tachikawa, K. (J. Less-Common Metals **33** [1973] 275/82).
[2] Jorda, J.L.; Graf, T.; Schellenberg, L.; Muller, J. (J. Less-Common Metals **136** [1988] 313/28).
[3] Wee, D.M.; Suzuki, T. (Trans. Japan Inst. Metals **20** [1979] 634/46).

Thermal Properties

Specific Heat. Determinations of the electronic specific heat (γ) were made at the zirconium-rich solid solution end of the system by [1]. The compound Zr_2Rh was examined by [2, 3]. Determinations on ZrRh and $ZrRh_3$ were also made by [3]. The high γ associated with Zr_2Rh is compared with those of the solid solution, ZrRh and $ZrRh_3$ [1, 3]. **Fig. 74** shows the low-temperature heat capacity of Zr_2Rh [2].

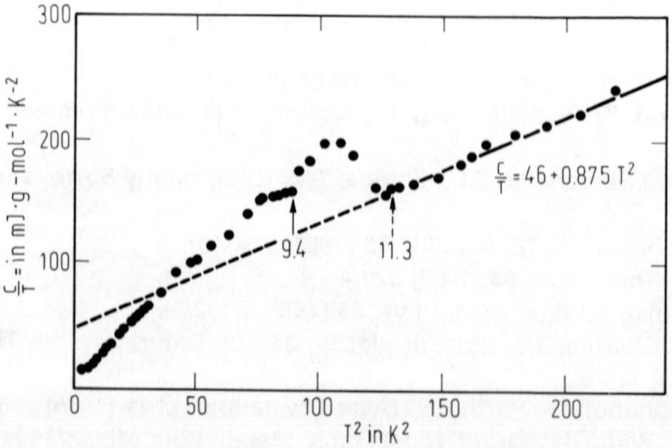

Fig. 74. Low-temperature heat capacity of Zr_2Rh (as-cast).

Debye Temperature. The following table shows the figures reported by [2, 3].

compound	θ_D in K	Ref.
Zr_2Rh	188	[2]
Zr_2Rh	212	[3]
ZrRh	300	[3]
$ZrRh_3$	436	[3]

Thermodynamic Data. The table below gives the available results, Δ_fH being expressed in kJ/mol and Δ_fS and ΔS (fusion) in $kJ \cdot mol^{-1} \cdot K^{-1}$ [4].

at% Rh	exp.	calculated Δ_fH	Δ_fS	ΔS (fusion)	T in K	Ref.
33	−55.4	—	—	—	1403	[5]
33	−55.4	−54.029	−15.1	12.8	1423″	[4]
50	−75.8	−78.807	−23.1	8.3	1700′	[4]
50	−75.8	—	—	—	1403	[5]
62.5	—	−86.277	−29.8	9.0	—	[4]
62.5	—	−88.000	—	—	300	[6]
75	−77.7	−68.644	−24.5	4.0	1423†	[4]
75	−71.8	—	—	—	1700*	[4]
75	—	−74.000	—	—	300	[6]
75	−72.0	—	—	—	1403	[5]

″σ=2000, ′σ=5000, †σ=4000, *σ=3000, see [4]

References:

[1] Dummer, G. (Z. Physik **185** [1965] 249).
[2] Fisk, Z.; Viswanathan, R.; Webb, G.W. (Solid State Commun. **15** [1974] 1797/9).
[3] Kuentzler, R.; Waterstrat, R.M. (Solid State Commun. **54** [1985] 517/24).
[4] Gachon, J.C.; Charles, J.; Hertz, J. (J. Calorim. Anal. Therm. Thermodyn. Chim. **17** [1986] 125/9).
[5] Jorda, J.L.; Graf, T.; Schellenberg, L.; Muller, J. (J. Less-Common Metals **136** [1988] 313/28).
[6] Kaufman, L.; Bernstein, H. (Computer Calculation of Phase Diagrams, Academic, New York 1970).

Electrical and Magnetic Properties

Superconductivity. This is by far the most important property of these alloys and the prime reason for most of the investigational work. It has been investigated by [1 to 5]; much further work has been done on amorphous alloys, but this will be dealt with in a separate section. Alloys containing >53 at% rhodium are not superconductive above 1 K, see table below.

at% Rh	phases	T_c in K	Ref.
1 to 7.5 cast	—	6.5	[6]
5	α-Zr	3.5 to 4	[2]
7.5	β-Zr	6.4	[2]
8	β-Zr	9.6	[2]
8	β-Zr	6.5	[4]
9	β-Zr	10.4	[2]
<10	—	9.0	[1]
12	Zr_3Rh	11.0	[2]
20	β-Zr; Zr_2Rh	11.3	[4]
25	Zr_3Rh	11.0	[2]
33	Zr_2Rh	11.3	[4]
33	Zr_2Rh	11.3	[5]
40	Zr_2Rh	11.4	[4]
52	ZrRh	2.4	[4]
57	Zr_3Rh_4	<1.1	[4]
63	Zr_3Rh_5	<1.1	[4]
65	$ZrRh_3$	<1.1	[4]
75	$ZrRh_3$	<1.1	[4]
85	$ZrRh_3$; Rh	<1.1	[4]

The effects of heat treatment on alloys containing up to 8 at% rhodium are shown in **Fig. 75**, p. 122 [3]. Measurements of the superconducting upper critical field H_{c2} of Zr_2Rh showed it to be as high as 80 kOe at 4.2 K; the anisotropy of H_{c2} may be caused by the crystalline anisotropy of Zr_2Rh. The critical current density of 14 and 33 at% alloys was found to be related to the metallographic structure [7].

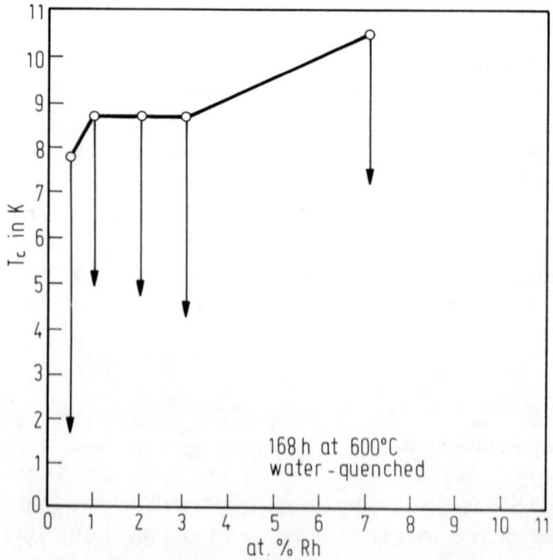

Fig. 75. Superconductivity of low-Rh alloys.

Magnetic Susceptibility. The susceptibility for some compounds is given below.

compound	χ in 10^{-6} emu/g-atom	Ref.
Zr_2Rh	151	[8]
Zr_2Rh	$\chi_{RT} \cdot 10^4$ emu/mol = 3.7	[9]
ZrRh	56.5	[8]
$ZrRh_3$	50.7	[8]

References:

[1] Matthias, B.T. (Phys. Rev. [2] **97** [1955] 74/6).
[2] Raub, C.J.; Andersen, C.A. (Z. Physik **175** [1963] 105/14).
[3] Zegler, S.T. (J. Phys. Chem. Solids **26** [1965] 1347/9).
[4] Jorda, J.L.; Graf, T.; Schellenberg, L.; Muller, J. (J. Less-Common Metals **136** [1988] 313/28).
[5] Fisk, Z.; McCarthy, S.L. (J. Low-Temp. Phys. **4** [1971] 489).
[6] Raub, C.J.; Hull, G.W. (Phys. Rev. [2] **133** [1964] A32/A34).
[7] Togano, K.; Tachikawa, K. (J. Less-Common Metals **33** [1973] 275/82).
[8] Kuentzler, R.; Waterstrat, R.M. (Solid State Commun. **54** [1985] 517/24).
[9] Fisk, Z.; Viswanathan, R.; Webb, G.W. (Solid State Commun. **15** [1974] 1797/9).

Amorphous Materials

Preparation. Amorphous materials are prepared by rapidly cooling from the liquid state by techniques such as melt spinning, splat cooling against a surface, or by sputtering. By such procedures, as much as 17 at% rhodium above the equilibrium limit of ~8 at% is retainable in solid solution [1]. Whilst this results generally in a lower T_c than in the corresponding equilibrium crystalline alloys, some of these have interesting resistivity properties and are of theoretical interest in progress in the understanding of superconductivity. Techniques are described by [1, 2, 3].

Transition Temperatures. The table below gives some of the data (A = annealed).

at% Rh	treatment	T_c in K	Ref.
20	—	3.72 ± 0.01	[4]
21	—	4.376	[5]
22	—	4.386	[5]
23	—	4	[1]
23	A	4 to 11	[1]
23	—	3.66 ± 0.02	[4]
24	—	4.355	[5]
24	—	3.67 ± 0.01	[4]
25	—	4.30	[6]
25	A	3.97	[6]
25	—	4.20 to 4.38	[7]
25	—	4.30	[8]
25	—	4.3	[9]
25	—	4.19 to 4.30	[3]
26	—	4.204	[5]
27	—	3.57 ± 0.01	[4]
30	—	4.4	[10]
30	—	4.010	[5]
37	—	2.86 ± 0.05	[4]
41	—	2.44 ± 0.23	[4]
43	—	2.18 ± 0.01	[4]

References:

[1] Togano, K.; Tachikawa, K. (J. Appl. Phys. **46** [1975] 3609/13).

[2] Missell, F.P.; Bergeron, R.; Keem, J.E.; Ovshinsky, S.R. (Solid State Commun. **47** [1983] 177/9).

[3] Poon, S.J.; Dunn, P.L.; Elmquist, R.E.; Smith, L.M. (Solid State Commun. **42** [1982] 267/70).

[4] Missell, F.P.; Frota-Pessoa, S.; Wood, J.; Tyler, J.; Keem, J.E. (Phys. Rev. [3] B **27** [1983] 1596/604).

[5] Eschner, W.; Gey, W. (Supercond. d-f-Band Metals Proc. 4th Conf., Karlsruhe 1982, pp. 359/62).

[6] Wong, K.M.; Cotts, E.J.; Poon, S.J. (Phys. Rev. [3] B **30** [1984] 1253/9).

[7] Drehman, A.J.; Johnson, W.L. (Phys. Status Solidi **52** [1979] 499/507).

[8] Elmquist, R.E.; Poon, S.J. (Solid State Commun. **41** [1982] 221/4).

[9] Garoche, P. (Solid State Commun. **39** [1981] 403/6).

[10] Tenhover, M.; Johnson, W.L. (Physica B+C **108** [1981] 1221/2).

Crystallography

Annealing. Annealing behaviour has been studied at temperatures below T_c (relaxation) and during crystallization; the latter is of particular interest since some metastable crystalline structures have been found, similar to those reported in early work on the equilibrium diagram. For example, annealing amorphous $Zr_{75}Rh_{25}$ in either the relaxed or unrelaxed state at 360 °C for 5 h produced an $E9_3$ structure found earlier by [1]. An $E9_3$ structure was also found on annealing Zr_3Rh under conditions probably free from oxygen contamination in samples of melt-sprayed material, and material hermetically sealed in a gold boat during DSC measurements [2]. Study of a liquid quenched amorphous 23 at% Rh alloy using transmission electron micrographs and electron diffraction patterns showed that whilst the expected equilibrium phases were obtained at annealing temperatures of 700 °C, after treatments for 10 min at 450 °C a hexagonal structure was found with a=5.17, c=2.91 Å [3].

Crystallization of Amorphous Alloys. The following table shows the crystallization temperature T_x for some alloys.

at% Rh	heating rate in K/m	T_x in K	determination method	Ref.
20	50	712	DSC	[4]
20	50	712	DSC	[5]
25	50	741	DSC	[4]
25	–	730	–	[6]
25	50	741	DSC	[5]
25	20	720 to 731$^{1)}$	DSC	[2]
25	20	716$^{2)}$	DSC	[2]
25	–	633$^{3)}$	X-ray	[7]
25	–	640	–	[8]
30	50	748	DSC	[4]
30	–	758	DSC	[9]
34	–	742	DSC	[9]

$^{1)}$=sample prepared by melt spinning; $^{2)}$=sample prepared by anvil/piston method; $^{3)}$=5 h anneal; similar specimen after 1 h at 873 K was in equilibrium and α-Zr and Zr_2Rh structure with $T_c=11.3\pm0.03$ K.

Surface Modification. The surface of glassy $Rh_{25}Zr_{75}$ can be modified by the scanning tunneling microscope by raising the surface temperature leading to enhanced diffusion of constituents [10].

References:

[1] Raub, C.J.; Andersen, C.A. (Z. Physik **175** [1963] 100/14).
[2] Cantrell, J.S.; Wagner, J.E.; Bowman, R.C. (J. Appl. Phys. **57** [1985] 545/53).
[3] Togano, K.; Tachikawa, K. (J. Appl. Phys. **46** [1975] 3609/13).
[4] Buschow, K.H.J. (J. Phys. Colloq. [Paris] **41** [1980] C8-559/C8-562).
[5] Eifert, H.-J.; Elschner, B.; Buschow, K.H.J. (Phys. Rev. [3] B **29** [1984] 2905/11).
[6] Buschow, K.H.J. (J. Appl. Phys. **52** [1981] 3319).
[7] Drehman, A.J.; Johnson, W.L. (Phys. Status Solidi **52** [1979] 499/507).
[8] Belotskii, A.V.; Gritskiv, Ya.P.; et al. (Metallofizika Otd. Fiz. **5** [1983] 111/3; Phys. Metals [USSR] **5** [1985] 816/20).
[9] Buschow, K.H.J. (Solid State Commun. **43** [1982] 171/4).
[10] Staufer, U.; Wiesendanger, R.; Eng, L.; Rosenthaler, L.; Hidber, H.-R.; Güntherodt, H.-J.; Garcia, N.J. (J. Vac. Sci. Technol. [2] A **6** [1988] 537/9).

Thermal Properties

Specific Heat. The specific heat of amorphous Zr_3Rh was determined with and without an applied field of 75 kG between 0.35 and 10 K; the applied high field allows accurate determination of the phonon term βT^3 for the lowest temperatures measured. **Fig. 76** shows the specific heat determinations and the effect of the applied field; an excess heat capacity was determined by subtraction of the phonon term and this is attributed to thermal excitations with the two-level configuration or tunneling systems [1].

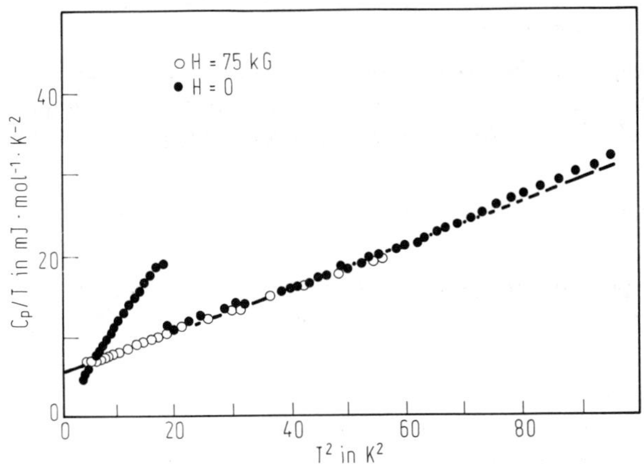

Fig. 76. Low-temperature specific heat of amorphous Zr_3Rh.

Fig. 77 shows similar determinations in a field of ~6 T between 2 and 7 K using amorphous and metastable crystalline 25 at% Rh material; determinations up to 700 K were also made and $\gamma(0)$, the coefficient of the electronic heat capacity for the amorphous and metastable crystalline forms were found to be 5.1 and $4.0 \pm 1\%$ $mJ \cdot g\text{-}atom^{-1} \cdot K^{-2}$ [2].

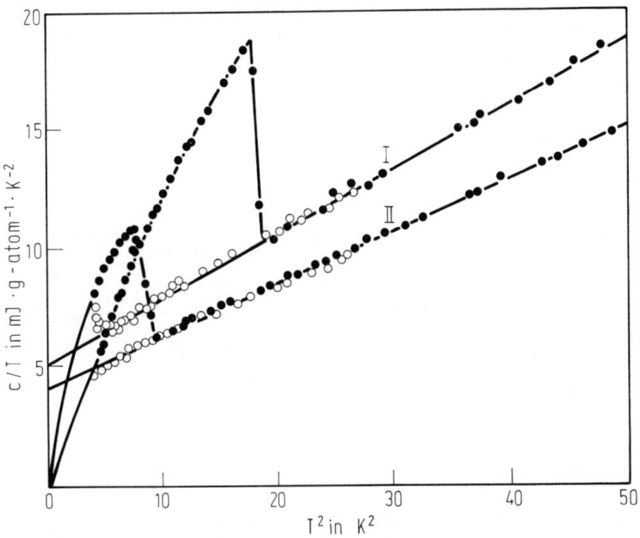

Fig. 77. Heat capacity of amorphous (I) and metastable crystalline (II) specimens of $Zr_{75}Rh_{25}$ in the absence of a field (full circles) and in a field of ~6T (open circles) in the temperature range from 2 to 7 K.

The table gives figures for γ by [3].

at% Rh	21	22	24	26	30
γ in $mJ \cdot mol^{-1} \cdot K^{-2}$ . .	5.11	4.39	4.02	4.07	3.86

Further figures for γ in $mJ \cdot g\text{-}atom^{-1} \cdot K^{-2}$ plotted against at% Rh for amorphous and crystalline alloys are given in [4].

Debye Temperature. Figures for θ_D plotted against at% Rh for material in the amorphous and crystalline state are given by [4]. Figures deduced from heat capacity measurements made between 2 and 7 K using 25 at% Rh material were 190 and 204 ± 1.5% K for amorphous and crystalline material, respectively [2].

Heat of Formation. ΔH for $Zr_{70}Rh_{30}$ was said to be −79 kJ/mol, for $Zr_{66}Rh_{34}$ −90 kJ/mol [5].

Activation Energy of Crystallization. Values of ΔE were determined as 3.64, 0.25 and 0.30 eV for alloys containing 20, 25 and 30 at% Rh [6].

References:

[1] Garoche, P. (Solid State Commun. **39** [1981] 403/6).
[2] Panova, G.Kh.; Chernoplekov, N.A.; Shikov, A.A.; Savel'ev, B.I. (Zh. Eksperim. Teor. Fiz. **82** [1982] 548/60; Soviet Phys.-JETP **55** [1982] 319/26).
[3] Eschner, W.; Gey, W. (Supercond. d-f-Band Metals Proc. 4th Conf., Karlsruhe 1982, pp. 359/62).
[4] Kuentzler, R. (Ann. Chim. [Paris] [15] **9** [1984] 967/70).
[5] Buschow, K.H.J. (Solid State Commun. **43** [1982] 171/4).
[6] Buschow, K.H.J. (J. Phys. Colloq. [Paris] **41** [1980] C8-559/C8-562).

Electrical and Magnetic Properties

The resistance of amorphous $Zr_{75}Rh_{25}$ and the same material in the metastable crystalline state was measured up to 300 K, and the results are shown in **Fig. 78**; the metastable crystalline state was examined by X-ray diffraction and was found to have the $E9_3$ structure [1].

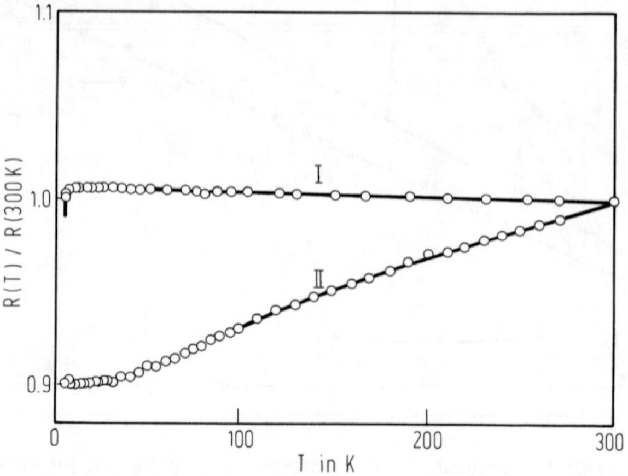

Fig. 78. Temperature dependence of the electric resistance of a specimen of $Zr_{75}Rh_{25}$ in the amorphous (I) and metastable crystalline (II) states.

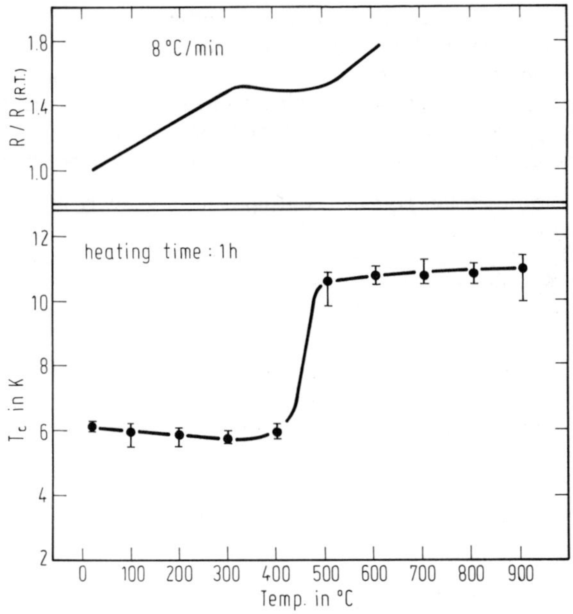

Fig. 79. Electric resistivity change and isochronal dependence of superconducting transition temperature for liquid-quenched Zr–11 at% Rh alloy.

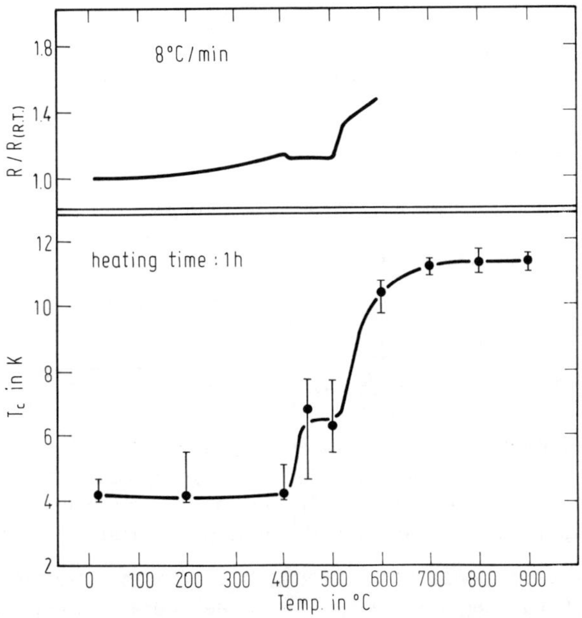

Fig. 80. Electric resistivity change and isochronal dependence of superconducting transition temperature for liquid-quenched Zr–23 at% Rh alloy.

The amorphous 25 at% Rh alloy in common with other transition metal alloys was found to have a hump in the resistivity/temperature curve at 2 to 3 T_c, 12.98 K in this case; there were also negative temperature coefficients, −1.48 for the zirconium alloy ($1/\rho_{250}\, d\rho/dT$) at 250 K; the hump was ascribed to the generation of superconducting fluctuations [2]. **Fig. 79**, p. 127, and **Fig. 80**, p. 127, show the electric resistivity and T_c values of amorphous 11 and 23 at% Rh material after isochronal annealing at increasing temperatures [3].

The following table gives the residual and specific resistivity of the 23 at% alloy (Q = quenched) [3].

treatment	ρ_0	ρ in μΩ·cm 77 K	300 K
liquid Q	359 (at 6 K)	351	356
1 h 700 °C (cryst.)	64 (at 12 K)	184	331
arc-melted	13 (at 12 K)	65	265

The table below gives resistivity results at 25 and 300 K for 20 to 41 at% Rh alloys [4].

at% Rh	ρ(25 K) in μΩ·cm	ρ(300 K)
20	184	171
23	175	165
24	185	180
27	177	166
37	194	182
41	200	187
43	214	205

Figures for ρ calculated at T_c [5].

at% Rh	ρ in μΩ·cm	T_c in K
21	217	4.376
22	223	4.386
24	240	4.355
26	223	4.204
30	248	4.010

The resistivity of amorphous 25 at% Rh alloy is given as 1.63×10^{-6} Ω/m with a temperature coefficient of -1.2×10^{-6} measured at 300 K; in the crystalline state the figures given are 0.8×10^{-6} and 15×10^{-6}, respectively [6]. Figures for amorphous 25 at% material up to 300 K shown in **Fig. 81**; again a negative temperature coefficient is shown [7].

The temperature dependence of the 25 at% Rh alloy was determined between 4 and 12 K, with and without a magnetic field of 7.4 T (see figure in original publication). The findings are found to be adequately described by the theory of weak localization of electron and electron-electron interaction in three-dimensional systems [8].

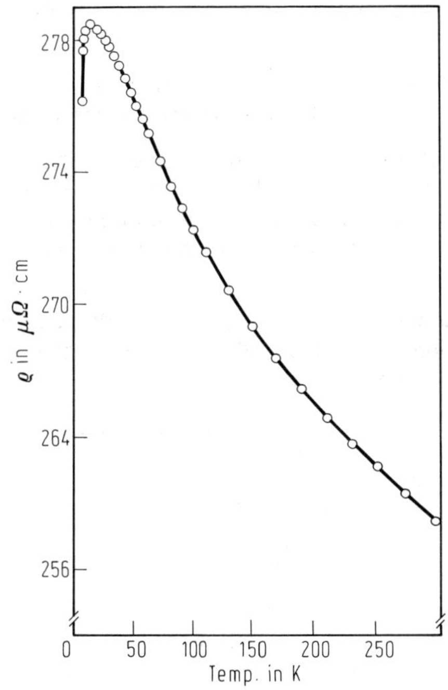

Fig. 81. Temperature dependence of the electric resistivity of amorphous $Zr_{75}Rh_{25}$.

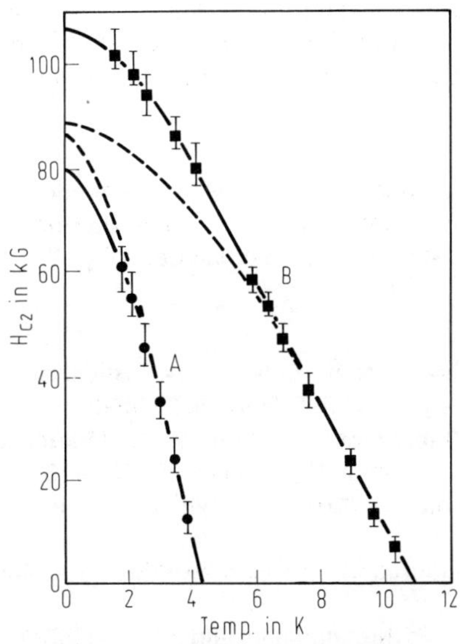

Fig. 82. Temperature dependences of upper critical field for the noncrystalline $Zr_{77}Rh_{23}$ alloy (A) and the Zr_2Rh compound (B). The dashed lines were calculated using WHH theory ignoring the paramagnetic limiting and spin-orbit scattering effects ($\alpha = 0$ and $\lambda_{SO} = 0$).

Hall Coefficient R_H. This was determined as $+2.7\,[10^{-11}\,m^3/As]$ for amorphous $Zr_{72}Rh_{28}$ [9].

Thermoelectric Power. This was found to be between -25 and $+35\,\mu V/K$ at 300 to 1050 K for amorphous $Zr_{75}Rh_{25}$; the μV/T curve is shown in the original paper [6].

Magnetic Measurements. Measurements of the upper critical field have been made by [4, 5, 10 to 13]. **Fig. 82**, p. 129, shows results for a 23 at% Rh amorphous alloy and the crystalline compound Zr_2Rh; the results are discussed applying the WHH (Werthamer-Helfand-Hohenberg) theory [10]. Similar results were obtained, and a good fit with WHH calculations was found by [5].

Fig. 83 gives results obtained by [11] and [4]. The solid lines represent the predictions of the GLAG (Ginzburg-Landau-Abrikosov-Gor'kov) theory for no paramagnetic limiting. Good fits with WHH theory were found by [12].

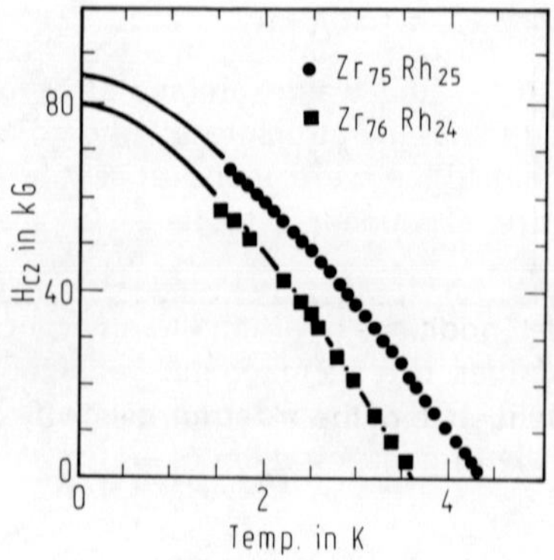

Fig. 83. Upper critical field H_{c2} vs. temperature T for $Zr_{76}Rh_{24}$ amorphous film of [4] (●) and $Zr_{75}Rh_{25}$ amorphous melt-spun ribbon of this work (▲). Solid lines represent predications of GLAG theory for no paramagnetic limiting.

Measurements of flux-flow resistivity and H_{c2} were made and the results compared with the predictions of the time-dependent microscopic theories for bulk superconductors in the dirty limit; the original paper should be consulted [13].

References:

[1] Panova, G.Kh.; Chernoplekov, N.A.; Shikov, A.A.; Savel'ev, B.I. (Zh. Eksperim. Teor. Fiz. **82** [1982] 548/60; Soviet Phys.-JETP **55** [1982] 319/26).

[2] Inoue, A.; Matsuzaki, K.; Masumoto, T.; Chen, H.S. (J. Mater. Sci. **21** [1986] 1258/68).

[3] Togano, K.; Tachikawa, K. (J. Appl. Phys. **46** [1975] 3609/13).

[4] Missell, F.P.; Frota-Pessoa, S.; Wood, J.; Tyler, J.; Keem, J.E. (Phys. Rev. [3] B **27** [1983] 1596/604).

[5] Eschner, W.; Gey, W. (Supercond. d-f-Band Metals Proc. 4th Conf., Karlsruhe 1982, pp. 359/62).

[6] Belotskii, A.V.; Gritskiv, Ya.P. (Metallofizika Otd. Fiz. **5** [1983] 111/3; Phys. Metals [USSR] **5** [1985] 816/20).

[7] Prekul, A.F.; Rassokhin, V.A.; Yartsev, S.V. (Pis'ma Zh. Eksperim. Teor. Fiz. **38** [1983] 340/3; JETP Letters **38** [1983] 408/12).

[8] Gumbatov, S.G.; Pashaev, Kh.M.; Panova, G.Kh.; Shikov, A.A. (Zh. Eksperim. Teor. Fiz. **93** [1987] 2122/8; Soviet Phys.-JETP **66** [1987] 1211/4).
[9] Tschumi, A.; Laubscher, T.; Jeker, R.; Schüpfer, E.; Künzi, H.-U.; Güntherodt, H.-J. (J. Non-Cryst. Solids **61/62** [1984] 1091/6).
[10] Togano, K.; Tachikawa, K. (Phys. Letters A **54** [1975] 205/6).
[11] Missell, F.P.; Bergeron, R.; Keem, J.E.; Ovshinsky, S.R. (Solid State Commun. **47** [1983] 177/9).
[12] Wong, K.M.; Cotts, E.J.; Poon, S.J. (Phys. Rev. [3] B **30** [1984] 1253/9).
[13] Poon, S.J.; Wong, K.M. (J. Low-Temp. Phys. **54** [1984] 15/25).

19.2 Ternary Alloys

Rh-Zr-Al. A new G-type phase with a = 12.32(3) Å was reported by [1].

Rh-Zr-Ga. The system has an $Mg_6Cu_{16}Si_7$-type phase with a = 12.43 Å [2]. An Fe_2P-type phase is also found with a = 7.207, c = 3.366 Å [3].

Rh-Zr-Gd. The alloy $Zr_{74.75}Gd_{0.25}Rh_{25}$ was examined by magnetic and resistance measurements; conclusions reached about the structure of inhomogeneities in superconductors [4]. Giant A.C. susceptibility anomalies were found at low temperatures in Rh-Zr alloys diluted with Gd; the results are explained in terms of an indirect exchange interaction via the d-band [5].

Rh-Zr-Er. The effect of small additions of erbium to amorphous Zr_3Rh has been studied; it was found that the Er impurities exhibit magnetic behaviour characteristic of the free Er^{3+} ion in a weak crystal field. Magnetic ordering of the Er moments occurred at T_M, a temperature proportional to the Er concentration [6].

References:

[1] Ganglberger, E.; Nowotny, H.; Benesovsky, F. (Monatsh. Chem. **97** [1966] 829/32).
[2] Dwight, A.E. (J. Less-Common Metals **34** [1974] 279/84).
[3] Markiv, V.Ya.; Storozhenko, A.I. (Dopov. Akad. Nauk Ukr. RSR Fiz. Tekh. Mat. Nauki A **35** [1973] 941/3).
[4] Zwicknagl, G.E.; Wilkins, J.W. (Phys. Rev. Letters **53** [1984] 1276/9).
[5] Dunn, P.L.; Poon, S.J. (J. Phys. F **12** [1982] 2273/8).
[6] Johnson, W.L. (Solid State Commun. **32** [1979] 981/7).

20 Alloys with Hafnium

20.1 The Rh-Hf System

Phase Diagram. An investigation of the complete compositional range has been made using X-ray diffraction, electron microprobe analysis and metallographic methods; no lattice parameter figures were included, it being considered that there was sufficient data available.

The terminal solid solubilities determined at 1100, 1400 and 1700 °C are based on the assumption of the linearity of the atomic volume with composition at the rhodium-rich end of the diagram; the hafnium-rich limits were based on the metallographic examination of quenched samples. The position of eutectic points was based on metallographic estimates of the respective volume fractions of eutectic to primary phase constituents, as no pure eutectic phases were seen. The authors suggest that some high-temperature X-ray diffraction studies should be made to decide whether a CsCl-type structure exists at the equiatomic composition. The diagram thus determined is shown in **Fig. 84** [1]. Two earlier papers included partial phase diagrams up to 50 at% rhodium [2, 3]. The diagram due to [2] shows the β-eutectoid at 880 °C and at ~2 at% rhodium, whilst the eutectic horizontal was found at the same temperature as in [1]. The position of a possible eutectic, estimated from the diagram, would be at ~10 at% rhodium. The other relevant paper again found the hafnium-rich eutectic temperature to be 1350 °C at 21 at% rhodium; the eutectic horizontal is shown to be at 1315 °C, and the peritectic at ~70 at% hafnium and 1450 °C [3].

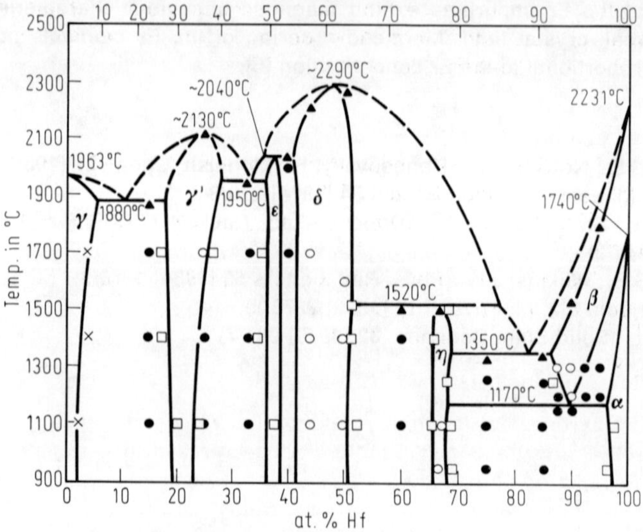

Fig. 84. Hf-Rh constitution diagram: ● two-phase alloys; ○ single-phase alloys; ▲ melting began on heating; □ boundary by electron microprobe analysis; × boundary by X-ray diffraction analysis.

Preparation. The samples used were prepared from iodide hafnium ~99.9% pure excluding 3% zirconium, and 99.999% rhodium sponge (excluding gaseous impurities); weight losses on degassing in vacuum suggested a high oxygen level in the sponge. The materials were arc-melted in a 50/50 argon/helium mixture and the resultant weight losses were generally <0.5%. Homogenization and equilibration of the as-cast samples was performed in a tungsten element resistance furnace at pressures between 5×10^{-6} and 5×10^{-7} mm mercury. The solid state phase boundaries in a series of two-phase samples were established by quantitative microprobe measurements with an estimated accuracy of ± 1 at%; the authors suggested that some segregation might have been present in samples near equiatomic proportions [1].

Crystallography. The compound Hf_2Rh was found at 33 at% rhodium at an oxygen level of 500 ppm. It had a Ti_2Ni-type structure and a lattice parameter of 12.3255 ± 0.0002 Å at nominal 30% rhodium when quenched from 1200 °C [4]. This compound was confirmed by several other authors [1, 2, 3, 5]; figures for the lattice parameter were 12.30 [5], 12.324 [2]. X-ray diffraction studies yielded data on HfRh, $Hf_{46}Rh_{54}$, $Hf_{45}Rh_{55}$, $Hf_{42}Rh_{58}$, $Hf_{42}Rh_{58}$, Hf_2Rh_3, Hf_3Rh_5, and $HfRh_3$; the lattice parameters are given in the following table [5].

compound	type	a	b in Å	c
Hf_2Rh	Ti_2Ni	12.30 12.3255	—	—
HfRh	ZrIr	—	—	—
HfRh	CsCl	3.227	—	—
$Hf_{46}Rh_{54}$	CsCl	3.268	—	3.150
$Hf_{45}Rh_{55}$	CsCl	3.12	—	3.41
$Hf_{42}Rh_{58}$	NbRu	4.392	4.306	3.470
$Hf_{42}Rh_{58}$	—	—	—	—
Hf_2Rh_3	CuAu	4.35	—	3.451
Hf_3Rh_5	Ti_3Rh_5	—	—	—
$HfRh_3$	Cu_3Au	3.911 3.912	— —	— —

A lattice parameter of a = 3.916 Å for $HfRh_3$ was found by [6]. Distorted CsCl structures which have an average electron concentration e/a = 6.5 are discussed and amongst those considered was $Hf_{48}Rh_{52}$ [7]. Hf_3Rh_5 was found to have the orthorhombic Ge_3Rh_5 structure with parameters a = 5.58, b = 10.73, c = 4.25 Å [8].

Thermodynamic Data. The standard enthalpy of formation of RhHf has been determined by high-temperature mixing calorimetry at 1400 K. $\Delta H_f^0 = -191.6 \pm 4.3$ kJ/mol [9].

Hardness. The hardness of hafnium-rhodium alloys rises steadily from 112 kp/mm^2 for pure hafnium to 600 kp/mm^2 at 33 at% rhodium, which is in the Hf_2Rh region [2]. Hot hardness tests were made on a number of $L1_2$ ordered alloys including Rh_3Hf. The hardness fell almost linearly with temperature between room temperature and 927 °C; at room temperature the hardness was 275×10 MPa and at 927 °C it had fallen to $\sim 200 \times 10$ MPa [6].

Specific Heat. This was measured at low temperatures on alloys containing 2, 5, 8 and 33 at% rhodium; the peak at ~1.7 K was associated with the compound Hf_2Rh, see **Fig. 85**, p. 134 [2]. C/T against T^2 plots for the compounds Hf_2Rh, HfRh, and $HfRh_3$ are shown in **Fig. 86**, p. 134. The electronic specific heat constant γ was determined for the three compounds as follows:

Hf_2Rh $\gamma = 2.72$ (as-cast), 2.79 $mJ \cdot g\text{-atom}^{-1} \cdot K^{-2}$ after 2 h at 1300 °C;
HfRh $\gamma = 1.25$ $mJ \cdot g\text{-atom}^{-1} \cdot K^{-2}$ after 2 h at 1300 °C;
$HfRh_3$ $\gamma = 2.54$ $mJ \cdot g\text{-atom}^{-1} \cdot K^{-2}$ after 3 h at 1700 °C [10].

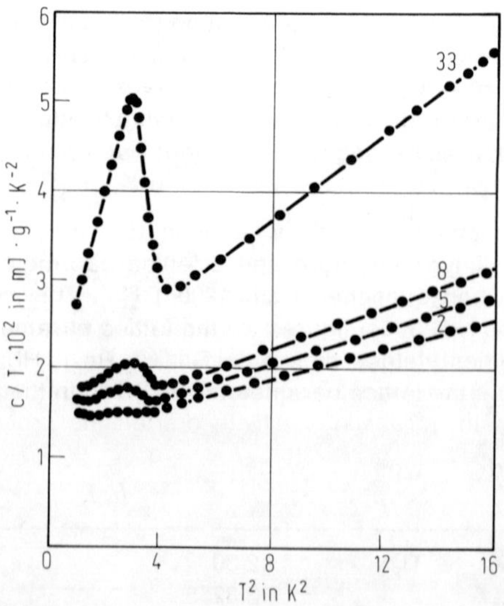

Fig. 85. Specific heat at 0.9 to 4 K. Figures at curves denote Rh contents in at%.

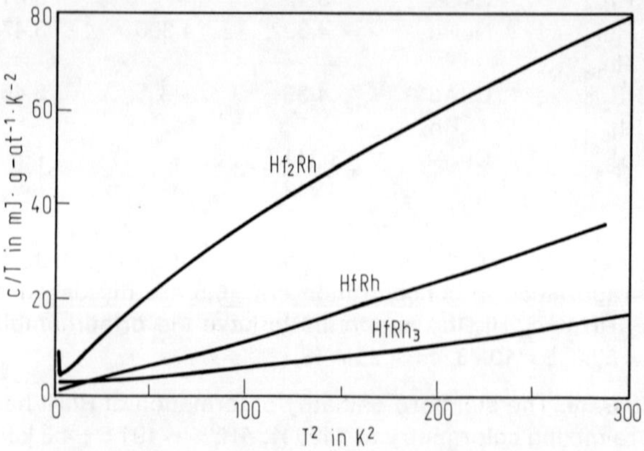

Fig. 86. Specific heat for Hf_2Rh, HfRh and $HfRh_3$.

Debye Temperature. This was found to be 176 K for as-cast Hf_2Rh and 178 K after heat treatment of 2 h at 1300 °C, $\theta_D = 270$ K for HfRh after 3 h at 1600 °C, and for $HfRh_3$ it was 400 K after 3 h at 1700 °C [10].

Electronic Structure. The electronic structure and stability of the ordered alloys in this system were investigated using low-temperature specific heat, magnetic susceptibility, and superconductivity measurements. $HfRh_3$ had high stability and no magnetic order, HfRh is more complex, possibly due to the intervention of the martensitic transformation at high temperatures; no magnetic or superconducting properties were observed, and the minimum value for γ is not reached. In Hf_2Rh the properties appeared to be strongly dependent on the crystallographic structures, and a structural instability is suggested [10].

Amorphous Materials. The crystallization behaviour of alloys of the type $Hf_{100-y}Rh_y$ (y = 18 to 35 at%) formed by the arc-melting hammer-anvil technique has been examined by differential scanning calorimetry (DSC). The exothermal curves for $Hf_{80}Rh_{20}$, $Hf_{75}Rh_{25}$ and $Hf_{70}Rh_{30}$ are shown in a figure (see original paper). The crystallization temperatures and kinetic and thermodynamic parameters are given in the table on p. 136. T_x is the crystallization temperature, T_p is the peak temperature and K_0 and K_p are kinetic parameters, MS I, II, III = metastable stages of crystallization [11].

Superconductivity. The superconducting transformation temperatures of alloys containing up to 20 at% rhodium are shown in **Fig. 87** and **Fig. 88**. There are important differences between results on the two sets of samples. A common feature was the tendency to approach a limit of $T_c = 1.98$ K coupled with a decreased scatter in the measurements, which was associated by the authors with the presence of Hf_2Rh. The rise in the T_c of quenched samples was considered to be associated with martensitic precipitation of the α' phase [2].

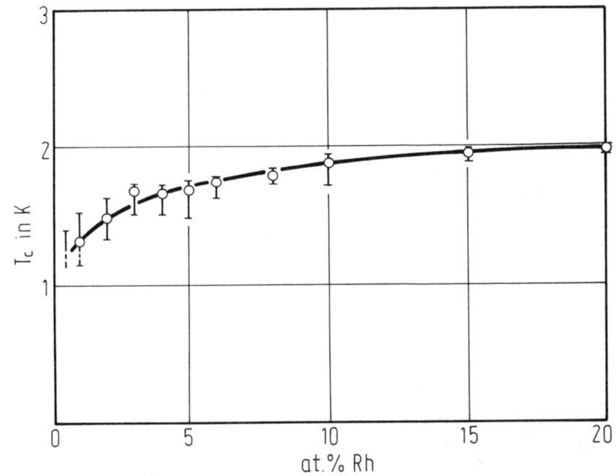

Fig. 87. Transformation temperatures of annealed Hf-Rh alloys.

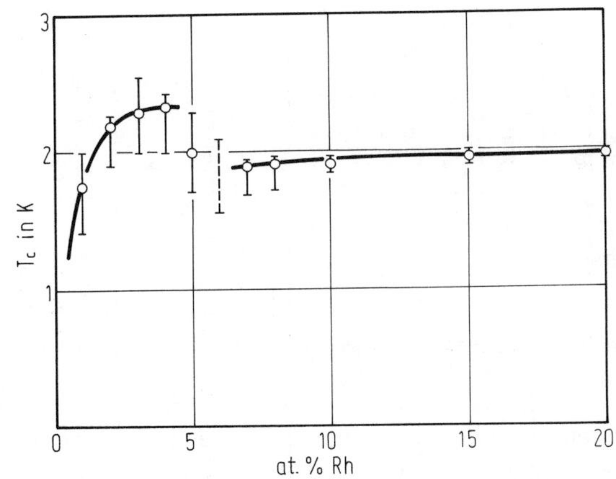

Fig. 88. Transformation temperatures of quenched Hf-Rh alloys.

Crystallization temperature, heat and other kinetic parameters of crystallization of metallic glasses $Hf_{70}Rh_{30}$, $Hf_{75}Rh_{25}$ and $HF_{80}Rh_{20}$.

alloys	B in K/min	MS I								MS II			
		T_x in K	T_p in K	ΔH in mcal/mg	$(dx/dt)_p$ (1/s)	E in kcal/mol	K_p (1/s)	K_0 (1/s)	n	T_x in K	T_p in K	ΔH in mcal/mg	$(dx/dt)_p$ (1/s)
$Hf_{70}Rh_{30}$	2.5	760	777	7.22	0.00392	74	0.00264	1.7×10^{18}	4.1		791	0.46	0.00660
	5	767	788		0.00813		0.00513		4.3		804		0.01220
	10	775	800		0.01560		0.00995		4.2		816		0.02395
	20	780	812		0.03099		0.01930		4.3		830		0.04769
$Hf_{75}Rh_{25}$	2.5	746	767	7.41	0.00310	64	0.00236	4.1×10^{15}	3.6	834	840	0.56	
	5	758	780		0.00627		0.00457		3.7	848	855		
	10	770	792		0.01252		0.00887		3.8	863	871		
	20	782	805		0.02415		0.01720		3.8	878	885		
$Hf_{80}Rh_{20}$	2.5		753	0.91	0.00330	77	0.00289	6.5×10^{19}	3.1		795	3.66	0.00550
	5		763		0.00619		0.00563		3.0		807		0.01105
	10		773		0.01180		0.0110		3.0		820		0.02060
	20		784		0.02480		0.0213		3.1		832		0.04181

E in kcal/mol	K_p (1/s)	K_0 (1/s)	n	MS III							ΣΔH
				T_p in K	ΔH in mcal/mg	$(dx/dt)_p$ (1/s)	E in kcal/mol	K_p (1/s)	K_0 (1/s)	n	
67	0.00226	7.6×10^{15}	7.9								7.68
	0.00437		7.5								
	0.00849		7.6								
	0.01642		7.8								
68 (from T_p)											7.97
71.5	0.00243	1.1×10^{17}	6.1	853	0.77	0.00536	71.5	0.00208	4.0×10^{15}	7.1	5.34
	0.00472		6.3	868		0.01121		0.00402		7.5	
	0.00915		6.1	883		0.02302		0.00777		8.0	
	0.01780		6.3	898		0.04397		0.0150		7.9	

MS = metastable state

The following figures for T_c are given by [10].

compound	heat treatment	T_c in K
Hf_2Rh	as-cast	1.54
Hf_2Rh	2 h 1300 °C	1.60
HfRh	3 h 1600 °C	<1.4
$HfRh_3$	3 h 1700 °C	<1.4

Magnetic Susceptibility. The following figures are given by [10].

compound	heat treatment	χ in emu/g-atom
Hf_2Rh	as-cast	67.5
Hf_2Rh	2 h 1300 °C	64.8
HfRh	3 h 1600 °C	40.0
$HfRh_3$	3 h 1700 °C	40.4

References:

[1] Waterstrat, R.M.; Giuseppetti, A.A. (J. Less-Common Metals **119** [1986] 327/35).
[2] Wühl, H. (Z. Physik **197** [1966] 276/87).
[3] Eremenko, V.N.; Shtepa, T.D.; Velikanova, T.A. (Dopov. Akad. Nauk Ukr. RSR A Fiz. Mat. Tekh. Nauki **1985** No. 10, pp. 70/2).
[4] Nevitt, M.V.; Schwartz, L.H. (Trans. Am. Inst. Mining Metall. Petrol. Eng. **208** [1958] 700/3).
[5] Raman, A.; Schubert, K. (Z. Metallk. **55** [1964] 704/10).
[6] Wee, D.M.; Suzuki, T. (Trans. Japan. Inst. Metals **20** [1979] 634/46).
[7] Dwight, A.E.; Beck, P.A. (Trans. Metall. Soc. AIME **245** [1969] 389/90).
[8] Giessen, B.C.; Wang, R.; Grant, N.J. (Trans. Metall. Soc. AIME **245** [1969] 1207/10).
[9] Topor, L.; Kleppa, O.J. (J. Less-Common Metals **135** [1987] 67/75).
[10] Kuentzler, R.; Waterstrat, R.M. (Solid State Commun. **68** [1988] 85/91).

[11] Mao, Y.; Gao, Yi-Q. (J. Non-Cryst. Solids **105** [1988] 134/8).

20.2 Ternary Alloys

Rh-Hf-Sb. The composition RhHfSb has been found to have an MgAgAs-type structure with a = 6.236 Å [1].

Rh-Hf-Al. A new G-phase alloy with a = 12.27(8) Å has been reported by [2].

A two-phase alloy was found at the composition RhHfAl with a = 5.23, c = 8.33 Å [3].

Rh-Hf-Ga. The composition RhHfGa was found to be of the Fe_2P-type with a = 7.237, c = 3.296 Å [1].

References:

[1] Dwight, A.E. (J. Less-Common Metals **34** [1974] 279/84).
[2] Ganglberger, E.; Nowotny, H.; Benesovsky, F. (Monatsh. Chem. **97** [1966] 829/32).
[3] Marazza, R.; Ferro, R.; Rambaldi, G.; Mazzone, D. (J. Less-Common Metals **37** [1974] 285/8).

21 Alloys with Thorium

21.1 The Rh-Th System

Phase Diagram. The complete range of composition was studied between 1000 and 1500 °C using metallographic and X-ray diffraction methods producing the diagram in **Fig. 89**. The solubility of rhodium in thorium was found to be <1 at%. The diagram is a complex one, with seven intermetallic compounds identified, Th_7Rh_3, ThRh, Th_3Rh_4, Th_3Rh_5, $ThRh_2$, $ThRh_3$, and $ThRh_5$. Compounds with congruent melting points are Th_7Rh_3, ThRh, and $ThRh_3$, the remainder is formed peritectically. $ThRh_2$ may have a β-form stable at high temperature and an α-form at low temperature [1].

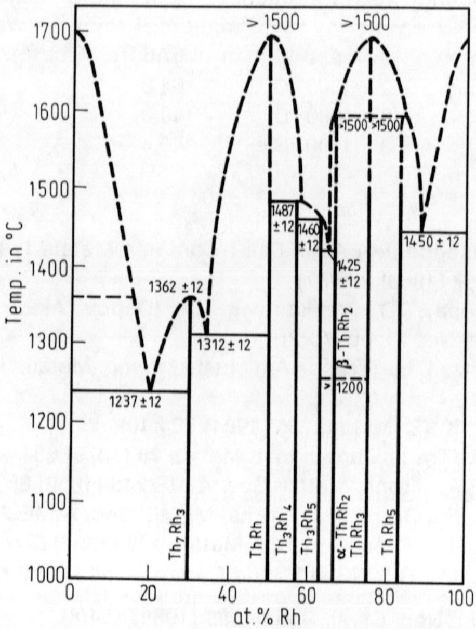

Fig. 89. Phase diagram Th-Rh.

Preparation. The specimens used were made by alloying sponge rhodium with electrolytic or iodide thorium as 1 g buttons molten in an arc covered with zirconium-gettered argon; compositional accuracy was estimated at ±1%. Heat treatments were carried out under a vacuum of better than 5×10^{-6} Torr in silicon tubes supported against collapse by tantalum capsules [1].

Crystallography. The following data is available:

compound	type	a in Å	b in Å	c in Å	Ref.
Th_7Rh_3	$D10_2$	10.031±0.003	—	6.287±0.002	[1, 2]
Th_7Rh_3	$D10_2$	10.02(8)	—	6.29(3)	[3]
β-$ThRh_2$	$B8_2$	4.629±0.002	—	5.849±0.003	[1, 4]
$ThRh_3$	$L1_2$	4.139	—	—	[5]
$ThRh_3$	$L1_2$	4.148±0.003	—	—	[1]
ThRh	CrB(Bf)	3.866±0.003	11.24±0.01	4.220±0.003	[1]

Density. The density of Th_7Rh_3 calculated from X-ray data was found to be 11.71 g/cm^3 [2, 3]. A pycnometric determination of 11.5 g/cm^3 was obtained by [3].

Hardness. Vickers hardness H_v according to [1]:

composition	t in °C	H_v in kg/mm^2
Th+1 at% Rh	600*)	65±3
Th_3Rh_4	—	155

*) ageing temperature

Free Energy. The Gibbs free energy of formation of the intermetallic compounds has been determined by electromotive force measurements [6]. The following table gives the figures.

compound	ΔG in cal/g-atom
$ThRh_5$	−13400+2.0 T
$ThRh_3$	−20000+2.9 T
$ThRh_2$	−23500+3.0 T
Th_3Rh_5	−25200+3.1 T
Th_3Rh_4	−26600+2.4 T
ThRh	−28600+3.7 T
Th_7Rh_3	−18900+3.2 T

The thermodynamics of platinum metal-actinide systems has been investigated using high-temperature galvanic cells with solid electrolytes and measuring continuously the oxygen partial pressure from the coupled reaction in highly purified hydrogen; no measurable reaction was found between thorium and rhodium at temperatures up to 1673 K [7].

Chemical Reactions. Th_7Rh_3 was found to tarnish slowly in air; ThRh tarnished more slowly and had a pleochroic response to polarized light. Th_3Rh_5 did not respond to polarized light [1].

References:

[1] Thomson, J.R. (J. Less-Common Metals **5** [1963] 437/42).
[2] Thomson, J.R. (Nature **189** [1961] 217).
[3] Ferro, R.; Rambaldi, G. (Acta Crystallogr. **14** [1961] 1094).
[4] Thomson, J.R. (Nature **194** [1962] 465).
[5] Dwight, A.E.; Downey, J.W.; Conner, R.A. (Acta Crystallogr. **14** [1961] 75/6).
[6] Murabayashi, M.; Kleykamp, H. (J. Less-Common Metals **39** [1975] 235/46).
[7] Möbius, S.; Hellwig, L.; Keller, C. (J. Less-Common Metals **121** [1986] 43/8).

21.2 Ternary Alloys

Rh-Th-Sb. The intermetallic compound RhThSb has an MgAgAs-type structure with a = 6.663 Å and V/M = 73.95 (V/M = volume per formula) [1].

Rh-Th-Al. The compound RhThAl was found to have an Fe_2P-type structure with a = 7.1828±0.0003 Å, c = 4.1198±0.0003 Å; **Fig. 90**, p. 140, shows the cell structure compared with that of $ThRh_2$ [2].

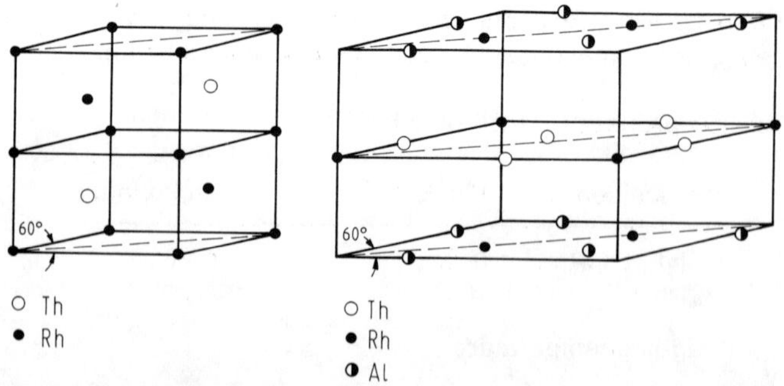

Fig. 90. Comparison of $ThRh_2$ (Ni_2In-type) with ThRhAl (Fe_2P-type).

Rh-Th-Ga. RhThGa has an Fe_2P-type structure with $a = 7.2143 \pm 0.0003$ Å, $c = 4.0441 \pm 0.0004$ Å [2].

Rh-Th-Ce. The effects of the substitution of thorium in the compound $CeRh_3$ to give alloys of the type $Ce_xTh_{1-x}Rh_3$ has been studied by specific heat, susceptibility and resistivity measurements. $CeRh_3$ is considered as a tetravalent saturated compound, and it was found that with the substitution of thorium Ce remained in a saturated valence state [3].

References:

[1] Dwight, A.E. (J. Less-Common Metals **34** [1974] 279/84).
[2] Dwight, A.E.; Mueller, M.H.; Conner, R.A.; Downey, J.W.; Knott, H. (Trans. Metall. Soc. AIME **242** [1968] 2075/80).
[3] Selim, R.; Michels, D.; Mihalisin, T. (J. Magn. Magn. Mater. **47/48** [1985] 99/101).

22 Alloys with Germanium

22.1 The Rh-Ge System

Phase Diagram. The diagram shown in **Fig. 91** is based on melting point, metallographic and X-ray diffraction data due to [1]. The alloys were prepared by high-frequency melting mixtures of high purity materials in quartz ampoules under argon. Difficulties in metallographic etching were found with alloys containing more than 30 wt% rhodium, and these were examined by relief polishing. Four intermediate compounds were found in this system, Rh_3Ge_4, RhGe, Rh_5Ge_3, and Rh_2Ge; Rh_3Ge_4 and Rh_5Ge_3 were formed by peritectic reaction, the other two crystallized congruently from the melt. Two eutectics were found but no terminal solid solubility.

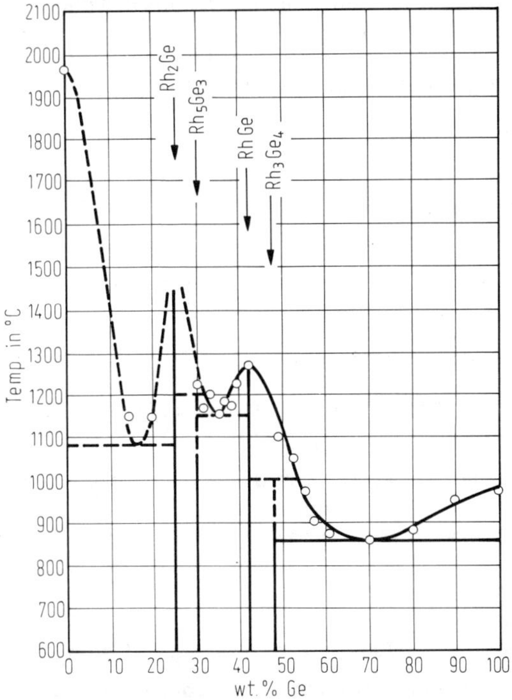

Fig. 91. Phase diagram Rh-Ge.

Crystallography. The lattice parameters and crystallography are given in the table below.

compound	space group	type	a	b in Å	c	Ref.
Rh_3Ge_4	—	tetragonal	5.7	—	10	[1]
Rh_2Ge	Pnam	—	5.44	7.57	4.00	[2]
Rh_5Ge_3	Pbam	—	5.42	10.32	3.96	[2]
RhGe	Pnam	—	5.70	6.48	3.25	[2]
$RhGe_4$ 1)	IrGe	hexagonal	6.19 ± 0.003	—	7.802 ± 0.004	[3]
$Rh_{17}Ge_{22}$ 2)	—	—	5.603	—	4.613	[4]
$Rh_{17}Ge_{22}$	—	tetragonal	5.604 ± 0.002	—	78.45 ± 0.04	[5]

[1] Formed under pressure of 2.5 GPa at temperatures of 1000 to 1300 K. This compound, after annealing in vacuum at 1200 K, decomposed to give $Rh_{17}Ge_{22}$ with the Nowotny chimney-ladder structure and pure germanium [3]. The threedimensional structural stability diagrams for 648 AB_3 and 389 A_3B_5 intermetalic compounds include data on Rh_5Ge_3 [6].

[2] Determined on single crystal specimen; accuracy ±0.004 Å included a factor of 17 indicating the formation of a multiple cell; see [5].

References:

[1] Zhuravlev, N.N.; Zhdanov, G.S. (Kristallografiya **1** [1956] 205/8; Soviet Phys. Crystallogr. **1** [1956] 158/60).
[2] Geller, S. (Acta Crystallogr. **8** [1955] 15/21).
[3] Larchev, V.I.; Popova, S.V. (J. Less-Common Metals **98** [1984] L1/L3).
[4] Völlenkle, H.; Wittmann, A.; Nowotny, H. (Monatsh. Chem. **97** [1966] 506/13).
[5] Jeitschko, W.; Parthé, E. (Acta Crystallogr. **22** [1967] 417/30).
[6] Villars, P. (J. Less-Common Metals **102** [1984] 199/211).

Mechanical Properties

Density. Some figures are given below.

compound	Rh_3Ge_4	RhGe			Rh_5Ge_3			Rh_2Ge			$RhGe_4$	
D in g/cm^3	8.5	9.8	9.7 [1]	9.7 [2]	10.7	11.0 [1]	10.6 [2]	11.0	11.2 [1]	11.4 [2]	7.56 [1]	7.5 ± 0.1 [2]
Ref.	[1]	[1]	[2]	[2]	[1]	[2]	[2]	[1]	[2]	[2]	[3]	[3]

[1] calculated value, [2] experimental value

Hardness. The figures in the table below are due to [1]. Hardness figures are microhardness determinations of VPN [1].

compound	average hardness in kg/mm^2	range of figures
Rh_3Ge_4	500	420 to 575
RhGe	580	470 to 660
Rh_5Ge_3	475	420 to 525
Rh_2Ge	700	500 to 900

The hardness of as-cast rhodium containing 1 wt% germanium was 237 VPN [4].

Stress-Rupture Strength. The time to failure at a loading of 345 bar for an alloy containing 1 wt% germanium was 3.0 h at 1200 °C and 0.6 h at 1400 °C [4].

References:

[1] Zhuravlev, N.N.; Zhdanov, G.S. (Kristallografiya **1** [1956] 205/8; Soviet Phys. Crystallogr. **1** [1956] 158/60).
[2] Geller, S. (Acta Crystallogr. **8** [1955] 15/21).
[3] Larchev, V.I.; Popova, S.V. (J. Less-Common Metals **98** [1984] L1/L3).
[4] Handley, J.R. (Platinum Metals Rev. **33** [1989] 64/72).

Electrical Properties

Superconductivity. The compound Rh_3Ge_2 was found to be superconducting with transition temperature $T_c=2.12$ K [1]. This compound was probably Rh_5Ge_3 according to [2]. The compound RhGe is said to be superconducting with $T_c = 0.96$ K [3]. $RhGe_4$ is superconducting with $T_c = 2.5 \pm 0.1$ K [4].

References:

[1] Matthias, B.T. (Phys. Rev. [2] **91** [1953] 413).
[2] Geller, S. (Acta Crystallogr. **8** [1955] 15/21).
[3] Braun, H.F.; Segre, C.U. (Solid State Commun. **35** [1980] 735/8).
[4] Larchev, V.I; Popova, S.V. (J. Less-Common Metals **98** [1984] L1/L3).

22.2 Ternary Alloys

Rh-Ge-Ga. The compound Rh_5Ge_4Ga has a B20-type structure with a = 4.822 kX. $Rh_{50}Ge_{42}Ga_8$, after annealing for 16 h at 550 °C, was shown by X-ray diffraction to be 80% B20 structure; $Rh_{50}Ge_{37.5}Ga_{12.5}$, after the same heat treatment, was also 80% B20 and $Rh_{50}Ge_{25}Ga_{25}$ was assumed to have the same structure [7]. New examples of a particular structural pattern shown in defect silicides have been found within the pseudo-homogeneous regions of $Rh(Ge, Ga)_{2-x}$ alloys; the lattice parameters are given below [8].

compound	Ga/Ge ratio	a (in Å)	c (in Å)
$Rh_{43}(Ga, Ge)_{57}$	10:90	6.623 ± 0.004	4.640 ± 0.004
$Rh_{23}(Ga, Ge)_{31}$	25:75	5.620	4.642
$Rh_{12}(Ga, Ge)_{17}$	35:65	5.680	4.655
$Rh_{39}(Ga, Ge)_{58}$	50:50	5.714	4.665
$Rh_{43}(Ga, Ge)_{69}$	75:25	5.770	4.695

Rh-Ge-Sc. The structure of $Rh_7Ge_6Sc_4$ has given the following crystallographic data: a = 8.1255_8 Å, atomic volume = 536.5_1 Å^3, Z = 2; density 8.269 g/cm^3 at 293 K; the crystal structure is of the $U_4Re_7Si_6$ type [6].

Rh-Ge-La. The compound $La_3Rh_4Ge_4$ has been found to be orthorhombic with a structure of the $U_3Ni_4Si_4$ type; the lattice parameters are a = 4.1746(3), b = 4.2412(2), c = 25.234(3) Å, Z = 2; density 8.32 g/cm^3 [5].

Rh-Ge-Ce. The magnetic structure of $CeRh_2Ge_2$ has been studied by magnetic and resistivity measurements as well as neutron diffraction. The compound has a $ThCr_2Si_2$-type structure with a = 4.150(1), c = 10.468(3) Å. The variation of the lattice parameters in rare earth-Rh_2Ge_2 compounds shows the cerium to be trivalent in this compound; Néel temperature $T_N = 15$ K, $\theta_P = -25$ K; effective magnetic moment 2.46 μ_B. The results are discussed in terms of the RKKY (Ruderman-Kittel-Kasuya-Yosida) exchange interaction and also in terms of the effects of 4f hybridization on ordering. **Fig. 92**, p. 144, shows the magnetic structure of $CeRh_2Ge_2$ [3].

Rh-Ge-Gd. The compound Rh_2Ge_2Gd was found to be related to the $Al_4Ba(ThCu_2Si_2)$ type, with a = 4.127, c = 10.152 Å [4].

Rh-Ge-Zr. The compound RhGeZr, prepared by arc-melting and heat-treated for 7 d at 840 °C followed by quenching, was found to have a TiNiSi-type structure with a = 6.5923(4), b = 3.9906(3), c = 7.5620(5) Å [2].

Rh-Ge-Th. Alloys of composition Rh_2ThGe_2 have a tetragonal $Al_4BaThCu_2Si_2$-type structure with a = 4.230 ± 0.002, c = 10.055 ± 0.005 Å [1]. The author discusses the geometrical characteristics in terms of reduced cell parameters (a/R_A).

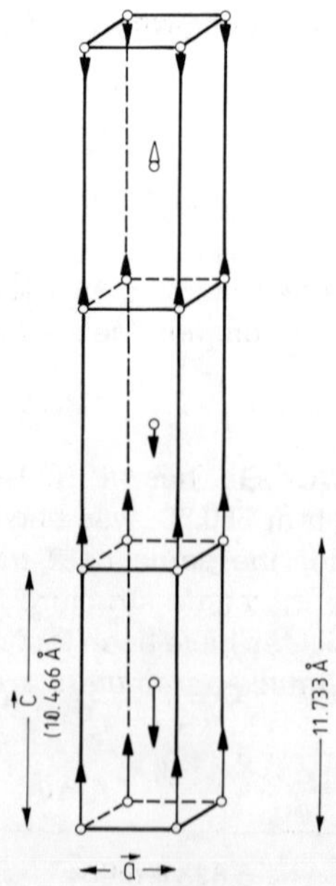

Fig. 92. Magnetic structure of $CeRh_2Ge_2$.

References:

[1] Marazza, R.; Ferro, R.; Rambaldi, G.; Zanicchi, G. (J. Less-Common Metals **53** [1977] 193/7).
[2] Johnson, V.; Jeitschko, W. (J. Solid State Chem. **4** [1972] 123/30).
[3] Venturini, G.; Malaman, B.; Pontonnier, L.; Fruchart, D. (Solid State Commun. **67** [1988] 193/7).
[4] Rossi, D.; Marazza, R.; Ferro, R. (J. Less-Common Metals **66** [1979] P17/P25).
[5] Hoverstreydt, E.; Klepp, K.; Parthé, E. (Acta Crystallogr. B**38** [1982] 1803/5).
[6] Engel, N.; Chabot, B.; Parthé, E. (J. Less-Common Metals **96** [1984] 291/6).
[7] Esslinger, P.; Schubert, K. (Z. Metallk. **48** [1957] 126/34).
[8] Völlenkle, H.; Wittmann, A.; Nowotny, H. (Monatsh. Chem. **97** [1966] 506/13).

23 Alloys with Tin

23.1 The Rh-Sn System

Phase Diagram. The diagram in **Fig. 93** is that given by [1].

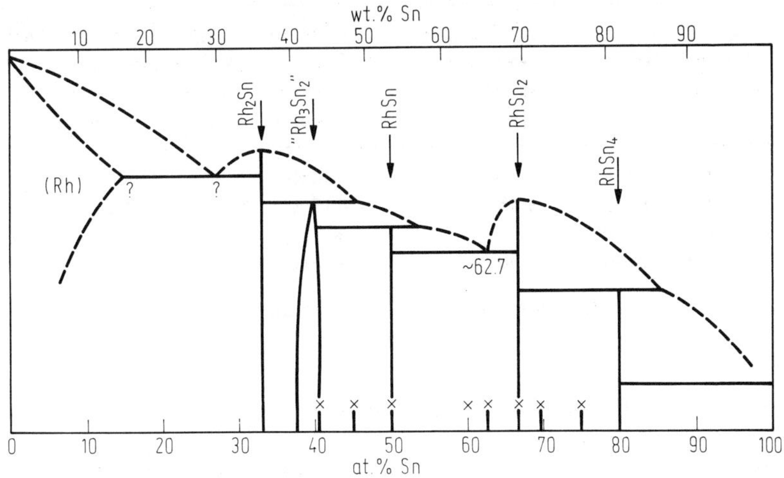

Fig. 93. Schematic phase diagram Rh-Sn.

Crystallography. Early work using X-ray diffraction and microscopical methods found the compound of approximate composition Rh_3Sn_2 existing in the composition range of 41 to 44 wt% tin; it had an NiAs-type structure (B8), with lattice parameters a=4.331, c=5.542 Å [2]. In another paper, this compound was found to have a homogeneous range of 56 to 59 wt% [3]. An NiAs-type phase at 40 wt% tin which had parameters a=4.331, c=5.544 kX and also three other phases corresponding to the compounds RhSn and $RhSn_2$ were found, the latter in the composition range 66.7 to 75 at% tin. The RhSn phase was cubic with an FeSi-type structure, its parameters being a=5.120 kX, $x_{Rh}=0.1$, $x_{Sn}=0.4$; the other phase was tetragonal, C16 type with a=6.397 to 6.401 kX, c=5.643 to 5.647 kX. The investigator found that an additional phase appeared after annealing above 475 °C [4]. The transformation of $RhSn_2$ has been further investigated, and on quenching from above 500 °C the C16 $CuAl_2$-type phase was found with a=6.398, c=5.643 Å, whilst at temperatures below 500 °C the tetragonal $CoGe_2$ phase was found with $a=6.31_{9\pm5}$, $c=11.9_{7\pm1}$ Å which had a density of 9.39 g/cm^3 [5]. The crystallography of the low-temperature form of $RhSn_2$ found by [4] has been further investigated and a mixed $CuAl_2$-CaF_2 structure proposed [6, 7]. The compound Rh_2Sn was found to have the Co_2Si-type structure with $a=8.19_2$, $b=5.50_9$, $c=4.21_2$ kX [8]. A previously unknown compound Rh_3Sn with Co_2Si-type structure was reported by [9]. An investigation of the partial molar heat of solution of rhodium in liquid tin and rhodium-tin alloys suggests that the solubility limit of rhodium in tin given in [1] should be 0.9 at% rather than 7 at% at 775 °C. Also, as a consequence of an unexplained increase in the heat of solution of rhodium between 725 and 750 °C a transformation in $RhSn_4$ is suggested [10].

Hardness. The as-cast hardness of rhodium containing 1 wt% tin is 205 VPN [11].

Stress-Rupture Properties. The stress-rupture time of rhodium containing 1 wt% tin under a loading of 345 bar is given as 37.2 h at 1200 °C and 2.5 h at 1400 °C [11].

Partial Molar Heat of Solution. This was determined by [10]. The results are given in the table below. It shows Δ_{Rh}, the partial molar heat of solution of rhodium in tin, the solubility, ΔH_i the heat of solution in the two-phase region, and the calculated heat of formation of $RhSn_4$.

T in K	$\overline{\Delta H}_{Rh}$ at infinite dilution in Sn in cal/g-atom	solubility of Rh in Sn in at%	$\overline{\Delta H}_i$ in two-phase region in cal/g-atom		$\Delta H^F_{RhSn_4}$ in cal/g-atom
			i = rhodium	i = tin	
700	−29950 ± 550 [a] (3) [b]	0.35	−48350 ± 600(5)	50	−9630 ± 120
725	−28790 ± 550(3)	0.48	−48350 ± 570(10)	80	−9610 ± 140
750	−28490 ± 430(5)	0.64	−43820 ± 600(9)	90	−8690 ± 120
775	−28260 ± 220(4)	0.90	−42560 ± 360(27)	115	−8420 ± 70

[a] 95% confidence limits of the mean, [b] number of determinations

References:

[1] Hansen, M.; Anderko, K. (Constitution of Binary Alloys, 2nd Ed., New York – Toronto – London1958, p. 1155).
[2] Notwotny, H.; Schubert, K.; Dettinger, U. (Metallforschung **1** [1946] 137/45).
[3] Schubert, K. (Z. Naturforsch. **2a** [1947] 120).
[4] Nial, O. (Svensk Kem. Tidskr. **59** [1947] 172/83).
[5] Schubert, K.; Pfisterer, H. (Z. Metallk. **41** [1950] 433/41).
[6] Hellner, E. (Z. Kristallogr. **107** [1956] 99/123).
[7] Jagodzinski, H.; Hellner, E. (Z. Krist. **107** [1956] 124/49).
[8] Schubert, K.; Breimer, H.; Burkhardt, W.; Günzel, E.; Haufler, R.; Lukas, H.L.; Vetter, H.; Wegst, J.; Wilkens, M. (Naturwissenschaften **44** [1957] 229/30).
[9] Schubert, K.; Lukas, H.L.; Meissner, H.G.; Bhan, S. (Z. Metallk. **50** [1959] 534/40).
[10] Miner, R.V.; Spencer, P.J.; Pool, M.J. (Trans. Metall. Soc. AIME **242** [1968] 1553/4).
[11] Handley, J.R. (Platinum Metals Rev. **33** [1989] 64/72).

23.2 Ternary Alloys

23.2.1 Miscellaneous Alloys

Rh-Sn-Ca. $Rh_{1.2}Sn_{4.5}Ca$ alloys have a phase I primitive cubic structure with a = 9.702 Å, T_c = 8.6 to 8.7 K, a critical field dH_{c2} = 4 kG/K, a resistivity of 110 μΩ·cm at 300 K and a resistance ratio $\varrho_{300K}/\varrho_{10K}$ = 8.8 [1]. Arrhenius behaviour was found between the reduction in superconducting transition temperature and the quenching temperature causing it in the compound $Ca_3Rh_4Sn_{13}$; this was used to determine the disordering energy which proved to be ~1.16 eV [3].

Rh-Sn-Sr. Rh_xSn_y alloys have a primitive cubic structure with a = 9.800 Å and a superconducting critical temperature of 4.3 to 4.0 K [1, 2].

Rh-Sn-Ti. RhSnTi has an MgAgAs-type structure with a = 6.11 Å [4].

Rh-Sn-Zr. RhSnZr has an Fe_2P-type structure with a = 7.338, c = 3.610 Å [4].

Rh-Sn-Hf. The structure of RhSnHf was found to be of the Fe_2P-type with a = 7.318 Å, c = 3.577 Å [4].

Rh-Sn-Th. The intermetallic compound RhThSn has an Fe_2P-type structure with a = 7.515, c = 4.088 Å [1]. The compound $Rh_4Sn_{12}SnTh_3$ has a critical superconducting temperature T_c = 1.9 K and a cubic structure with a ≈ 9.7 Å [2, 3].

References:

[1] Remeika, J.P.; Espinosa, G.P.; Cooper, A.S.; Barz, H.; Rowell, J.M.; McWhan, D.B.; Vandenberg, J.M.; Moncton, D.E.; Fisk, Z.; Woolf, L.D.; Hamaker, H.C.; Maple, M.B.; Shirane, G.; Thomlinson, W. (Solid State Commun. **34** [1980] 923/6).
[2] Hodeau, J.L.; Marezio, M.; Remeika, J.P.; Chen, C.H. (Solid State Commun. **42** [1982] 97/102).
[3] Westerveld, J.P.A.; Lo Cascio, D.M.R.; Bakker, H. (J. Phys. F **17** [1987] 1963/71).
[4] Dwight, A.E. (J. Less-Common Metals **34** [1974] 279/84).

23.2.2 Rh-Sn-RE Alloys

Crystallography

RERhSn (RE = rare-earth element)

The ternary alloys of general formula RERhSn with Fe_2P-type structures have been the subject of X-ray diffraction and Mössbauer studies; the results are given in the following two tables [4].

compound	a (in Å)	c (in Å)
YRhSn	7.54	3.77
SmRhSn	7.46	3.98
GdRhSn	7.527 ± 0.001	3.862 ± 0.001
TbRhSn	7.542 ± 0.002	3.796 ± 0.001
DyRhSn	7.529 ± 0.002	3.771 ± 0.001
HoRhSn	7.527 ± 0.002	3.748 ± 0.001
ErRhSn	7.519 ± 0.001	3.730 ± 0.001
TmRhSn	7.534 ± 0.001	3.690 ± 0.001
YbRhSn	7.52	3.67
LuRhSn	7.529 ± 0.001	3.646 ± 0.001

Mössbauer Parameters. Values according to [4]:

compound	T in K	S in mm/s*)	E_Q in mm/s**)	Γ1 in mm/s
GdRhSn	295.0	1.85	0.48	1.01
TbRhSn	295.0	1.79	0.63	0.99
DyRhSn	295.0	1.79	0.65	1.00
HoRhSn	295.0	1.78	0.68	1.01

*) relative to $Ba^{119}SnO_3$ at 295 K, **) total splitting eqQ/2

For further details see original paper [4].

$RERh_xSn_y$. A most important paper on a new family of ternary intermetallic compounds of the general formula RE-Rh_x-Sn_y which showed either superconducting or magnetic properties appeared in 1980. In general, compounds of the heavier rare earths were superconducting, and those containing the lighter elements were found to be magnetic. The compound $ErRh_{1.1}Sn_{3.6}$ showed re-entrant superconductivity, i.e., it was superconducting ($T_c = 0.97$ K) but became magnetic with higher electric resistivity at 0.57 K [3].

The table below gives the lattice parameters and other physical properties determined.

Properties of $RERh_xSn_y$:

compound	structure	lattice constant in Å	density in g/cm³ at 25 °C	superconductivity T_c in K	magnetic T_m in K	critical field $dH_{c2}/dT (T=T_c)$ in kG/K	resistance ϱ_{300K} in μΩ·cm	resistance $\varrho_{300K}/\varrho_{10K}$
$ScRh_xSn_y$	II			4.5 to 4.1 a)				
YRh_xSn_y	II			3.2 to 3.1 a)		21.5	290	1.03
$LaRh_xSn_y$	I	9.745		3.2 to 3.0 a)				
$CeRh_{1.2}Sn_{4.0}$	I	9.710	8.3					
$PrRh_xSn_y$	I	9.693					114	21
$NdRh_{1.2}Sn_{4.1}$	I	9.676	8.4					
$SmRh_{1.2}Sn_{4.3}$	I	9.657	8.7					
$EuRh_xSn_y$	I	9.749			~11 c)			
$GdRh_xSn_y$	I	9.638			11.2 b)			
$TbRh_{1.1}Sn_{3.6}$	III	13.774	8.7		e) {3.8 a), 2.8}			
$DyRh_{1.1}Sn_{3.6}$	III	13.750	8.8		2.08 a)			
$HoRh_{1.2}Sn_{3.9}$ } $HoRh_{1.2}Sn_{3.9}$	II, III	13.750	8.9		1.68 a)		310	0.84
$ErRh_{1.1}Sn_{3.6}$ } $ErRh_{1.1}Sn_{3.6}$	II, III	13.714	9.1	{1.31 a), 1.22 b), 0.97 c)}	{0.34 b), 0.57 c), 0.61 d)}		340	0.85
$TmRh_{1.3}Sn_{4.0}$ } $TmRh_{1.3}Sn_{4.0}$	II, III	13.701	9.2	2.3 to 2.2 a)			286	0.94
$YbRh_{1.4}Sn_{4.6}$	I		8.9	8.6 to 8.2 a)		3.5		
$LuRh_{1.2}Sn_{4.0}$	II		9.6	4.0 to 3.9 a)		20.3		

I = primitive cubic, $a \approx 9.7$ Å, II = pseudotetragonal, $a \approx 13.7$, $c \approx 9.7$ Å, III = face-centred cubic, $a \approx 13.7$ Å

a) measured inductively, b) measured resistively, c) susceptibility measurement, d) neutron scattering; e) 2 T_m observed on same crystal

The compounds fell into three crystal types called phase I, II, and III (primitive cubic, pseudotetragonal and face-centred cubic structures, respectively) [3]. The crystallography of these compounds was investigated using X-ray methods, in particular a phase I-compound $PrRh_{1.2}Sn_{4.2}$ with $a = 9.693$ Å and density 8.3 g/cm³, and a phase III-compound $TbRh_{1.1}Sn_{3.6}$ with $a = 13.774$ Å and density 8.7 g/cm³. Structural models for both compounds are given; **Fig. 94** and **Fig. 95** show one octant of the cubic cell of the proposed phase I and III structures [5].

Mössbauer studies on $ErRh_{1.1}Sn_{3.6}$, $DyRh_{1.1}Sn_{3.6}$, and $TmRh_{1.3}Sn_{4.0}$ have been carried out; **Fig. 96**, p. 150, shows the spectrum for $ErRh_{1.1}Sn_{3.6}$ measured at 77 K with two quadru-

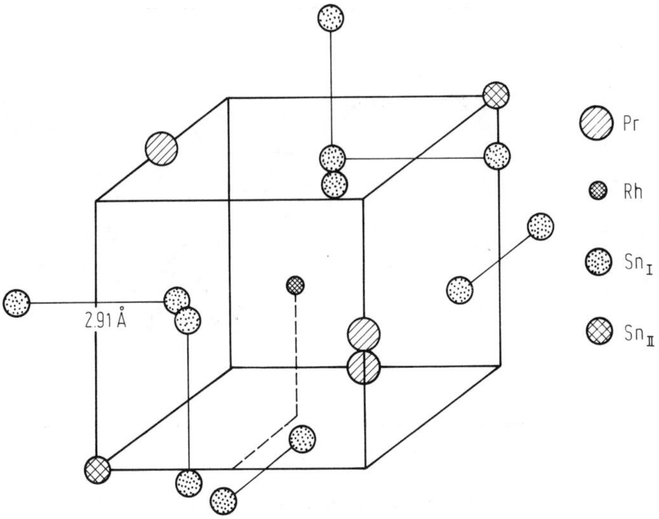

Fig. 94. One octant of the cubic cell of phase I.

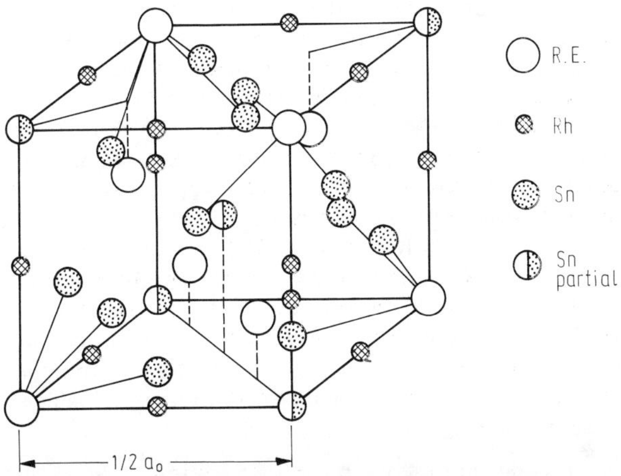

Fig. 95. One octant of trial structure of phase III.

pole doublets indicating two crystallographically different tin sites. The isomer shifts for these were distinct, suggesting large chemical dissimilarity between the two types of atom, and hence support the crystal chemistry of this complex structure with respect to tin clusters [6].

A structural distortion of the primitive cubic phase (I) of the compounds $SnRE_3Rh_4Sn_{12}$ where RE = La, Ce, Pr, Nd, Sm, and Gd was observed using electron diffraction and X-ray methods; with RE = Eu, Yb, no such distortion was found. The La, Yb, Ca and Sr compounds were superconducting at temperatures between 8.7 and 1.9 K, whereas the Eu and Gd compounds had a magnetic transition at ~11 K. The structural implications are discussed [2]. The structure of phase II in the Er-Rh-Sn system has been determined from single-

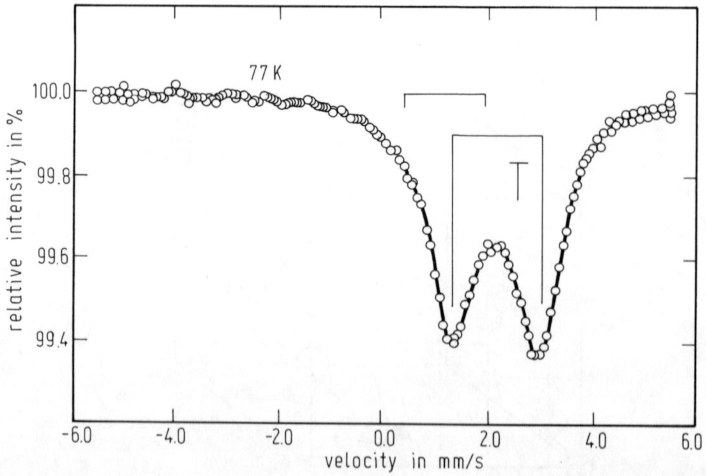

Fig. 96. Mössbauer spectra of $ErRh_{1.1}Sn_{3.6}$ measured using the 23.9 keV resonance in ^{119}Sn at 77 K.

crystal X-ray diffraction data; the chemical formula was found to be $[Er(1)_{1-x}Sn(1)_x]$-$Er(2)_4Rh_6Sn(2)_4Sn(3)_{12}Sn(4)_2$ where Rh and Sn(3) correspond to two and four different crystallographic sites, respectively (the numbers in parentheses refer to atoms occupying crystallographically independent sites and having different chemical environments); the original paper should be consulted [7]. The relationship between the crystal structure and the re-entrant superconducting properties of the compound $(Sn_{1-x}Er_x)Er_4Rh_6Sn_{18}$ has been investigated using electric resistance, magnetic and X-ray diffraction methods on single-crystal specimens; these compounds are superconducting with $x \approx 0$ $(T_c) = 1.3$ K, re-entrant superconducting with $x \approx 0.3$ $(T_c) = 1.24$ K and $T_M = 0.34$ K, and undergo a single magnetic transformation at $T_M = 0.68$ K with $x = \sim 0.75$. Their results clearly indicate that the occupancy of the $[Sn(1)_{1-x}Er(1)_x]Er(2)_4$ sublattice is responsible for the variation of the low-temperature properties of the $ErRh_xSn_y$ compounds; **Fig. 97** shows a threedimensional representation of the $[Sn(1)_{1-x}Er(1)_x]$ and $Er(2)_4$ sublattices. The authors indicate that this is the first system of re-entrant superconductors where stoichiometry within a sublattice controls both magnetic ordering and superconductivity [8].

The magnetic structure of $[Sn(1)_{1-x}Er(1)_x]Er(2)_4Rh_6Sn_{18}$ with $x = 0.42$ has been examined by neutron diffraction at 2.5 K (above T_c) and 70 mK (below T_M). In the magnetic state, all the reflections could be indexed in the crystallographic cell. The magnetic order is not collinear, and only one model deduced from Bertaut's macroscopic theory is in good agreement with the observed data [9]. The rare-earth ternary stannides can crystallize with four different structures I, I′, II, II′ and III (face-centred cubic); I′ is a distorted I structure whilst II′ is a disordered II structure. The valence fluctuations of Yb in the superconducting ytterbium-rhodium stannides has been examined by Yb L_{III} X-ray absorption edge measurements. Large single crystals of $(Yb_xSn_{1-x})(1)Yb_4Rh_6Sn(2)_{18}$ were used and these were found by X-ray diffraction and transmission electron microscopy to be of face-centred cubic structure phase II′ type with a = 13.740 Å. Comparison was made between the Yb L_{III} X-ray absorption edge spectra of this compound and the phase I structured $Sn(1)Yb_3Rh_4Sn(2)_{12}$ material with a = 9.676 Å. The results gave +2.20 and +2.50 for the Yb valence in the phase I and phase II′ compounds, respectively. The former is superconducting ($T_c \approx 8$ K)

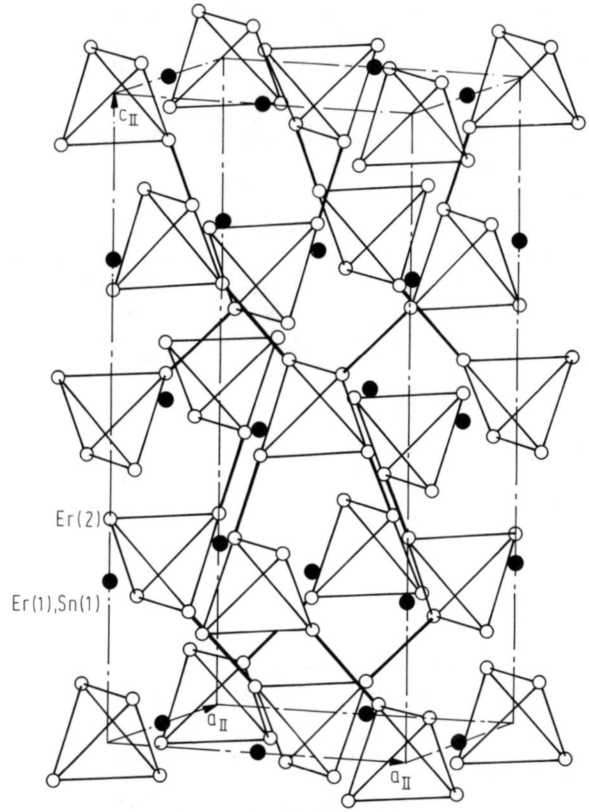

Fig. 97. Threedimensional representation of the $[Sn(1)_{1-x}Er(1)_x]$ and Er(2) sublattices. The $Er(2)_4$ tetrahedra are represented by a thin line and the short Er–Er doublets between tetrahedra by a thick line.

and contains only one crystallographic site for Yb, which suggests that this atom is in a valence fluctuation state [11].

$Rh_3Sn_5RE_2$. The lattice parameters of these compounds containing Y, Gd, Tb, Dy and Ho are given below [11]. RE = rare-earth element.

RE	a	b in Å	c
Y	4.387(2)	26.212(4)	7.1550(8)
Gd	4.419(6)	26.28(4)	7.191(18)
Tb	4.408(2)	26.22(1)	7.172(6)
Dy	4.394(2)	26.18(1)	7.142(7)
Ho	4.378(2)	26.12(2)	7.120(9)

The structure of $Rh_3Sn_5Y_2$ was analysed and a three-dimensional model of the "covalent" arrangement of Rh(1), Rh(2) and Rh(3) atoms with those of Y(1) and Y(2) derived [11].

Electrical and Magnetic Properties. Measurements on cast samples of $Rh_3Sn_5Y_2$ do not show superconductivity down to 4.2 K. $Rh_3Sn_5Y_2$ showed little increase of resistivity with temperature; it was $\sim 2.7 \times 10^{-5}$ at 4.2 K and 4.8×10^{-5} $\Omega \cdot cm$ at 290 K. $Rh_3Sn_5Y_2$ is diamagnetic [11].

$Rh_2Sn_{4-x}RE_{1+x}$, $0 \leqq x \leqq 0.5$; RE = La, Ce, Pr, Nd, Sm. The lattice parameters are given below for compounds where x = 0 [12].

RE	a	b in Å	c
La	18.692(7)	4.536(3)	7.259(4)
Ce	18.591(6)	4.494(4)	7.252(4)
Pr	18.571(6)	4.477(3)	7.238(3)
Nd	18.535(3)	4.463(1)	7.229(1)
Sm	18.450(7)	4.421(3)	7.210(4)

The structure of Rh_2Sn_4Nd was analysed and found to be of a new type with space group Pnma and z = 4; a projection on the (010) plane is shown. This structure is closely related to that of $Rh_3Sn_5Y_2$ [11].

Rh_2Sn_4La was not superconducting down to 1.35 K; the resistivity ratio $\varrho_{290K}/\varrho_{1.35K} = 2.8$ [12].

References:

[1] Dwight, A.E. (J. Less-Common Metals **34** [1974] 279/84).

[2] Hodeau, J.L.; Marezio, M.; Remeika, J.P.; Chen, C.H. (Solid State Commun. **42** [1982] 97/102).

[3] Remeika, J.P.; Espinosa, G.P.; Cooper, A.S.; Barz, H.; Rowell, J.M.; McWhan, D.B.; Vandenberg, J.M.; Moncton, D.E.; Fisk, Z.; Woolf, L.D.; Hamaker, H.C.; Maple, M.B.; Shirane, G.; Thomlinson, W. (Solid State Commun. **34** [1980] 923/6).

[4] Dwight, A.E.; Harper, W.C.; Kimball, C.W. (J. Less-Common Metals **30** [1973] 1/8).

[5] Vandenberg, J.M. (Mater. Res. Bull. **15** [1980] 835/47).

[6] Shenoy, G.K.; Viccaro, P.J.; Cashion, J.D.; Niarchos, D.; Dunlap, B.D.; Pröbst, F.; Remeika, J.P. (Ternary Supercond. Proc. Intern. Conf., Lake Geneva, Wis., 1980 [1981], pp. 233/7).

[7] Hodeau, J.L.; Marezio, M.; Remeika, J.P. (Acta Crystallogr. B**40** [1984] 26/38).

[8] Miraglia, S.; Hodeau, J.L.; Marezio, M.; Ott, H.R.; Remeika, J.P. (Solid State Commun. **52** [1984] 135/7).

[9] Hodeau, J.L.; Bordet, P.; Wolfers, P.; Marezio, M.; Remeika, J.P. (J. Magn. Magn. Mater. **54/57** [1986] 1527/8).

[10] Bordet, P.; Hodeau, J.L.; Wolfers, P.; Krill, G.; Weiss, F.; Marezio, M. (J. Magn. Magn. Mater. **63/64** [1987] 524/6).

[11] Meot-Meyer, M.; Venturini, G.; Malaman, B.; Steinmetz, J.; Roques, B. (Mater. Res. Bull. **19** [1984] 1181/6).

[12] Meot-Meyer, M.; Venturini, G.; Malaman, B.; Roques, B. (Mater. Res. Bull. **20** [1985] 913/9).

Electrical and Magnetic Properties

$ErRh_{1.1}Sn_{3.6}$. The reentrant superconducting compound $ErRh_{1.1}Sn_{3.6}$ in single-crystal form was examined by low field absolute static magnetization measurements; it showed a clear Meissner effect at the superconducting transition. In the superconducting state the magnetization shows reversible type-II behaviour in a field >6 Oe and a thermodynamic critical field of $H_0 = 20 \pm 5$ Oe can be deduced from the data. A search for a region of superconductivity and ferromagnetism below T_c showed that this must be <0.03 K [1].

The electric resistance, magnetic susceptibility and superconducting properties of this compound have been measured and contrasted with those of a similar compound containing Lu instead of Er. Resistance measurements plotted against temperature show the transition curves and the reentrant nature of the superconductivity, see **Fig. 98**; it is suggested that the large localized magnetic moment of the Er^{3+} ion is responsible for the magnetic properties, low superconducting parameters and destruction of superconductivity at T_{c2} [2].

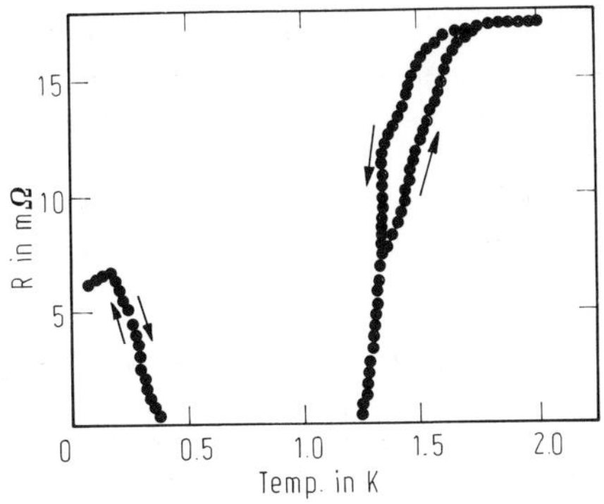

Fig. 98. Transition curves for reentrant superconductor $ErRh_{1.1}Sn_{3.6}$.

$YbRh_{1.4}Sn_{4.6}$. This compound with the simple cubic phase I-structure was found to transform at a pressure of 40 kbar and a temperature of 800 °C to the face-centred cubic phase III-structure with composition **$YbRh_{1.1}Sn_3$** with a = 13.735 Å, suggesting that Yb is in an intermediate valence state. The phase I-compound is superconducting at 8.6 K, but the phase III-compound is not superconducting at 0.9 K; it is suggested that mixed valency and superconductivity are incompatible [3]. The electric resistivity, magnetoresistance and upper critical field (H_{c2}) of single-crystal superconducting material have been studied. The coefficient of electronic specific heat was estimated as $\gamma = 11.7$ $mJ \cdot mol^{-1} \cdot K^{-2}$. It was concluded that the Yb ion is in a divalent state in the compound [4].

The degradation of the superconducting transition temperature in **$Yb_3Rh_4Sn_{13}$** produced by quenching from 650 to 990 °C is attributed to site-antisite disorder; an Arrhenius relation was derived between quenching temperature and the reduction in T_c which was used to derive the disordering energy = 1.15 ± 0.08 eV [5].

$GdRh_{1.07}Sn_{4.21}$. The electric resistivity and magnetoresistance of single-crystal material was measured between 4.2 and 25 K in a magnetic field up to 30 kOe. The compound orders antiferromagnetically below $T_N \approx 11.6$ K. A positive value for the magnetoresistance at temperatures $< T_N$ was attributed to field-induced enhancement of the spin fluctuations [6].

$LaRh_{1.1}Sn_4$. Mössbauer studies suggested that the tin atoms occupy two chemically inequivalent sites; electric resistance measurements showed that the superconducting transition temperature was between 2.8 and 3.1 K and that there was an unexpected temperature dependence of the resistivity in the normal state; this is attributed to the scattering of conduction electrons on localized spin fluctuations of the d electrons of rhodium [7].

$LuRh_{1.2}Sn_4$. Electrical, magnetic and superconductivity measurements have been made on this compound, and these contrasted with similar ones made on $ErRh_{1.1}Sn_{3.6}$; the high superconducting temperature and low magnetic susceptibility ($\chi \approx 1 \times 10^{-6}$/g) is attributed to the very low magnetic moment of the Lu^{3+} ion [11]. Single-crystal specimens were used for resistivity, magnetoresistance and upper critical field measurements on this compound; the resistivity measurements showed that it was superconducting below 4.1 K and that it had a Kondo anomaly at higher temperatures with a resistivity minimum at 210 K. The resistance and magnetoresistance data suggested weak itinerant antiferromagnetic ordering below 5.2 K; the electronic specific heat coefficient is estimated as 62.4 $mJ \cdot mol^{-1} \cdot K^{-2}$ [8]. The superconductivity of thin films produced by cathode sputtering has been reported; some of the data are given below, where RRR = residual resistivity ratio, T_c is the superconducting critical temperature and the temperature column gives the annealing temperature in °C [9].

sample	T_c in K	RRR	t in °C
1	3.75	43	600
2	3.80	11	525
3	3.96	34	800
4	2.25	1.4	—

References:

[1] Andres, K.; Remeika, J.P.; Espinosa, G.P.; Cooper, A.S. (Phys. Rev. [3] B **23** [1981] 1179/84).
[2] Rojek, A.; Sulkowski, C. (Phys. Status Solidi B **134** [1986] K133/K137).
[3] Jayaraman, A.; Remeika, J.P.; Espinosa, G.P.; Cooper, A.S.; Barz, H.; Maines, R.G. (Solid State Commun. **39** [1981] 1049/51).
[4] Ali, N.; Woods, S.B.; Kozlowski, G.; Rojek, A. (J. Phys. F **15** [1985] 1547/54).
[5] Westerveld, J.P.A.; Baaker, H. (Philos. Mag. [8] B **54** [1986] L15/L19).
[6] Ali, N.; Woods, S.B.; Kozlowski, G.; Rojek, A. (J. Phys. F **15** [1985] 155/60).
[7] Olejniczak, J.; Rojek, A.; Sulkowski, C. (Phys. Status Solidi B **130** [1985] K73/K76).
[8] Ali, N.; Datars, W.R. (J. Less-Common Metals **127** [1987] 49/54).
[9] Cendlewska, B.; Rojek, A. (Japan. J. Appl. Phys. **26** Pt. 1 [1987] 1295/6).

23.3 Quaternary Alloys

Alloys of general composition $\mathbf{La_{1-x}Pr_xRh_4Sn_{13}}$ with x = 0 to 0.15 have been investigated by NMR (nuclear magnetic resonance); the NMR spectrum revealed satellite absorption lines of the tin nuclei located in the nearest coordination spheres of the rare-earth impurity. The spatial distribution of the electron polarization near the Van Vleck ion was determined and estimates were obtained of the hyperfine and exchange constants. Analysis of the experimental results suggests that the compound $Pr_3Rh_4Sn_{13}$ may be a very effective working substance for obtaining ultralow temperatures by adiabatic demagnetization [1].

Alloys of composition $\mathbf{Er_xY_{1-x}Rh_{1.1}Sn_{3.6}}$ and $\mathbf{Ho_xY_{1-x}Rh_{1.1}Sn_{3.6}}$ have had their superconducting and magnetic ordering critical temperatures determined by magnetic measurements. For the erbium alloys, superconductivity appears for any value of x, and re-entry to a magnetically ordered state for $x \geqq 0.6$, whilst for the holmium alloys the compound becomes either superconducting or magnetically ordered, and is expected to remain paramagnetic for $x \approx 0.5$. Some possible theoretical models are discussed [2].

References:

[1] Garifullin, I.A.; Kataev, V.E.; Sadykov, I.I.; Tagirov, L.R. (Soviet Phys.-JETP **66** [1987] 761/5).
[2] van de Pasch, A.W.M.; Lazaro, F.J.; Blank, D.H.A.; Houwman, E.P.; Flokstra, J. (Japan. J. Appl. Phys. **26** Pt. 1 [1987] 1293/4).

24 Alloys with Lead

The Rh-Pb System

Phase Diagram. The partial phase diagram shown in **Fig. 99** was established by thermal analysis, metallography, and X-ray diffraction techniques [1]. The rhodium-rich end of the diagram is uncertain; the horizontal line at 1230 °C could not be explained satisfactorily, although it was suggested that it might be connected with a transformation in rhodium. Difficulties were also experienced with oxidation, and thermal analysis gave no clear indication of the expected Rh_3Pb phase. At the other end of the diagram, the horizontals at 300 and 320 °C are in doubt.

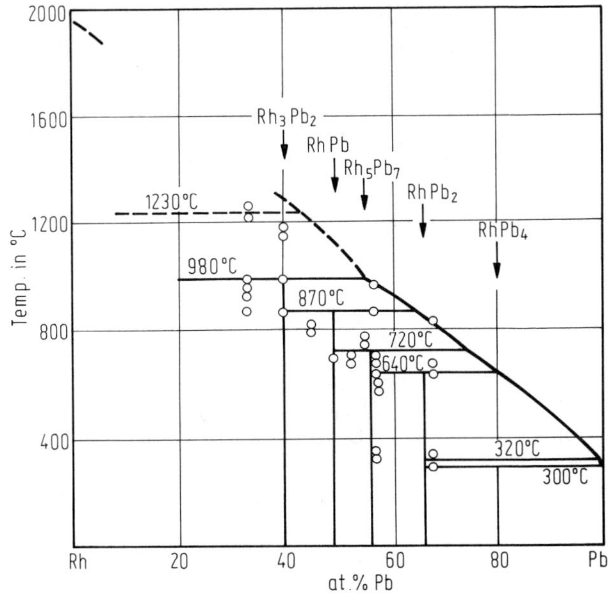

Fig. 99. Partial phase diagram Rh-Pb.

Preparation. The alloys were prepared from materials of >99.5% purity and were pre-alloyed by heating in quartz tubes for 1 h at 1000 °C under an atmosphere of 0.3 bar of argon; alloys were finally melted for a short time at 1500 °C using high-frequency methods. Lead from vapour condensation was redissolved by heating for 0.2 h at 1400 °C in a silica tube furnace arranged so that the tubes could be shaken [1].

Crystallography. The structure of the compound $RhPb_2$ was established as being of the $CuAl_2$ type with $a=6.651\pm0.003$, $c=5.853\pm0.003$ Å by [2]. What are considered to be the stable compounds of the diagram are shown in the table below [1].

phase	structure	a in Å	b in Å	c in Å
$Rh_3Pb(0)$	Cu_3Au	4.02	—	—
Rh_3Pb_2	NiAs	4.33	—	5.64
RhPb	CoSn	5.678	—	4.428
Rh_5Pb_7	Rh_5Pb_7	$9.84_{4\pm3}$	$5.71_{2\pm1}$	$26.53_{8\pm7}$
$RhPb_2$	$CuAl_2$	6.674	—	5.831
$RhPb_4$	—	—	—	—

The crystal structure of another compound, Rh_4Pb_5, has been examined by X-ray diffraction and a complex structure was revealed composed of mixed layers of the $CuAl_2$- and CoSn-type [3]. The lattice parameters of the $CuAl_2$-type compound $RhPb_2$ were determined as a = 6.674 ± 0.003, c = 5.831 ± 0.003 Å [4]. The stability of $CuAl_2$ (C16) compounds including $RhPb_2$ is discussed by [5].

Superconductivity. The compound $RhPb_2$ was found to be superconducting with a transition temperature of 2.66 K [6]. $RhPb_{1.9}$ was found to have a T_c of 1.32 ± 0.02 K [7].

References:

[1] El-Boragy, M.; Jain, K.C.; Mayer, H.W.; Schubert, K. (Z. Metallk. **63** [1972] 751/3).
[2] Wallbaum, H.J. (Z. Metallk. **35** [1943] 218/21).
[3] Mayer, H.W.; Schubert, K. (J. Less-Common Metals **33** [1973] 91/8).
[4] Havinga, E.E.; Damsma, H.; Hokkeling, P. (J. Less-Common Metals **27** [1972] 169/86).
[5] Havinga, E.E.; Damsma, H. (J. Less-Common Metals **27** [1972] 269/80).
[6] Gendron, M.F.; Jones, R.E. (Phys. Chem. Solids **23** [1962] 405/6).
[7] Havinga, E.E.; Damsma, H.; Kanis, J.M. (J. Less-Common Metals **27** [1972] 281/91).

25 Alloys with Vanadium

25.1 The Rh–V System

Phase Diagram, Crystallography

The **phase diagram** shown in **Fig. 100** is due to [1]. A diagram covering the whole composition range was produced at about the same time by [2]. Although there is good agreement in some respects, there are also considerable differences, especially with regard to melting points, and in view of the authors' own comments on the small number of samples examined in some cases, the diagram due to [1] is presented; this does not conflict with earlier work by [3, 4]. The diagrams by [1] and [2] have been discussed by [10].

The solubility of vanadium in rhodium is given as 19.5 at% at 1740 °C and that of rhodium in vanadium as 18 at% at 1730 °C. A number of intermediate phases were found: $Rh_3V(\gamma')$, $Rh_5V_3(\varepsilon)$, $Rh_5V_4(\alpha_2)$, RhV(α_3 at low temperature, α_1 at high temperature); α_2 has the AuCu-type structure [1].

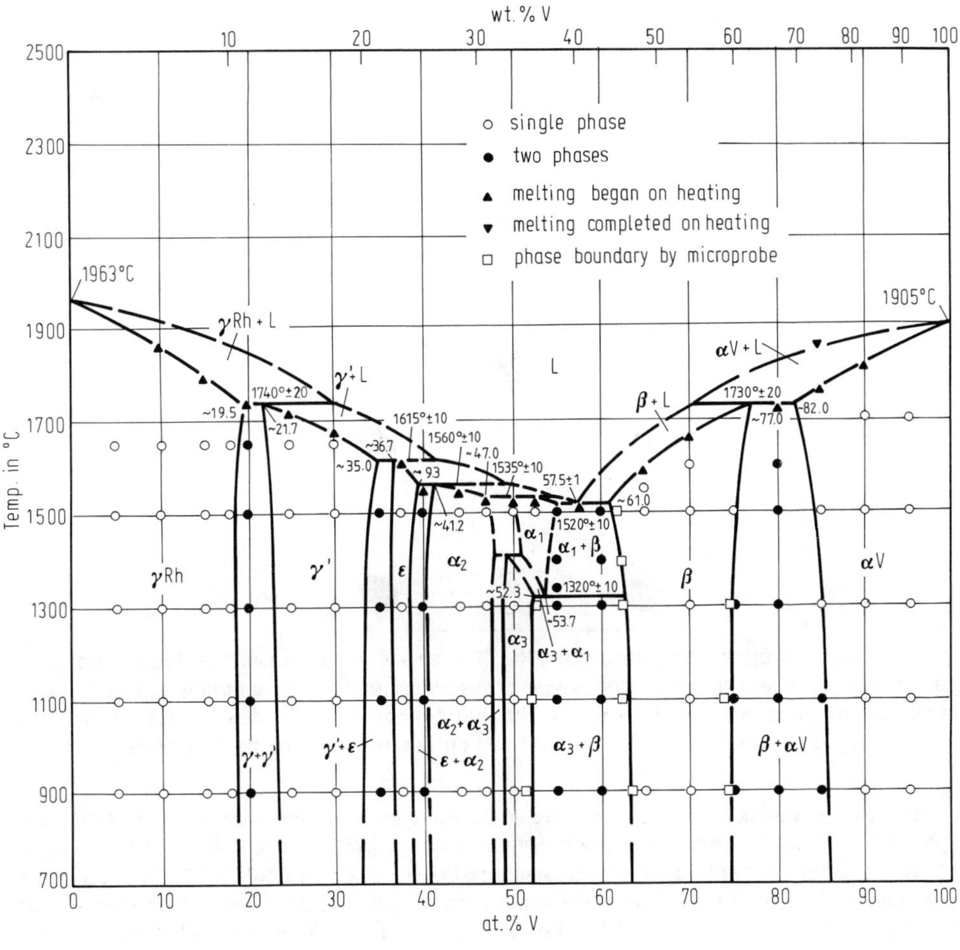

Fig. 100. Phase diagram Rh–V.

Crystallography. A considerable amount of information on the intermediate phases is given in the following table [2].

phase	structure	a	b in Å	c
β	cubic	4.767 ± 0.01	—	—
α	tetragonal	3.887 ± 0.04	—	3.593 ± 0.04
ε	orthorhombic	5.433 ± 0.04	9.268 ± 0.04	4.311 ± 0.04
γ′	cubic	3.784 ± 0.03	—	—

In an earlier paper which examined only four alloys annealed at 1200 °C containing 25, 42, 50 and 60 at% rhodium, hexagonal and cubic phases were found in the 60 at% rhodium alloy; the β-V_3Rh phase was identified as an A15-type structure with a = 4.767 Å [4]. A very detailed investigation of the V_3Rh_5 phase showed this to be orthorhombic with a = 5.420, b = 9.276, c = 4.320 Å, Z = 2; it was found to have a structure intermediate between the Cu_3Au- and CuAu-types except that it had a two-layer stacking sequence. The proposed structure is shown in **Fig. 101** [5].

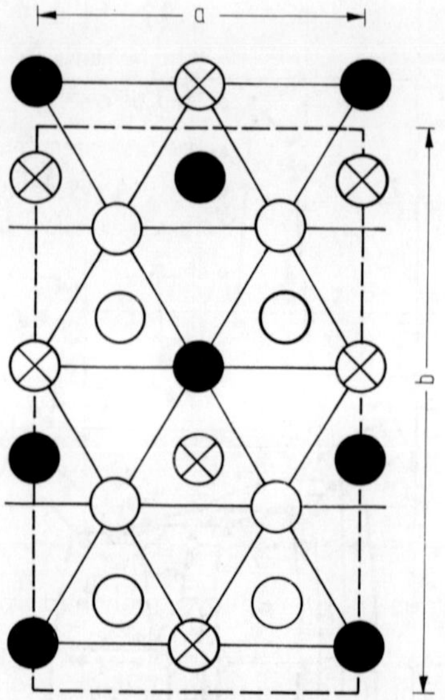

Fig. 101. Proposed crystal structure of V_3Rh_5. The space group is Cm2m. Filled atom sites: open circles represent rhodium atom sites; crossed circles represent atomic sites populated equally by vanadium and rhodium atoms. The interconnected atom sites are at z = 0, unpopulated sites are at z = $^1/_2$. The unit cell outline is shown by dashed lines.

The VRh_3 γ′-phase was found to have a Cu_3Au-type structure with a = 3.795 Å [6]. The β-phase V_3Rh was found to be of the A15 type with a = 4.7582 ± 0.0001 Å; it was also found to have marked long-range order, being assigned a value of 0.83 ± 0.04 for S, the long-range order parameter (S = 1.0 for complete order and a value of zero for complete disorder) [7]. A figure of a = 4.784 Å is given for the β-phase by [8]. A lattice parameter of a = 4.777 Å has been calculated for the A15 compound by [9].

References:

[1] Waterstrat, R.M.; Manuszewski, R.C. (J. Less-Common Metals **52** [1977] 293/305).
[2] Aksenova, O.V.; Kuprina, V.V.; Bernard, V.B.; Skolozdra, R.V. (Vestn. Mosk. Univ. Khim. **32** [1977] 429/32; Moscow Univ. Chem. Bull. **32** No. 4 [1977] 40/2).
[3] Raub, E.; Walter, P. (Heraeus Festschrift **1951** 124).
[4] Greenfield, P.; Beck, P.A. (Trans. Am. Inst. Mining Metall. Petrol. Eng. **206** [1956] 265/76).
[5] Waterstrat, R.M.; Dickens, B. (J. Less-Common Metals **31** [1973] 61/7).
[6] Dwight, A.E.; Beck, P.A. (Trans. Am. Inst. Mining Metall. Petrol. Eng. **215** [1959] 976/9).
[7] van Reuth, E.C.; Waterstrat, R.M. (Acta Crystallogr. B**24** [1968] 186/96).
[8] von Philipsborn, H. (Z. Kristallogr. **131** [1970] 73/87).
[9] Tarutani, Y.; Kudo, M. (J. Less-Common Metals **55** [1977] 221/9).
[10] Smith, J.F. (Trans. Indian Inst. Metals **40** [1987] 143/7).

Miscellaneous Physical Properties

Hardness. Figures determined over the whole composition range on samples quenched from 1400 °C are shown in **Fig. 102** (diamond pyramid hardness, load 10 kg/mm^2); this reflects the phases present, showing minima at intermetallic compounds and maxima corresponding to eutectic mixtures [1]. The hardness of as-cast rhodium containing 1 wt% vanadium is 164 VPN [2].

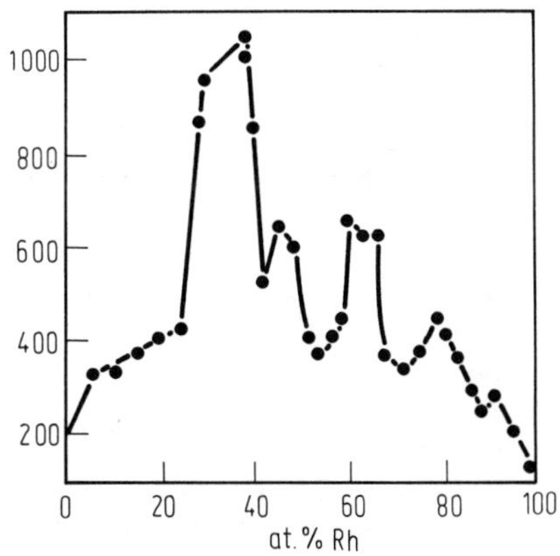

Fig. 102. Hardness of Rh-V alloys quenched from 1400 °C.

Stress-Rupture Properties. The 1 wt% vanadium alloy was said to have a stress-rupture life of 47.2 h at 1200 °C and 5.7 h at 1400 °C at a loading of 345 bar [2].

Specific Heat. The electronic specific heat for the compound V_3Rh is 2.51 ± 0.05 $mJ \cdot g-atom^{-1} \cdot K^{-2}$ [3].

Diffusion Coefficients. The diffusion between V-Rh bimetallic couples obtained by shock impulse and diffusion welding has been examined; the coefficients of mutual diffusion (D) in various phases of the bimetal at 1000 °C are given on page 160 [4].

X-ray Spectrum. X-ray $K_{\beta}5$ and M II, III emission bands showed that the electronic structure is characterized by the distribution of valence bands resulting from various components; the high-energy bands are due to the 3d states of vanadium and the low-energy ones are due to the d states of rhodium [5].

phase	at% V	$D^{1)}$ in cm^2/s	$D^{2)}$
V_3Rh	70	2.79×10^{-11}	5.45×10^{-12}
VRh	50	1.90×10^{-11}	4.10×10^{-12}
$\alpha 1$	45	1.80×10^{-11}	3.85×10^{-12}
VRh_3	30	2.50×10^{-11}	4.89×10^{-12}

$^{1)}$ = couple made by shock impulse, $^{2)}$ = couple made by diffusion welding

Superconductivity. A transition temperature $T_c < 0.015$ K for V_3Rh is given by [6]. $T_c < 0.3$ K for the same compound was given by [7]. Measurements of the A15 Rh-V compound showed that long-range order induced by low-temperature annealing may cause T_c to decrease rather than increase [8].

Susceptibility. This was determined as $59 \pm 0.5 \times 10^{-6}$ emu/mol for the compound VRh_3 [9].

Nuclear Magnetic Resonance Studies. The Knight shift and relaxation times for VRh_3 have been determined as $K_v = 0.48 \pm 0.02\%$ and $(T_1T)_v(s\,K) = 3.1 \pm 0.2\%$ [9]. The spin-echo patterns of vanadium in a series of A15 intermetallic compounds including V_3Rh were determined and the nuclear spin-lattice relaxation times of vanadium were measured; it was found that the contribution of the s electrons to the total relaxation rate R could be neglected; the paper should be consulted for more details [10].

References:

[1] Aksenova, O.V.; Kuprina, V.V.; Bernard, V.B.; Skolozdra, R.V. (Vestn. Mosk. Univ. Khim. **32** [1977] 429/32; Moscow Univ. Chem. Bull. **32** No. 4 [1977] 40/2).
[2] Handley, J.R. (Platinum Metals Rev. **33** [1989] 64/72).
[3] Spitzli, P.; Flükiger, R.; Heiniger, F.; Junod, A.; Muller, J.; Staudenmann, J.L. (J. Phys. Chem. Solids **31** [1970] 1531/7).
[4] Aksenova, O.V.; Arkhangel'skaya, O.V.; Bernard, V.B.; Kuprina, V.V. (Vestn. Mosk. Univ. Khim. **28** [1987] 74/8).
[5] Nemnonov, S.A.; Kurmaev, E.Z.; Formichev, V.A.; Ishmukhametov, B.Kh.; Belash, V.P.; Rudnev, A.V. (Izv. Akad. Nauk SSSR Ser. Fiz. **36** [1972] 317/24).
[6] Blaugher, R.D.; Hulm, J.K. (J. Phys. Chem. Solids **19** [1961] 134/8).
[7] Schoep, G.K.; Poulis, N.J. (Physica B+C **82** [1976] 227/38).
[8] Cox, J.E.; Waterstrat, R.M. (Phys. Letters A **46** [1973] 21/2).
[9] Panissod, P. (J. Phys. F **4** [1974] 484/96).
[10] Schoep, G.K.; van der Valk, H.J.; Frijters, G.A.M.; Kok, H.B.; Poulis, N.J. (Physica **77** [1974] 449/68).

25.2 Ternary Alloys

Rh-V-As. Alloys of composition RhVAs were investigated by Mössbauer spectroscopy, X-ray diffraction and magnetic measurements to study 3d and 4d transitional metal ordering. The crystal structure was found to be orthorhombic of the type Co_2P with an atomic volume of 43.11 Å^3 [3]. The atomic volume figure was confirmed, and the lattice parameters determined as a = 6.138, b = 3.870, c = 7.259 Å [4].

Rh-V-Ga. The effect of adding ~5 at% rhodium to the compound V_3Ga has been studied; the alloy was examined by differential thermal analysis, X-ray diffraction, microhardness and metallographically; an isothermal section of the ternary diagram at 100 °C is given [2].

Rh-V-Sn. The compound Rh_2VSn has been found to have unusual properties: a sample slowly cooled to room temperature is tetragonal resulting from a large tetragonal distortion of the Heusler structure, but a quenched sample is a two-phase mixture of the Heusler ($L2_1$) and tetragonal structure. The lattice parameters are shown below [1].

	a in Å	c in Å	c/a
Heusler	6.192	6.192	1
tetragonal	4.062	7.291	1.27

References:

[1] Suits, J.C. (Solid State Commun. **18** [1976] 423/5).

[2] Svechnikov, V.N.; Pan, V.M.; Spektor, A.Ts. (Izv. Akad. Nauk SSSR Met. **1971** No. 1, p. 195; Russ. Metall. **1973** No. 1, pp. 133/7).

[3] Sénateur, J.P.; Fruchart, D.; Boursier, D.; Rouault, A. (J. Phys. [Paris] **38** [1977] 61/6).

[4] Deyris, B.; Roy-Montreuil, J.; Rouault, A.; Krumbügel-Nylund, A.; Sénateur, J.P.; Fruchart, R.; Michel, A. (Compt. Rend. C **278** [1974] 237/9).

26 Alloys with Niobium

26.1 The Rh-Nb System

Phase Diagram. The constitutional diagram of this system is complex, there being nine intermediate phases; there is considerable solid solubility at both ends of the diagram, the maximum solubility of rhodium in niobium being 16.5 at% and that of niobium in rhodium 20.7 at%. The invariant reactions in the system are shown in the following table, and the complete phase diagram in **Fig. 103** [1].

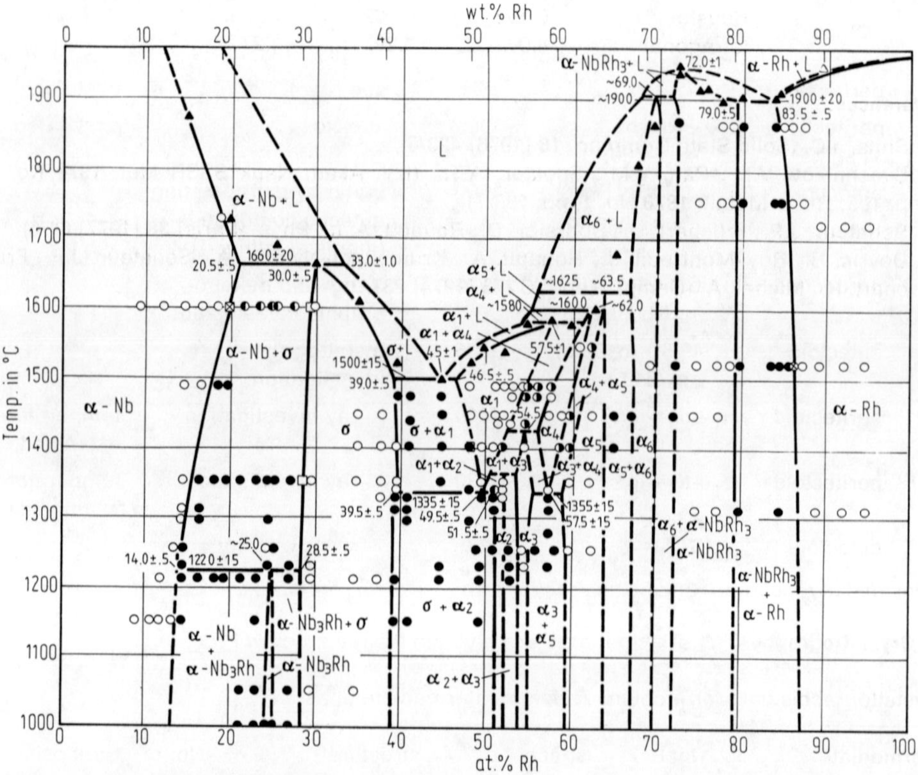

Fig. 103. Nb-Rh constitution diagram. Legend: ○ one phase; ◑ ● two phases; ▲ melting observed; □ metallographic boundary determination; × boundary determined by X-ray parametric method.

Invariant reactions in the niobium-rhodium system:

No.	type	reaction, at% Rh	temp. in °C	method of determination	remarks
1	peritectic	α Nb (20.5) + L (33.0) → σ (30.0)	1660	liquid by disappearing primary α Nb	–
2	eutectic	L (45) → σ (39.0) + α_1 (46.5)	1500	liquid by extrapolating eutectic component	–

No.	type	reaction, at% Rh	temp. in °C	method of determination	remarks
3	peritectic	$L+\alpha_4\rightarrow\alpha_1$ (57.5)	~1580	α_1 by extrapolation of solidus line and boundary $\alpha_1/(\alpha_1+\alpha_4)$	postulated
4	peritectic	$L+\alpha_5\rightarrow\alpha_4$	~1600	temperature estimated from intersection of solvus line with solidus line	postulated *)
5	peritectic	$L+\alpha_6\rightarrow\alpha_5$	~1625	see No. 4	postulated *)
6	peritectic	$L+\alpha$ $NbRh_3$ $\rightarrow\alpha_6$ (−69)	~1900	see No. 4	postulated
7	eutectic	$L\rightarrow\alpha$ $NbRh_3$ (79.0) $+\alpha$ Rh (83.5)	1900	liquid by extrapolating eutectic component (identical with α-Rh)	–
8	peritectoid	α Nb (14.0) + σ (28.5) $\rightarrow\alpha$ Nb_3Rh (25)	1220	X-ray and metallographic investigation	–
9	eutectoid	α_1 (49.5) → σ (39.5) $+\alpha_2$ (51.5)	1335	metallographic investigation	–
10	peritectoid	$\alpha_1+\alpha_3\rightarrow\alpha_2$	~1360	X-ray investigation	temperature estimated
11	peritectoid	$\alpha_1+\alpha_4\rightarrow\alpha_3$	~1430	X-ray investigation	temperature estimated
12	eutectoid	α_4 (57.5) $\rightarrow\alpha_3+\alpha_5$	1335	X-ray investigation	–

*) Instead of reactions 4 and 5, the reactions 4a, $L+\alpha_6\rightarrow\alpha_4$, and 5a, $\alpha_4+\alpha_6\rightarrow\alpha_5$, are possible.

Crystallography. The principal data are given in the table below [1].

Crystallographic data for niobium-rhodium intermediate phases:

intermediate phase	crystal system	space group	structure type	No. of atoms per unit cell	No. of close-packed layers	composition limits, at% Rh
α Nb_3Rh	cubic	O_h^3 − Pm3n	A15, Cr_3O	8	–	~25, 1200 °C
σ	tetragonal	D_{4h}^{14} − P4/mnm	$D8_b$, σ(Fe-Cr)	30	–	28.5 to 39, 1200 °C
α_1 NbRh	unknown	–	unknown	–	–	46.5 to 54, 1500 °C
α_2 $(Nb_{0.96}Rh_{0.04})Rh$	tetragonal	D_{4h}^1 − P4/mmm	$L1_0$, AuCu	4	3	51.5 to 52.5, 1200 °C
α_3 $(Nb_{0.90}Rh_{0.10})Rh$	ortho-rhombic	D_{2h}^5 − Pmcm	α_1(Ta-Rh)	12	6	~54.0 to 55.5, 1200 °C

intermediate phase	crystal system	space group	structure type	No. of atoms per unit cell	No. of close-packed layers	composition limits, at% Rh
α_4 $(Nb_{0.85}Rh_{0.15})Rh$	ortho-rhombic	D_{2h}^5 − Pmcm	B19, AuCd	4	2	~56 to 58.5, 1400 °C
α_5 $(Nb_{0.75}Rh_{0.25})Rh$	monoclinic	C_{2h}^1 − P2/m	distorted Sm	18 [1)]	9	~59.5 to 63.5,
or (Nb_3Rh_5)	monoclinic	C_{2h}^1 − P2/m		72 [2)]	9	1200 °C
α_6 $Nb(Rh_{0.90}Nb_{0.10})_3$	hexagonal	D_{3h}^1 − P$\bar{6}$m2	VCo_3	24	6	~67 to 70, 1200 °C
α $NbRh_3$		cubic	O_h^1 − Pm3m	$L1_2$, $AuCu_3$	4	3	73 to 79, 1200 °C

[1)] 18 atoms per cell assuming random substitution of niobium by rhodium in NbRh,
[2)] 72 atoms per cell assuming ordered Nb_3Rh_5.

Lattice parameter measurements for compositions between 12.0 and 94.7 at% rhodium are given in the following table [1].

Lattice parameters of intermediate phases and terminal solid solutions for niobium-rhodium:

at% Rh	phase	lattice parameter(s) in Å a	b	c	c/a	accuracy a	b	c
12.0	α Nb	3.265	—	—	—	±0.002	—	—
14.5	α Nb	3.257	—	—	—	±0.002	—	—
16.6	α Nb	3.250	—	—	—	±0.002	—	—
18.5	α Nb	3.245	—	—	—	±0.002	—	—
24.8	α Nb_3Rh	5.120	—	—	—	±0.003	—	—
29.7	σ	9.869	—	5.106	0.517_4	±0.004	—	±0.003
34.9	σ	9.835	—	5.083	0.516_8	±0.004	—	±0.003
37.8	σ	9.810	—	5.068	0.516_6	±0.004	—	±0.003
51.3	α_2	4.019	—	3.809	0.947_7	±0.004	—	±0.004
55.9	α_3	2.827	4.770	13.587	—	±0.002	±0.005	±0.010
58.8	α_4	2.813	4.808	4.510	—	±0.002	±0.005	±0.005
62.3	α_5	2.806	4.772	20.25	α = 90°31.5′	±0.002	±0.003	±0.010
69.2	α_6	5.483	—	13.405	2.445	±0.003	—	±0.005
74.5	α $NbRh_3$	3.857	—	—	—	±0.002	—	—
77.2	α $NbRh_3$	3.844	—	—	—	±0.002	—	—
89.1	α Rh	3.835	—	—	—	±0.002	—	—
92.0	α Rh	3.827	—	—	—	±0.002	—	—
94.7	α Rh	3.818	—	—	—	±0.002	—	—

The crystallography of the individual phases is discussed in more detail in [2]. The lattice parameters of Nb_3Rh_2 are given as a = 9.774, c = 5.054 Å, c/a = 0.517; the structure being σ-type [3]. Earlier work on AB type alloys indicated that the equiatomic rhodium-niobium alloy did not form a CsCl-type structure; it was found that in the 39 AB systems (including RhNb) where CsCl structures did not occur, the ratio of the Goldschmidt diameters of the elements for coordination number 8 D_A/D_B was between 0.985 and 1.439, that for RhNb being 1.009 [4]. A later paper examines the binding energy of the compounds in the rhodium-niobium system using a new method of assessing the generated shear stresses, these being considered as more due to spatial factors in the peripheral core electrons rather than to the arrangement of valence ones. As a consequence of this approach, there exists the possible generation of electric dipoles in the direction of the stacking normals, and the energetic interaction of these determining the stacking sequence of the structure [5]. A detailed theory has been developed for radiation-induced order-disorder transformations in Nb-Rh alloys amongst others. The calculated partial phase diagram as modified by irradiation at the rates indicated is shown in a figure. The increase in free energy is insufficient to remove the compound $NbRh_3$ from the phase diagram, and instead, the boundary of the terminal solid solution was progressively shifted to higher solubilities [6].

Thermodynamics. The thermodynamics of the rhodium-niobium and rhodium-niobium-oxygen systems have been determined; the results on the former system can be expressed in the following relationships:

$^f\Delta G^0(Nb_{0.13}Rh_{0.87}) = -29.7 + 0.0027\,T$ kJ/mol, $^f\Delta G^0(NbRh_{3.55}) = -223.0 + 0.0202\,T$ kJ/mol, both at 1100 to 1300 K.

The relative partial excess Gibbs energy of niobium in rhodium at infinite dilution is $\Delta_{Nb} = -200$ kJ/mol at 1200 K [7]. Solid-state galvanic cell measurements on the system have been made by [8].

Hardness. The effect of the addition of rhodium to niobium on the hardness at two temperatures, 77 and 300 K, was measured by [9]. The figures are given below.

Hardness H_v in kg/mm^2:

at% Rh	at 77 K	at 300 K
0.53	221	106
2.0	297	143
5.0	334	211
7.2	395	254

Hot hardness measurements between 300 and 1250 K were made on a large number of $L1_2$ alloys including the compound Rh_3Nb [10].

Diffusion. The interdiffusion of Rh-Nb couples made from 0.3/1.0 mm sheet material heated at ≦1400 °C was examined using electron microscopy and microprobe analyses; the growth characteristics of Rh_3Nb and other intermetallics were studied [11].

Superconductivity. The compound Nb_3Rh was found to be superconducting with a transition temperature $T_c = 2.5$ K [12]. The T_c of a number of A15-type compounds has been examined as a function of pressures up to 25 kbar. Under atmospheric pressure the T_c of Nb_3Rh was 2.797 K, a figure higher than that reported by [12]; the transition width ΔT_c was 0.012 K and $\delta T/\delta P = -0.32 \pm 0.02 \times 10^{-5}$ $kbar^{-1}$ [13]. The superconductivity of the composition $Nb_{58}Rh_{42}$ has been studied in the amorphous and microcrystalline states; the microcrystalline alloy had a $T_c \approx 4.7$ K, whilst annealed amorphous samples had a figure of 10.3 K [14].

Amorphous Materials. A large number of binary systems (127) were tested for the formation of amorphous behaviour after cooling at a rate of $\leqq 10^8$ K/s; $Nb_{60}Rh_{40}$ was found to be amorphous after this treatment. Criteria were developed for formation of the amorphous state [15]. The electronic structure of $Nb_{55}Rh_{45}$ has been examined by photoelectron spectroscopy; the calculated charge transferred from the Nb to Rh atoms was relatively small (0.1 electron per atom) compared with other glass forming transition metal alloys [16].

Most work on amorphous material has been carried out on $Nb_{55}Rh_{45}$. Arc-furnace quenched material of this composition was found to have the following properties:

crystallization temperature $T_x = 973 \pm 10$ to 980 ± 2 K, hardness $H_v = 780 \pm 40$ kg/mm², Young's modulus $E = 2.30 \times 10^6$ kg/cm² [17].

The superconducting critical temperature $T_c = 5.13$ K and the Debye temperature $\theta_D = 273$ K, this being determined from experimental determinations of Young's modulus [18]. T_c was found to be 4.75 K, during an investigation of pressure on T_c. The following properties were determined:

B (bulk modulus) $= 2.043 \times 10^2$ GPa, Debye temperature $\theta_D = 273$ K, γ_G (Grüneisen parameter) $= 1.57$, electron-phonon coupling constant $\lambda = 0.66$, $\delta T_c/\delta p = 2.40 \pm 0.50 \times 10^{-2}$ K/GPa [19].

Magnetic Properties. The calculated magnetic susceptibility of the alloy $Nb_{60}Rh_{40}$ which has the $D8_b$ structure is +79 emu/g-atom according to [20].

References:

[1] Ritter, D.L.; Giessen, B.C.; Grant, N.J. (Trans. Metall. Soc. AIME **230** [1964] 1250/9).
[2] Ritter, D.L.; Giessen, B.C.; Grant, N.J. (Trans. Metall. Soc. AIME **230** [1964] 1259/67).
[3] Hansen, R.C.; Raman, A. (Z. Metallk. **61** [1970] 115/20).
[4] Dwight, A.E. (Trans. Metall. Soc. AIME **215** [1959] 283/6).
[5] Schubert, K. (Z. Metallk. **76** [1985] 326/9).
[6] Liou, K.-Y.; Wilkes, P. (J. Nucl. Mater. **87** [1979] 317/30).
[7] Kleykamp, H. (J. Less-Common Metals **83** [1982] 105/13).
[8] Cima, M.J. (LBL-21951 [1986] 220 pp.; C.A. **106** [1987] No. 183555).
[9] Stephens, J.R.; Witzke, W.R. (J. Less-Common Metals **40** [1975] 195/205)
[10] Wee, D.-M.; Suzuki, T. (Trans. Japan Inst. Metals **20** [1979] 634/46).

[11] Khokhlova, L.M. (Deposited Doc. VINITI-575-82 [1981] 587/91).
[12] Matthias, B.T.; Wood, E.A.; Corenzwit, E.; Bala, V.B. (Phys. Chem. Solids **1** [1956] 188/90).
[13] Smith, T.F. (J. Low Temp. Phys. **6** [1972] 171/95).
[14] Johnson, W.L.; Poon, S.J. (J. Appl. Phys. **46** [1975] 1787/92).
[15] Ning, Y.; Zhou, X. (Jinshu Xuebao **17** [1981] 278/84).
[16] DasGupta, A.; Oelhafen, P.; Gubler, U.; Lapka, R.; Güntherodt, H.-J.; Moruzzi, V.L.; Williams, A.R. (Phys. Rev. [3] B **25** [1982] 2160/82).
[17] Davis, S.; Fischer, M.; Giessen, B.C.; Polk, D.E. (Rapidly Quenched Metals Proc. 3rd Intern. Conf., Brighton, Engl., 1978, Vol. 2, pp. 425/30).
[18] Koch, C.C.; Kroeger, D.M.; Scarbrough, J.O.; Giessen, B.C. (Phys. Rev. [3] **22** [1980] 5213/24).
[19] Müller, R.; Shelton, R.N.; Koch, C.C.; Kroeger, D.M. (Solid State Commun. **45** [1983] 327/30).
[20] Bucher, E.; Heiniger, F.; Muheim, J.; Muller, J. (Rev. Mod. Phys. **36** [1964] 146/9).

26.2 Ternary Alloys

Rh-Nb-Al. The structures of $Rh_{10}Nb_{65}Al_{25}$ and $Rh_{30}Nb_{62}Al_8$ are of σ-type; the lattice parameters are:

$Rh_{10}Nb_{65}Al_{25}$: a = 9.854, c = 5.162 Å, $Rh_{30}Nb_{62}Al_8$: a = 9.830, c = 5.122 Å [1].

Rh-Nb-Ti. The alloy $Ti_{0.875}Nb_{0.1}Rh_{0.025}$ was compared with $Ti_{80}Nb_{20}$ with respect to workability and electric resistivity at low temperatures in the annealed and progressively cold-worked conditions. The as-cast hardness was 256 VPN (10 kg) which fell slightly on annealing; the resistance results are summarized in **Fig. 104** which shows the specific resistance plotted against % cold-working of the specimen when tested at the indicated temperatures (77, 197, 293, and 373 K) [4].

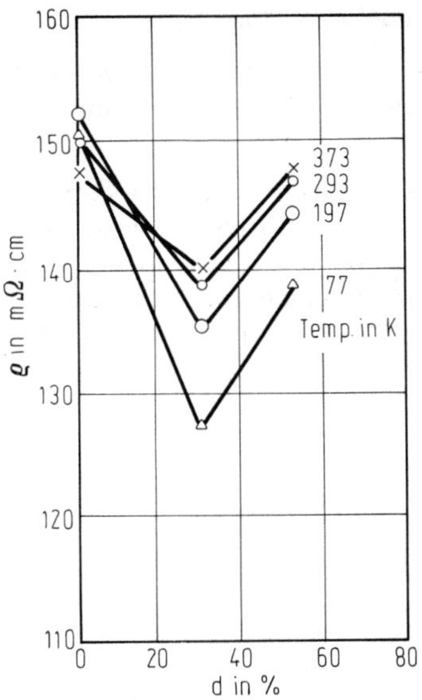

Fig. 104. Influence of deformation d on the resistivity ϱ of $Ti_{0.875}Nb_{0.1}Rh_{0.025}$.

Rh-Nb-Ge. The structure of NbRhGe arc-melted, heat-treated at 840 °C for 7d and then quenched was of TiNiSi type with a = $6.439_{5\pm4}$, b = $3.842_{7\pm3}$, c = $7.430_{7\pm4}$ Å [3].

Rh-Nb-Sn. The structure of arc-melted RhNbSn was found to be of the MgAgAs type, with a = 6.132 ± 1 Å; the density was 9.06 g/cm² [2].

References:

[1] Hansen, R.C.; Raman, A. (Z. Metallk. **61** [1970] 115/20).
[2] Jeitschko, W. (Metall. Trans. **1** [1970] 3159/62).
[3] Johnson, V.; Jeitschko, W. (J. Solid State Chem. **4** [1972] 123/30).
[4] Zwicker, U. (Z. Metallk. **56** [1965] 222/8).

27 Alloys with Tantalum

27.1 The Rh–Ta System

Phase Diagram. The rhodium-tantalum system was carefully investigated by thermal analysis, metallographic and X-ray techniques to give the diagram shown in **Fig. 105.** The diagram is a complex one, having five intermediate phases:

σ between 23.5 and 61 at% rhodium at 1300 °C;
α_2($TaRh_2$) with 66.5 at% rhodium at 1500 °C;
α_3 with 50 to 58 at% rhodium at 1700 °C;
α-$TaRh_3$ with 70 to 76.5 at% rhodium at 1500 °C.

$TaRh_3$ forms congruently from the melt and $TaRh_2(\alpha_2)$, α_1, α_3 and σ form peritectically. α_1 decomposes at 1375 ± 20 °C and 52 ± 0.5 at% rhodium eutectoidally to form $\sigma + \alpha$. The diagram shows a single eutectic at 45 ± 1 at% rhodium and 1740 ± 15 °C [1].

Preparation. The alloys were prepared as arc-melted buttons weighing 5 to 10 g, in most cases the melting process being carried out three times; melting was done under titanium-gettered welding grade argon at 400 Torr pressure. At compositions below 25 at% rhodium, these buttons frequently contained inhomogeneities, hence all buttons were sectioned and only those showing <1 at% of inhomogeneity were used, this being estimated from the amounts of phases present. In addition, only the centre portions of such buttons were used; all alloys were homogenized to remove coring [1].

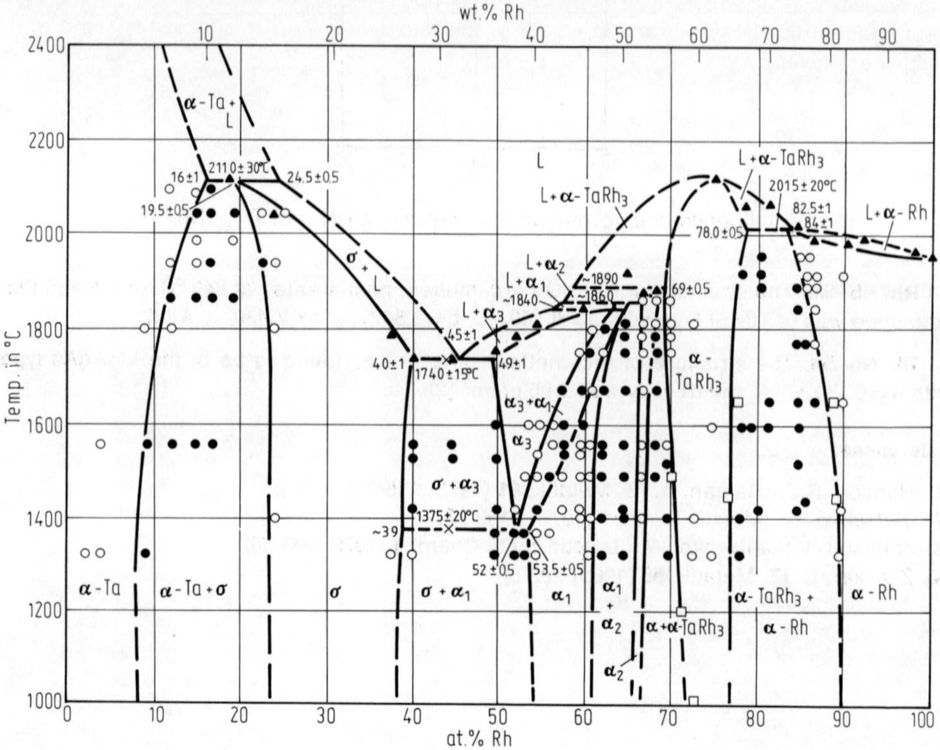

Fig. 105. Ta-Rh constitution diagram: ○ one phase; ● two phase; ▲ melting observed; × by thermal analysis; □ by X-ray analysis.

Crystallography. Some data is given in the following table from [1, 2, 3].

phase	type	at% Rh	a	b in Å	c	accuracy in Å	Ref.
tetr.		25	9.904	—	5.160	±0.002	[1]
		30	9.863	—	5.131	±0.002	[1]
		35	9.822	—	5.102	±0.002	[1]
α_1 rh.		60.5	2.822	4.742	13.551	a, b ±0.002, c [1)]	[1]
α_1 rh.		60.5	2.822	4.742	13.551	—	[3]
α_2 rh.		66.5	8.179	5.453	4.027	b, c ±0.001, a [2)]	[1]
α_2 rh.	Co_2Si	66.5	8.179	5.454	4.027	—	[3]
$TaRh_3$	cubic	70.0	3.884	—	—	±0.001	[1]
		75.0		3.860	—	±0.001	[1]
$TaRh_3$				3.86	—	—	[2]

rh. = orthorhombic, tetr. = tetragonal, c [1)] = ±0.006 Å, a [2)] = ±0.002 Å.

The structure of α_3 was not determined, but it was suggested that it was a structurally related high-temperature modification of α_1 stabilized by high tantalum contents [1]. Electron microscopy and electron microprobe analysis were used to study interdiffusion between rhodium-tantalum sheets 0.3 to 1 mm thick during annealing at ≦1400 °C; the growth kinetics of the compound Rh_3Ta was studied. The interdiffusion was significantly changed by introducing nickel [4]. The structure of alloys in the σ-phase range of 22 to 39 at% rhodium has been studied by X-ray diffraction, some of the alloys being splat-quenched; some of the results are given below [5].

alloy	$Ta_{76.5}Rh_{23.5}$		$Ta_{70.0}Rh_{30.0}$	$Ta_{68.6}Rh_{31.4}$	
sample No.	1	1a	2	3	3a
treatment . .	1600; 4 h	s.q.	1600; 4 h	1600; 4 h	s.q.
phase . . .	σ+ext.	σ+ext.	σ+ext.	σ+ext.	σ+ext.
lattice parameters in Å:					
a	9.931 [3)]	9.923 [1)]	9.906 [1)]	9.868 [2)]	9.862 [1)]
c	5.177 [1)]	5.200 [1)]	5.160 [1)]	5.147 [2)]	5.139 [2)]

Accuracies: [1)] = ±0.003, [2)] = ±0.004, [3)] = ±0.0025, s.q. = splat-quenched, temperatures are in °C, ext. = extra phase apart from σ.

Hardness. The hot hardness of a number of $L1_2$ compounds has been determined, including that of Rh_3Ta. The results at temperatures up to 1220 K are given [6].

Superconductivity. The superconductivity of various close-packed low-symmetry intermediate phases in a number of binary systems, including Rh-Ta, has been investigated; in each case, the σ-phase exhibited the highest superconducting temperature, that for the Rh-Ta σ-phase being 2.35 K [7]. The critical temperature for superconductivity (T_c) given in [7] was determined by [8]. A determination of T_c on the Rh-Ta σ-phase gave 2.0 K [9]. However, according to [5] the Rh-Ta alloys do not become superconducting above 1.2 K; measurements of the resistance ratio plotted against temperature for $Ta_{70.0}Rh_{30.0}$ and $Nb_{65.2}Rh_{34.8}$ are given in **Fig. 106**, p. 170 [5].

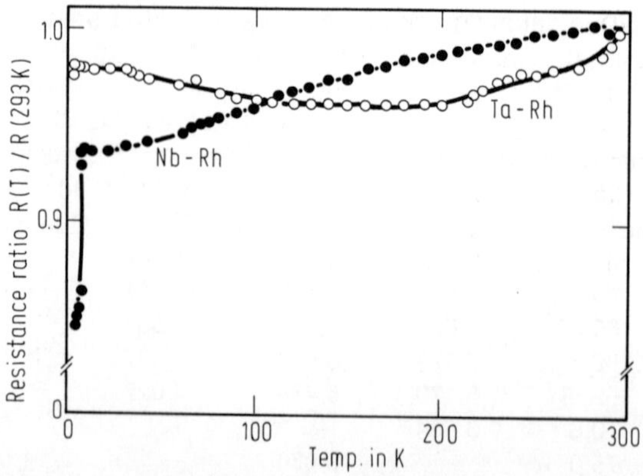

Fig. 106. Resistance ratio for $Nb_{65.2}Rh_{34.8}$ and $Ta_{70.0}Rh_{30.0}$ as a function of temperature.

Amorphous Alloys. Much interest has been shown in amorphous materials in recent years. Some properties are summarized below.

T_c	T_x in K	θ_D	H_v in kg/mm²	E	B	γ_G	Ref.
—	1118±10	—	890±5	2.72	—	—	[10]
<1.12	—	—	—	—	—	—	[5]
3.37	1151±10	247	—	—	—	—	[11]
—	—	—	9.022 1)	14.85 2)	—	—	[12]
3.19	—	247	—	—	2.475 3)	1.72	[13]
—	547/677	—	805±40	—		—	[14]

E is Young's modulus, B is the bulk modulus and γ_G is the Grüneisen parameter, θ_D is the Debye temperature, T_x is the crystallization temperature and T_c the superconducting temperature. H_v is the Vickers hardness.

1) $= 9.022 \times 10^9$ N/M² (919 kg/mm²), 2) $= 14.85 \times 10^{10}$ N/M² (15100 kg/mm²), 3) = bulk modulus $\times 10^2$ GPa (25200 kg/mm²).

Examination of 127 binary systems quenched at a cooling rate of $\leqq 1 \times 10^9$ K/s have had their non-crystallinity assessed; a criterion for the formation of the amorphous state was developed [15].

The pressure dependence of the transition temperature for superconductivity T_c was investigated. For $Ta_{55}Rh_{45}$, $T_c = 3.19$ K; $\delta T_c/\delta p = -3.10 \pm 0.30 \times 10^{-2}$ K/GPa (102 kg/mm²) [13].

Photoelectron Spectroscopy. The electronic structure of glassy $Ta_{55}Rh_{45}$ has been studied by this means. The charge transfer from the Ta to Rh atoms on alloying is relatively small (<0.1 electron per atom) compared with other glass-forming transition-metal alloys [16].

References:

[1] Giessen, B.C.; Ibach, H.; Grant, N.J. (Trans. Metall. Soc. AIME **230** [1964] 113/22).
[2] Dwight, A.E.; Beck, P.A. (Trans. Metall. Soc. AIME **215** [1959] 976).
[3] Giessen, B.C.; Grant, N.J. (Acta Crystallogr. **17** [1964] 615/6).

[4] Khokhlova, L.M. (Deposited Doc. VINITI-575-82 [1981] 587/91).
[5] Khan, H.R.; Lüders, K.; Raub, Ch.J.; Roth, G. (Z. Physik B **38** [1980] 27/33).
[6] Wee, D.-M.; Suzuki, T. (Trans. Japan Inst. Metals **20** [1979] 634/46).
[7] Sadagopan, V.; Gatos, H.C. (Phys. Status Solidi **13** [1966] 423/7).
[8] Bucher, E.; Heiniger, F.; Müller, J. (Helv. Phys. Acta **34** [1961] 843).
[9] Blaugher, R.D.; Hulm, J.K. (J. Phys. Chem. Solids **19** [1961] 134/8).
[10] Davis, S.; Fischer, M.; Giessen, B.C.; Polk, D.E. (Rapidly Quenched Metals Proc. 3rd Intern. Conf., Brighton, Engl., 1978, Vol. 2, pp. 425/30).
[11] Koch, C.C.; Kroeger, D.M.; Scarbrough, J.O. (Phys. Rev. [3] B **22** [1980] 5213/24).
[12] Whang, S.H.; Polk, D.E.; Giessen, B.C. (Proc. 4th Intern. Conf. Rapidly Quenched Metals, Sendai, Japan, 1981 [1982], Vol. 2, pp. 1365/8).
[13] Müller, R.; Shelton, R.N.; Koch, C.C.; Kroeger, D.M. (Solid State Commun. **45** [1983] 327/30).
[14] Fischer, M.; Polk, D.E.; Giessen, B.C. (Proc. 1st Conf. Rapid Solidif. Process. Princ. Technol. Proc. 1st Intern. Conf., Reston, Va., 1977 [1978], pp. 140/8).
[15] Ning, Y.; Zhou, X. (Jinshu Xuebao **17** [1981] 278/84).
[16] DasGupta, A.; Oelhafen, P.; Gubler, U.; Lapka, R.; Güntherodt, H.-J.; Moruzzi, V.L.; Williams, A.R. (Phys. Rev. [3] B **25** [1982] 2160/4).

27.2 Ternary Alloys

Rh-Ta-Nb. The pressure dependence of the superconducting transition temperature has been determined on alloys of the formula $(Nb_{1-x}Ta_x)_{55}Rh_{45}$; the dependence of the superconducting critical temperature on pressure is shown in **Fig. 107**. Figures for T_c, $\delta T_c/\delta p$, B (bulk

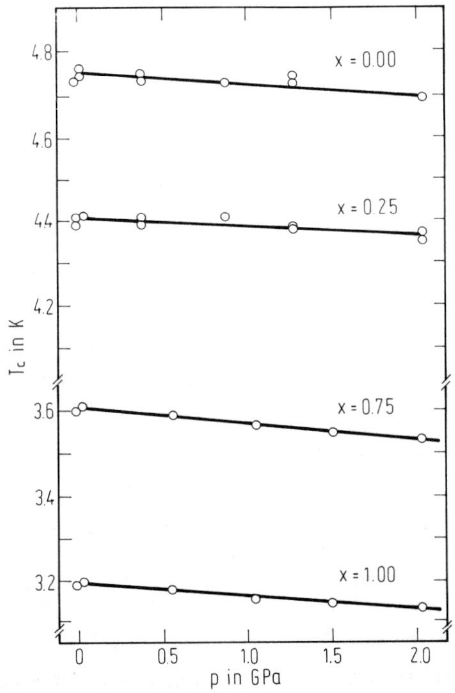

Fig. 107. Dependence of the onset of superconductivity on pressure for $(Nb_{1-x}Ta_x)_{55}Rh_{45}$ metallic glasses.

modulus), γ_G (Grüneisen constant), λ (the electron-phonon coupling constant) and θ_D (Debye temperature) are given below.

x	T_c in K	$\delta T_c/\delta p$	B	θ_D in K	λ	γ_G
0.00	4.75	-2.40 ± 0.50	2.043	273	0.66	1.57
0.25	4.41	-1.94 ± 0.50	2.151	266	0.65	1.55
0.75	3.60	-3.61 ± 0.30	2.367	253	0.62	1.75
1.00	3.19	-3.10 ± 0.30	2.475	247	0.60	1.72

$\delta T_c/\delta p$ is in 10^{-2} K/GPa, B is in 10^2 GPa.

Müller, R.; Shelton, R.N.; Koch, C.C.; Kroeger, D.M. (Solid State Commun. **45** [1983] 327/30).

28 Alloys With Chromium

28.1 The Rh-Cr System

Phase Diagram. The diagram in **Fig. 108** is due to [1]. It is based on the work of [2]. This diagram was compiled from data obtained over the whole composition range by metallographic, X-ray diffraction and electron microprobe investigations. The two intermediate phases, the ε close-packed hexagonal (hcp) phase and the β Cr_3Si (A15) phase have composition ranges of ~20 at% at 900 °C to 68 at% at 1475 °C and 77 to 78 at% below ~1200 °C, respectively. The γ(Rh) ε-phase boundary slopes strongly with temperature. Peritectic and eutectic reactions were found at 1700±10 and 1475±10 °C; a peritectoid reaction takes place at 1265±12 °C [2]. Earlier work of an exploratory nature using X-ray diffraction, and metallographic methods had been carried out by [3, 4]. The former investigation showed the existence of the hcp ε-phase but not of the β A15 phase; the maximum solid solubility of chromium in rhodium was reported to be only ~8 at%. The latter paper, which carried out limited work on 29 other binary systems as well, looked at six alloys between 25 and 75 at% chromium; the alloys were arc-melted under helium, the operation being carried out twice to obtain adequate homogeneity. After annealing for up to 120 h at 1200 °C the hcp ε-phase was found between 50 and 65 at% chromium and the A15 $RhCr_3$ β phase at 75 at% chromium [4].

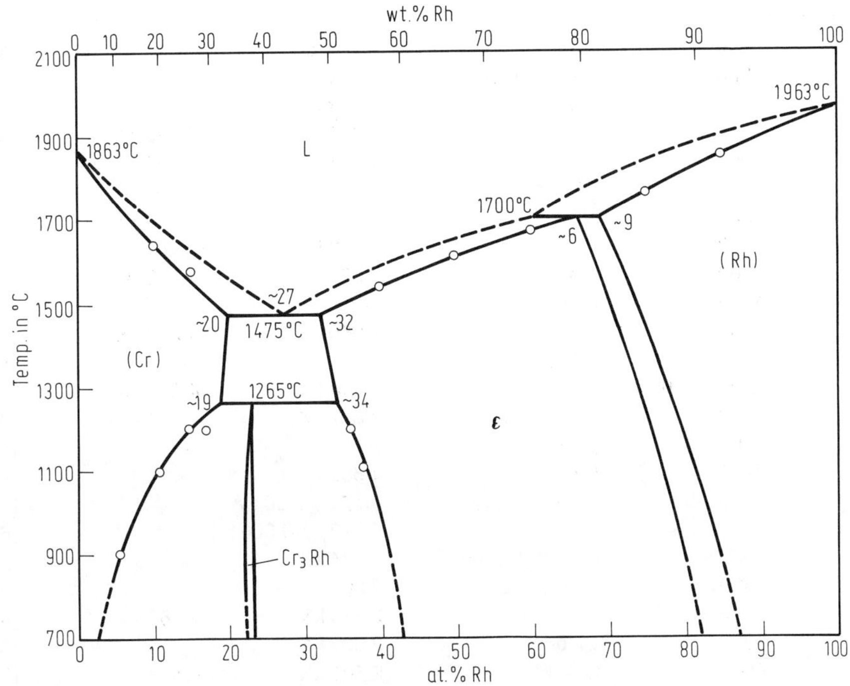

Fig. 108. Phase diagram Rh-Cr.

Preparation. Alloys were prepared from 99.999% chromium and rhodium. To avoid the occurrence of chromium oxide particles in the samples, it was found necessary to deoxidise the chromium by heating at 1400 °C for 3 d in dry hydrogen. The alloys were then arc-melted in an atmosphere of 50/50 argon/helium. Errors in the synthetic compositions achieved were estimated to be ±1 at% from the weight losses on melting. Annealing treatments were carried out in a tantalum strip resistance furnace under a vacuum of 5×10^{-6} to 5×10^{-7} Torr; typical annealing treatments are given in the table below [2].

t in °C	time	cooling mode
1600	1 h	rapid
1400*)	3 d	rapid
1200	66 h	rapid
1100	3 weeks	quenched
900	4 months	quenched

*) This treatment was carried out in pure dry hydrogen, all others were carried out in high vacuum.

Crystallography. The available lattice parameters and phases present are given in the following table according to [3 to 6].

at% Cr	quenching temp. in °C	phases	a	c	Ref.
5.8	900	α fcc	3.784 kX	—	[3]
10.0	800	α+ε (weak)	3.773 kX	—	[3]
19.7	800	α+ε	3.773 kX	—	[3]
			2.678 kX	4.285 kX	[3]
24.4	900	ε	2.686 kX	4.300 kX	[3]
25	1200	Rh fcc	—	—	[4]
30.0	900	ε	2.682 kX	4.278 kX	[3]
50.0	900	ε	2.660 kX	4.261 kX	[3]
50.0	1200	ε hcp	2.668 Å	4.249 Å	[4]
60	1200	ε hcp	—	—	[4]
61.0	800	ε	2.660 kX	4.272 kX	[3]
65	1200	ε hcp	—	—	[4]
70.0	1300 to 900	ε	2.644 kX	4.265 kX	[3]
	450	ε+Cr oxide	2.667 kX	4.283 kX	[3]
70	1200	β+ε	—	—	[4]
75	1200	β A15	4.656 Å	—	[4]
75	1020 (450 h)	β A15≈80 at%	4.674±0.001 Å		[5]
75	—	A15	4.6731±0.0001 Å		[6]
80.4	1300	Cr bcc+oxide	2.914 kX	—	[3]
	800	Cr bcc+ε	2.893 kX	—	[3]
			2.658 kX	4.266 kX	[3]
89.7	1300	Cr bcc	2.903 kX	—	[3]
	900	Cr+ε	2.893 kX	weak	[3]

The lattice parameters for the ε-phase between 20 and 60 at% chromium are shown in **Fig. 109** [2]. The degree of long-range order in 20 binary A15-type phases including $RhCr_3$ has been examined from X-ray diffractometer measurements, and a figure of S= 0.83±0.04 assigned to the chromium-rhodium alloy. The long-range order parameter S is defined as representing the maximum degree of order possible when equal to 1 [6].

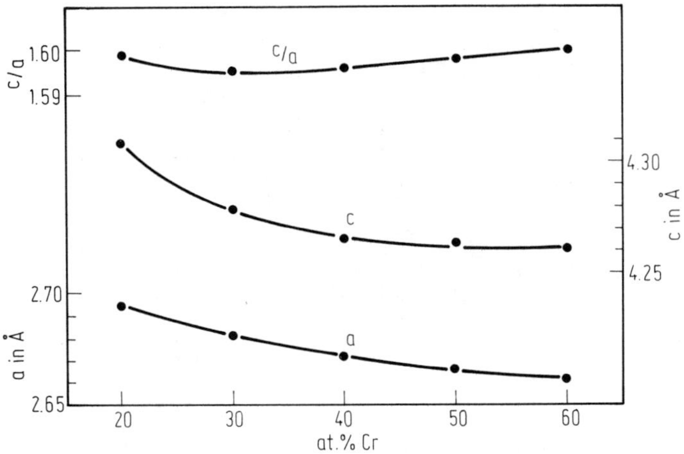

Fig. 109. Lattice parameters of ε-phase Rh-Cr alloys.

Hardness. Fig. 110 shows Vickers microhardness plotted against chromium content [3]. The Vickers hardness of as-cast chromium was found to increase from ~120 to ~220 kg/ mm^2 with a 1 at% addition of rhodium [7].

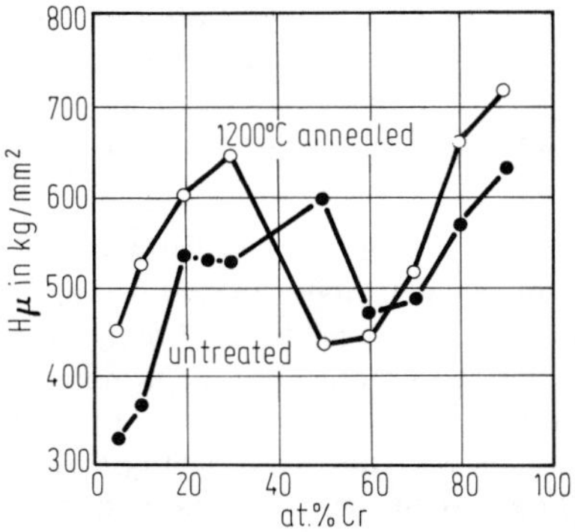

Fig. 110. Microhardness of Rh-Cr alloys.

Ductility. Additions of rhodium of up to 3 wt% to pure chromium were found to considerably raise the temperature of the ductile-brittle transition [8].

Young's Modulus. This was measured dynamically using alloys containing between 0.12 and 1.14 at% rhodium at temperatures between 90 and 500 K, and compared with Re, Os and Ru alloys. These alloys showed a deep dip in their modulus near the Néel temperature (~340 K) which is shown in **Fig. 111**, p. 176, according to [10]. An explanation is offered based on the theories of [9, 10].

Debye Temperature. A figure of 460±20 K was determined for the A15 β-phase $RhCr_3$ by [5]. A Debye temperature of 267±1 K for the same phase is reported by [11].

Specific Heat. The electronic specific heat was determined as 8.65±0.05 $mJ \cdot g-atom^{-1} \cdot K^{-2}$ by [5].

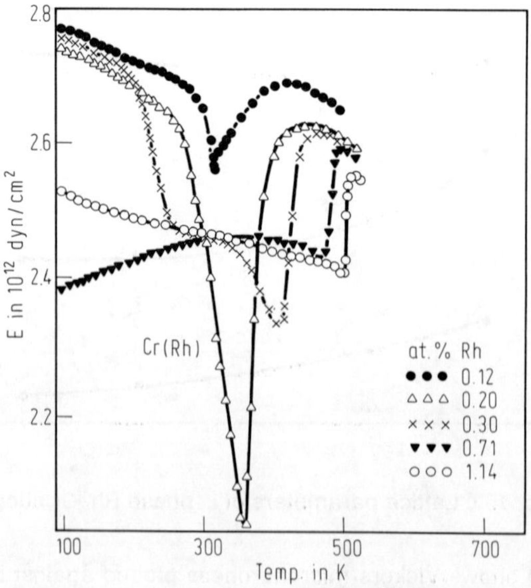

Fig. 111. Temperature dependence of Young's modulus of Rh-Cr alloys.

Superconductivity. Some rhodium-chromium alloys have been found to be superconductive. The superconducting transition temperatures of the ε-phase are plotted against composition and valence-electron concentration (see figure in original paper). The A15 Cr_3Rh β-phase is not superconducting above 0.3 K [12]. The transition temperature is quoted as 0.073 K by [13].

Magnetic Susceptibility. The hcp ε-phase was said to behave ferrogmagnetically by [14]. This was not confirmed in as-cast material containing 60 at% chromium, but cold-working was found to increase the magnetic susceptibility and weak ferromagnetism was then observed [12]. Magnetic measurements on alloys containing up to 14 at% rhodium showed that there was an unexpected change in the Néel temperature/concentration curve at the higher rhodium contents; this was attributed to the proximity to the ε-phase boundary. Quenching in oil from 1500 °C gave a single-phase sample which showed fine precipitation after tempering. This treatment caused the Néel temperature to rise to the maximum of 285 °C. Similar behaviour was found in a 9 at% rhodium specimen [15]. Further investigations on the effects of ageing were interpreted in the light of the composition/phase relationships of [3]; the antiferromagnetic-weak ferromagnetic behaviour was interpreted as possibly due to strain generated during precipitation [16]. The temperature dependence of the magnetic susceptibility of the A15 $RhCr_3$ phase was measured between room temperature and 3 K; χ at room temperature was found to be 3.63×10^6 emu/g [13]. The magnetic properties related to the phases present are shown in **Fig. 112**; the hatched areas are those showing weak ferromagnetism. Investigation of two alloys at the chromium-rich end of the diagram with single-phase and two-phase structures showed the single-phase alloy (98 at% Cr) to be antiferromagnetic, whilst the two-phase sample (80.4 and 89.5 at% Cr) shows weak ferromagnetism (see figure in original paper) [17].

Susceptibility measurements were made on alloys containing 0.55 to 14.5 at% rhodium in the paramagnetic region. The temperature dependence of χ is shown in Fig. 112 indicating some previously unreported features. The decrease in χ in the paramagnetic region, and

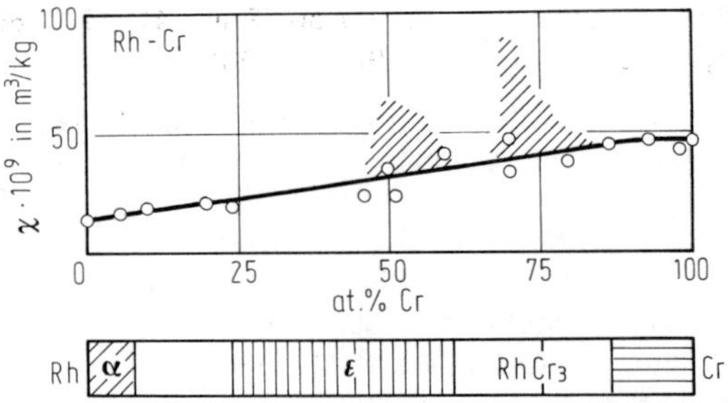

Fig. 112. Magnetic susceptibility in the Rh-Cr system, hatched areas denote weak ferromagnetism.

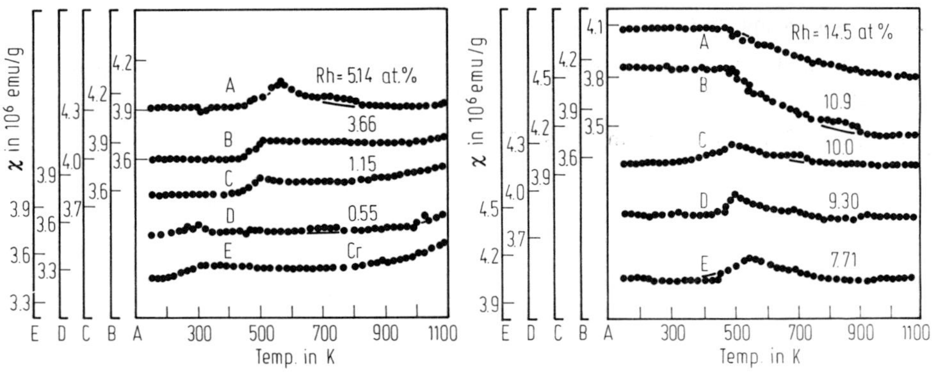

Fig. 113. Temperature dependence of χ in 0.55 to 14.5 at% Rh alloys.

the temperature-independent behaviour of the ordered phase between 10.9 and 14.5 at% Rh are apparent. A model is proposed in which rhodium induces a slight localization ($\mu_{eff} < 0.6\ \mu_B$) in chromium atoms, the interactions between the localized magnetic moments and SDW (spin density waves) being responsible for the non-conventional magnetic behaviour in the ordered phase; see **Fig. 113** [18].

Corrosion. The effect of small additions of rhodium in reducing corrosion rates of chromium in boiling acids is reported; 0.5 wt% of rhodium reducing the attack in 30% H_2SO_4 from 23 to 16 mils/year [19].

References:

[1] Venkatraman, R.; Neumann, J.P. (Bull. Alloy Phase Diagrams **8** [1987] 128/31).
[2] Waterstrat, R.M.; Manuzewski, R.C. (J. Less-Common Metals **32** [1973] 331/43).
[3] Raub, E.; Mahler, W. (Z. Metallk. **46** [1955] 210/5).
[4] Greenfield, P.; Beck, P.A. (Trans. Inst. Mining Metall. **206** [1956] 265/76).
[5] Flükiger, R.; Heiniger, F.; Junod, J.; Muller, P.; Spitzli, P.; Staudenmann, J.L. (J. Phys. Chem. Solids **32** [1970] 459/63).
[6] van Reuth, E.C.; Waterstrat, R.M. (Acta Crystallogr. B **24** [1968] 186/96).
[7] Allen, B.C.; Jaffee, R.I. (Am. Soc. Metals Trans. Quart. **56** [1963] 387/402).
[8] Carlson, O.N.; Sherwood, L.L.; Schmidt, F.A. (J. Less-Common Metals **6** [1964] 439/50).

[9] Jayaraman, A.; Rice, T.M.; Bucher, E. (J. Appl. Phys. **41** [1970] 869).
[10] Munday, B.C. (Phys. Status Solidi A **8** [1971] K129/K131).
[11] Reddy, S.V.; Suranarayana, S.V. (J. Less-Common Metals **99** [1984] L1/L3).
[12] Matthias, B.T.; Geballe, T.H.; Compton, V.B.; Corenzwit, E.; Hull, G.W. (Phys. Rev. [2] **128** [1962] 588/90).
[13] Bardos, D.I.; Waterstrat, R.M.; Rowland, T.J.; Darby, J.B. (J. Low Temp. Phys. **3** [1970] 509/18).
[14] Raub, E. (J. Less-Common Metals **1** [1959] 1/18).
[15] Booth, J.G. (Phys. Status Solidi **7** [1964] K157/K160).
[16] Booth, J.G. (J. Appl. Phys. **37** [1966] 1332/3).
[17] Kussman, A.; Müller, K.; Raub, E. (Z. Metallk. **59** [1968] 859/63).
[18] Dadarlat, D.; Petrisor, T.; Giurgiu, A.; Pop, I. (Phys. Status Solidi B **117** [1983] 155/9).
[19] Greene, N.D.; Bishop, C.R.; Stern, M. (J. Electrochem. Soc. **108** [1961] 836/41).

28.2 Ternary Alloy Rh_2CrSn

This magnetic alloy was found to have an exceptionally large tetragonal distortion of the Heusler structure; a = 4.093, c = 7.291 Å, c/a = 1.27.

Suits, J.C. (Solid State Commun. **18** [1976] 423/5).

29 Alloys with Molybdenum

29.1 The Rh-Mo System

Phase Diagram. The phase diagram given in **Fig. 114** is that from thermal, microscopical, hardness and X-ray investigations of [1, 2], presented by [3]. Some earlier work on the system using X-ray diffraction, microhardness, and microscopical methods had established the phases present and generally the later, more detailed work showed good agreement with the earlier findings [4]. The diagram shows the solid solubility of rhodium in molybdenum to be small at low temperatures, but this rises to 20 at% at the eutectic temperature of 1940±15 °C, which occurs at ~89 at% rhodium. Rhodium can dissolve ~15 at% of molybdenum at 2000±10 °C. The ε-phase extends roughly between 45 and 85 at% rhodium; the axial ratio of this phase passes through a minimum in the region of 75 at% rhodium. The melting point of the ε-phase rises to a maximum of 2075±10 °C at ~67 at% rhodium. Later work using X-ray powder diffraction methods on alloys in the ε-area containing 50 and 75 at% rhodium suggests that ordering of the ε-phase may take place after heat treatment at 950 °C for 36 h [5].

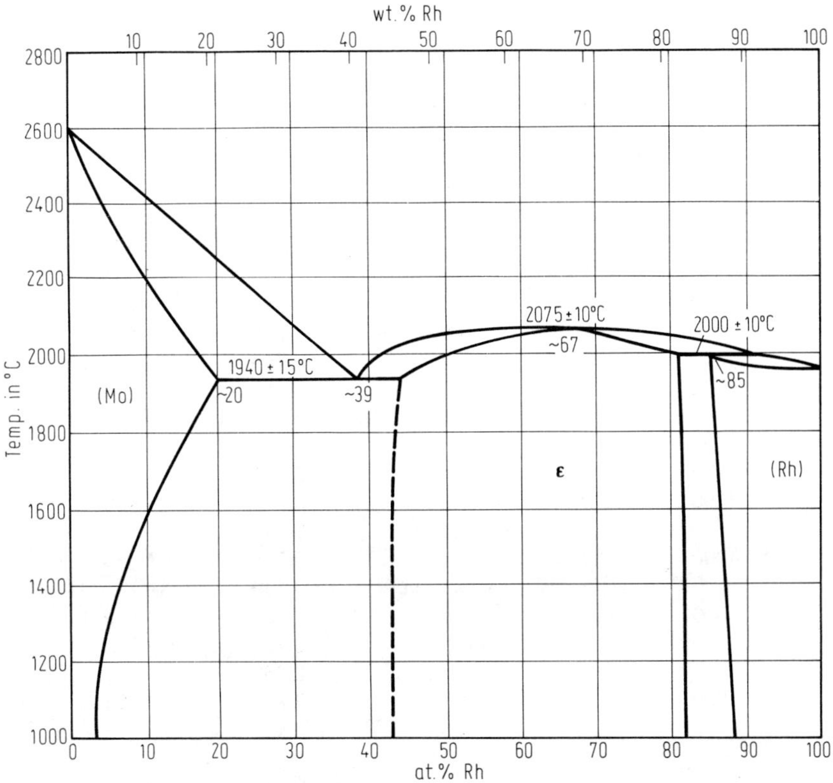

Fig. 114. Phase diagram Rh-Mo.

Preparation. The samples used were prepred by mixing and pressing high purity powders which were sintered in hydrogen at temperatures between 1680 and 1760 °C [2]. Difficulty had been experienced in obtaining homogeneous material by argon-arc melting [1].

Crystallography. The lattice parameters determined by [1, 2, 4] are given in the following table.

at% Rh	t in °C	annealing time in h	phases	a (in kX)	c (in kX)	Ref.
100	—	—	—	3.7954	—	[2]
97.2	1300	2	fcc	3.804	—	[4]
94.7	1800	1	fcc	3.8010	—	[2]
94.6	1300	2	fcc	3.810	—	[4]
91.5	1810	1	fcc	3.8046	—	[2]
89.4	1800	1	fcc	3.8066	—	[2]
89.35	1300	2	fcc	3.814	—	[4]
			+hcp	2.712	4.347	[4]
89.4	1800	1	fcc	3.8066	—	[2]
85.1	1750	1	fcc	3.8108	—	[2]
			hcp	2.7146	4.3415	[2]
85.1	1510	2	fcc	3.8100	—	[2]
			hcp	2.7144	4.3414	[2]
83.89	1300	2	hcp	2.714	4.345	[4]
80.9	1760	1	hcp	2.7144	4.3376	[2]
78.86	1200	2	hcp	2.720	4.349	[4]
78.8	1670	1	hcp	2.7185	4.3387	[2]
73.9	1650	1	hcp	2.7226	4.3412	[2]
69.2	1700	1	hcp	2.7278	4.3549	[2]
68.51	1200	2	hcp	2.737	4.376	[4]
58.7	1750	1	hcp	2.7400	4.3818	[2]
55.5	1680	1	hcp	2.7426	4.3895	[2]
48.25	1400	2	hcp	2.756	4.423	[4]
48.25	1200	2	hcp	2.754	4.423	[4]
48.25	800	25	hcp	2.754	4.424	[4]
			+Mo bcc	3.139		
48.1	1700	1	hcp	2.7502	4.4119	[2]
45.2	1700	1	hcp	2.7541	4.4258	[2]
38.33	1400	2	hcp	2.754	4.426	[4]
			+Mo bcc	3.141	—	[4]
38.33	1200	2	hcp	2.756	4.418	[4]
			+Mo bcc	3.139	—	[4]
38.33	800	25	hcp	2.754	4.409	[4]
			+Mo bcc	3.139		
38.3	1505	2	Mo hcp	2.7617	4.4499	[2]
28.53	1200	2	bcp	2.756	4.423	[4]
			+Mo bcc	3.141	—	[4]

The phase diagrams of Rh-Mo, Rh-W, Ir-W, and Mo-Ir, which all have in common an intermediate close-packed hexagonal, disordered ε-phase of broad stoichiometry have been examined for ordering using X-ray powder diffraction techniques on argon-arc melted material. The AB and AB_3 compositions were generally used, except in the case of Rh-Mo where only the composition RhMo was examined; however, deductions were made by analogy using data from [2] on the composition Rh_3Mo. The RhMo composition was found to be ordered after treatment at 950 °C for 36 h; after the original treatment at 1200 °C for 24 h, the material had been still disordered. Ordered MoRh was found to have a B19-$MgCd_3$ type structure, with a = 2.745, b = 1.785, c = 4.413 Å, b/a = 1.743, c/a = 1.607. The calculated data for ordered Rh_3Mo was: structure type D19-$MgCd_3$, a = 5.45_6, c = 4.35_0 Å, 2c/a = 1.595 [5]. The phase diagrams of 21 binary systems including Rh-Mo have been analysed for stability by minimizing the residual interaction parameters by successive approximation and the results compared with conventional computer phase analysis. Some deviations from experimental values were noted for this system [6].

Thermodynamics. Electromotive force cells have been used to determine the activity of molybdenum in the system over the temperature range 1200 to 1300 K and thermodynamic functions derived from the results. Solid ZrO_2 + 11 mol% CaO was used as the electrolyte. The results are shown in **Fig. 115** [7].

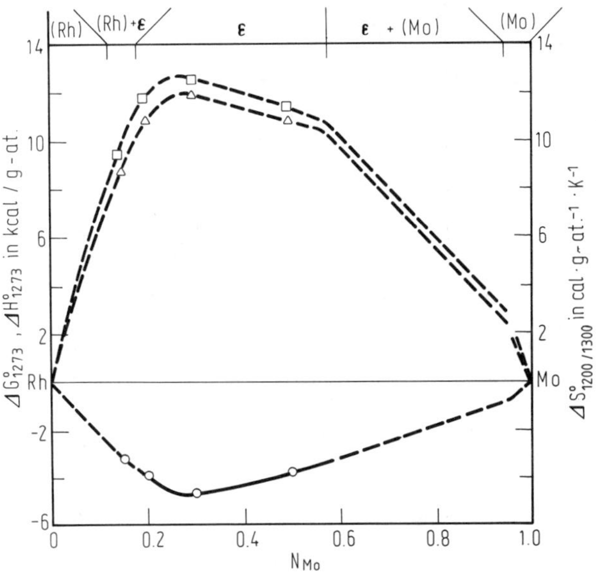

Fig. 115. Relative integral molar Gibbs free energies, enthalpies and entropies of formation of Rh-Mo alloys.

Thermal Expansion. This was determined on an Mo-67 at% rhodium alloy (ε phase) between 293 and 1173 K as 8.60×10^{-6}/K [8].

Hardness. VPN figures are reported in the composition range 45 to 100 at% Rh by [2]. Microhardness (Vickers) figures are given for alloys containing up to 70 at% Mo by [4]. The data are shown in **Fig. 116**, p. 182 [2].

Superconductivity. Theoretical grounds suggested that some Rh-Mo alloys might show superconductivity, and this has been found to be true in the case of RhMo which has a transition temperature of 1.75 K [9]. The superconducting transition temperature (T_c) of a number of alloys with different electrons/atom ratios was determined down to 0.31 K; the decrease with increasing electrons/atom is shown in **Fig. 117**, p. 182. This was explained on the basis of the theory of spin fluctuations [10]. The superconducting transition tempera-

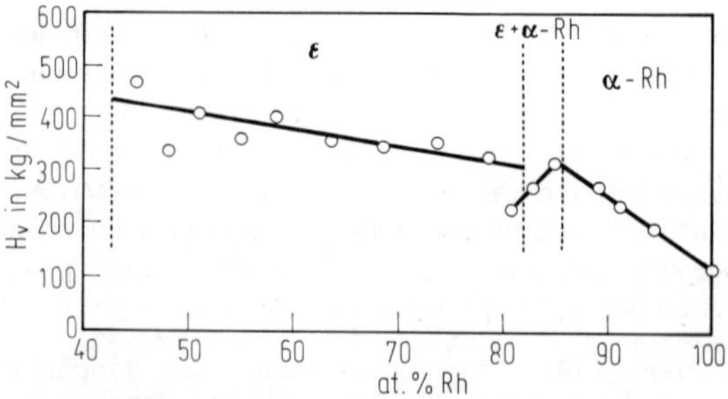

Fig. 116. Vickers pyramid hardness of Rh-Mo alloys.

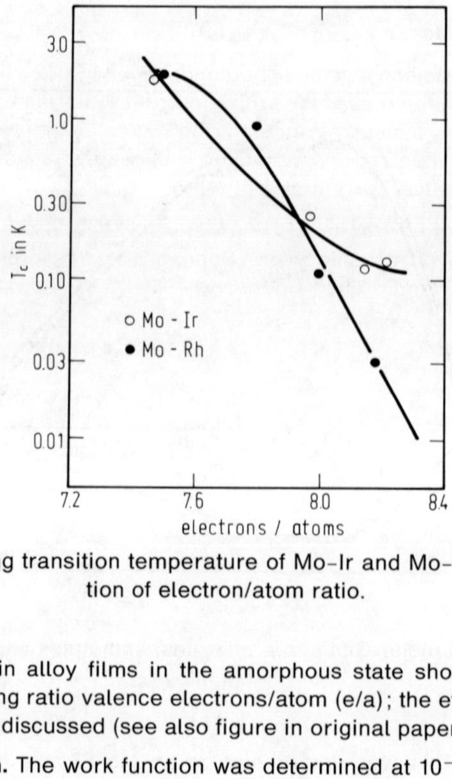

Fig. 117. Superconducting transition temperature of Mo-Ir and Mo-Rh hcp alloys as a function of electron/atom ratio.

tures determined on thin alloy films in the amorphous state showed a peak and then a steady fall with increasing ratio valence electrons/atom (e/a); the effects of spin fluctuations and charge transfer are discussed (see also figure in original paper) [11].

Thermionic Emission. The work function was determined at 10^{-9} Torr for an Mo 65 at% Rh alloy (ε-phase), in single-crystal form, for the (0001), (10$\bar{1}$2) and (11$\bar{2}$4) faces. The slopes of the Richardson lines gave $\phi_{(0001)}=4.87$ eV, $A_{(0001)}=230\ A\cdot cm^{-2}\cdot K^{-2}$, $\phi_{(10\bar{1}2)}=4.46$ eV, $A_{(10\bar{1}2)}=110\ A\cdot cm^{-2}\cdot K^{-2}$, $\phi_{(11\bar{2}4)}=4.30$ eV, $A_{(11\bar{2}4)}=170\ A\cdot cm^{-2}\cdot K^{-2}$ [12].

References:

[1] Hawarth, C.W. (J. Inst. Metals **87** [1958/59] 265/9).

[2] Anderson, E.; Hume-Rothery, W. (J. Less-Common Metals **2** [1960] 19/28).

[3] Elliott, R.P. (Constitution of Binary Alloys, 1st Suppl., McGraw-Hill, New York-Toronto-London-Sydney 1965, p. 630).

[4] Raub, E. (Z. Metallk. **45** [1954] 23/30).

[5] Giessen, B.C.; Jaehnigen, U.; Grant, N.J. (J. Less-Common Metals **10** [1966] 147/50).
[6] Lesnik, A.G.; Nemoshkalenko, V.V.; Ovcharenko, A.A. (Metallofizika Inst. Metallofiz. No. 75 [1979] 3/19).
[7] Yamawaki, M.; Nagai, Y.; Kogai, T.; Kanno, M. (Thermodyn. Nucl. Matter Proc. Intern. Symp. 1979 [1980], Vol. 1, pp. 1249/61).
[8] Savitskii, E.M.; Burov, I.V.; Tomilin, N.A. (Dokl. Akad. Nauk SSSR **271** [1983] 1370/2; Soviet Phys.-Dokl. **28** [1983] 676/7).
[9] Matthias, B.T.; Corenzwit, B.T. (Phys. Rev. [2] **94** [1954] 1069).
[10] Riblet, G.; Jensen, M.A. (Physica **55** [1971] 622/5).
[11] Collver, M.M.; Hammond, R.H. (Solid State Commun. **22** [1977] 55/7).
[12] Savitskii, E.M.; Burov, I.V.; Litvak, L.N.; Savel'ev, Yu.A. (Zh. Tekh. Fiz. **43** [1973] 818/20; Soviet Phys.-Tech. Phys. **18** [1973] 513/4).

29.2 Ternary Alloys Rh-Mo-Zr

The effect of rhodium additions on the superconductivity of zirconium-molybdenum alloys has been studied by measurements of the critical current. After annealing the ternary alloys, at 450 °C, when only the α-phase precipitated, the critical current increased significantly; after annealing at 500 °C, when Zr_2Rh precipitated with coarse α-phase, the critical temperature was higher and the critical current lower.

Baranov, I.A.; Bychkov, Yu.F.; Korzhov, V.P.; Mal'tsev, V.A.; Slavgorodskii, M.P.; Shamalevich, R.S. (Sverkhprovodyashchie Splavy Soedin. Dokl. 6th Vses. Soveshch. Metalloved. Fiz. Khim. Metallofiz. Sverkhprovodn., Moscow 1969 [1972], pp. 140/7).

30 Alloys with Tungsten

The Rh-W System

Phase Diagram. The diagram shown in **Fig. 118** is based on the microscopic, thermal analysis, X-ray diffraction and microprobe analysis results of [1]. The diagram shows only one intermediate phase (ε), which has a wide homogeneity range and a hexagonal structure; the range of homogeneity shown decreases markedly with falling temperature. At 52 at% tungsten the melting point is 2250 °C. The results confirm earlier but less detailed work by [2, 3].

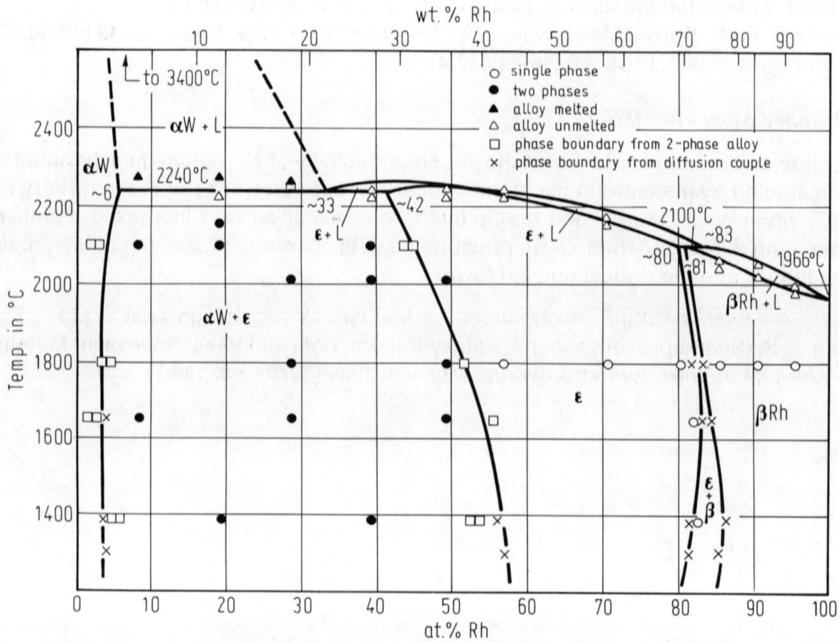

Fig. 118. Phase diagram Rh-W.

Crystallography. The close-packed hexagonal ε-phase found at 19.2 at% tungsten had the following lattice parameters: a = 2.708, c = 4.328 Å, c/a = 1.598 [2].

Ordering was found in WRh_3 after homogenizing for 24 h at 1600 °C to remove coring, followed by an ordering anneal for 24 h at 1200 °C. The alloy has the Do-$MgCd_3$ structure, with a = 5.453, c = 4.350 Å [4].

Hardness. Figures determined during the course of an investigation of diffusion barrier materials are shown in **Fig. 119**; the microhardness determinations were made across a rhodium-tungsten couple annealed at 1700 ± 15 °C for 1 h [5]. As-cast rhodium containing 1 wt% tungsten has a hardness of 116 VPN according to [6].

Stress-Rupture Properties. Figures are given in the table below for rhodium alloys containing 1, 3 and 6 wt% tungsten, tested at 1200 and 1400 °C [6].

wt% W	1	1	3	3	6	6
test temperature in °C . .	1200	1400	1200	1400	1200	1400
time to rupture in h	11.1	0.3	21.7	1.6	13.2	0.8

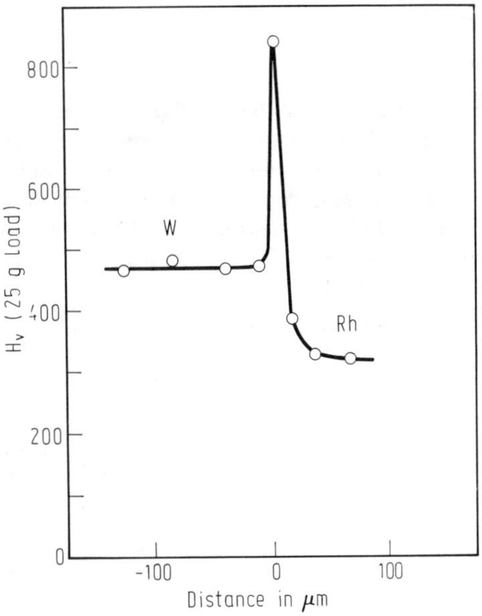

Fig. 119. Microhardness of W-Rh diffusion couples.

Vapourization Rates. These have been determined for a rhodium-50% tungsten alloy by [7].

Electrochemistry. The high power density performance of rhodium-tungsten alloy electrodes in alkali metal thermoelectric converters was investigated using a β″-alumina electrolyte; electrical measurements are given by [8].

References:

[1] Rapperport, E.J.; Smith, M.F (WADD-TR-60-132-Pt. II AD-287-540 [1962] 8/27).
[2] Greenfield, P.; Beck, P.A. (Trans. Inst. Mining Metall. **206** [1956] 265/76).
[3] Raub, E.; Walter, P. (Alloys of The Platinum Group Metals with Tungsten, Heraeus Festschrift 1951, p. 124).
[4] Giessen, B.C.; Jaehnigen, U.; Grant, N.J. (J. Less-Common Metals **10** [1965] 147/50).
[5] Passmore, E.M.; Boyd, J.E.; Neal, L.P.; Andersson, C.A.; Lement, B.S. (WADD-TR-60-343 [1960] pp. 49; N.S.A. **15** [1961] No. 11607).
[6] Handley, J.R. (Platinum Metals Rev. **33** [1989] 64/72).
[7] Rytvin, E.I.; Malashkin, V.V. (Tr. Inst. Fiz. Metall. Akad. Nauk SSSR Ural. Nauchn. Tsentr No. 28 [1971] 250/3).
[8] Williams, R.M.; Jeffries-Nakamura, M.L.; Underwood, B.L.; Wheeler, M.E.; Loveland, S.J.; Kikkert, J.L.; Lamb, T.; Kummer, J.T.; Bankston, C.P. (J. Electrochem. Soc. **136** [1989] 893/4).

31 Alloys with Uranium

31.1 The Rh-U System

Phase Diagram. The phase diagram shown in **Fig. 120** was produced from thermal analysis, metallographic and X-ray diffraction data. The system has four intermetallic compounds: U_4Rh_3, U_3Rh_4, U_3Rh_5, and URh_3. One eutectic occurs at ~865 °C and 24.5 at% rhodium and a second at 1393 °C and 87 at% rhodium. The maximum solid solubility of rhodium in uranium is ~8 at% and ~3 at% of uranium in rhodium. A solid-state transformation occurs in the compound U_3Rh_4 at ~720 °C; peritectics occur at 1450 and 1550 °C and the melting point of URh_3 is ~1700 °C [1].

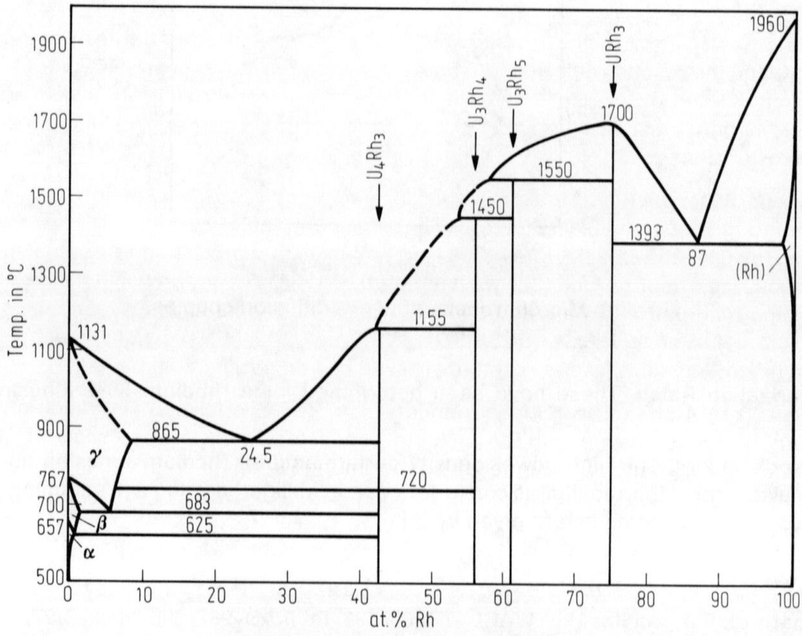

Fig. 120. Phase diagram Rh-U.

Preparation. Arc or induction melting in beryllia crucibles was used; generally induction melting was used for the lower rhodium content alloys. Impurity levels and details of experimental technique are given in the paper [1].

Crystallography. X-ray diffraction data gave the following figures for the lattice parameter of URh_3 which has a $CuAu_3$-type structure.

composition	a in Å	Ref.
60.7 to 74.9 at% Rh	~4.0	[1]
	3.992	[2]
	3.991	[3]
	3.988 ± 0.001	[4, 5]
	3.9915	[6]

The interatomic distance is given as 2.82 Å by [7]. For X-ray diffraction data and interplanar spacings for U_4Rh_3, U_3Rh_4 and U_3Rh_5 see table in original paper [1].

Density. This is given for URh_3 as 14.27 g/cm^3 by [2].

Thermodynamic Data. The free energy of formation of URh_3 is $\Delta G_{1000K} = -62.4$ kcal/mol (-261.26 kJ) [8]. The expression $\Delta G = -316.369 + 0.009257\,T$ kJ/mol between 1090 and 1180 K is derived by [9]. $\Delta_F G$ at 1500 K was determined as 306 kJ/mol [12].

Heat Capacity. That of URh_3 was measured by adiabatic calorimetry in the temperature range 5 to 350 K, and by enthalpy-increment drop calorimetry from 300 to 850 K. Phase transitions were not observed over the whole temperature range. The results are shown in a figure where C_P/R is plotted against temperature; $R = 8.3143\ J \cdot K^{-1} \cdot mol^{-1}$. The electronic coefficient γ was found to be 0.0016/RK at 0 K, and the Debye temperature at 0 K, $\theta_D(0) = 297$ K [6].

Electric Resistivity. Electric resistivities have been measured in the temperature range 2 to 300 K, and the results can be computed from the expression $\varrho = \varrho_0 + AT^3$ where $A = 2.53 \times 10^{-5}\ \mu\Omega \cdot cm/K^3$ and $\varrho_0 = 1.26\ \mu\Omega \cdot cm$. The T^3 dependence is attributed to interband (s-d or s-f) electron-phonon scattering expected from the occupation of a broad 6 d-5 f band with a high density of states [4].

Magnetic Susceptibility. Measurements in the range 2.5 to 300 K show little variation with temperature. An almost temperature-independent value $\chi = 1.10 \times 10^{-3}$ emu/mol was measured at 5 K; the spacial distribution of induced magnetization was also measured using a polarized neutron beam of wavelength 0.980 Å and a diffractometer, the measurements being made at 5 K in a field of 48 kOe. Expressed as atomic susceptibilities, $\chi(U) = (0.42 \pm 0.05) \times 10^{-3}$ emu, $\chi(Rh) = (0.25 \pm 0.04) \times 10^{-3}$ emu. Results suggested that the rhodium 4 d band is partially filled as the result of strong 6 d-4 d hybridization [4].

de Haas-van Alphen Effect. Measurements have been performed and much of the data were found to be consistent with the multiple-connected Fermi surface shown in **Fig. 121**; band-structure calculations confirmed the experimental findings [10].

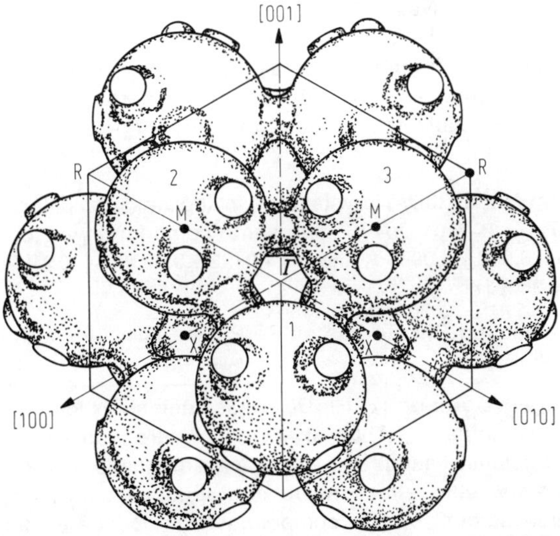

Fig. 121. Proposed topology for the large Fermi-surface piece in URh_3. Large ball-like surfaces centred at the M points in the Brillouin zone are connected by necks in ⟨110⟩ directions which would be intersected by the line S between the points X and R.

Photoemission. The resonant photoemission of URh_3 was measured and the 5 f emission lineshape is shown in a figure [11].

Nuclear Magnetic Resonance. This has been observed in URh_3 and the Knight shift at 4.2 K found to be $0.00 \pm 0.02\%$ [11].

References:

[1] Park, J.J. (J. Res. Natl. Bur. Std. A **72** [1968] 11/7).
[2] Ferro, R. (Atti Accad. Nazl. Lincei Classe Sci. Fis. Mat. Nat. Rend. [8] **25** [1958] 189/91).
[3] Dwight, A.E.; Downey, J.W.; Conner, R.A. (Acta Crystallogr. **14** [1961] 75/6).
[4] Nellis, W.J.; Harvey, A.R.; Brodsky, M.B. (AIP Conf. Proc. No. 10 Pt. 2 [1973] 1076/80).
[5] Delapalme, A.; Lander, G.H.; Brown, P.J. (J. Phys. C **11** [1978] 1441/50).
[6] Cordfunke, E.H.P.; Muis, R.P.; Wijbenga, G. (J. Chem. Thermodyn. **17** [1985] 1035/44).
[7] Ferro, R. (Papers Sect. Inorg. Chem. 16th Intern. Congr. Pure Appl. Chem., Paris 1957 [1958], pp. 353/8).
[8] Berndt, U.; Erdmann, B.; Keller, C. (Platinum Metals Rev. **18** [1974] 29/34).
[9] Wijbenga, G.; Cordfunke, E.H.P. (J. Chem. Thermodyn. **14** [1982] 409/17).
[10] Arko, A.J.; Brodsky, M.B.; Crabtree, G.W.; Karim, D.; Koelling, D.D.; Windmiller, L.R. (Phys. Rev. [3] B**12** [1975] 4102/12).
[11] Zolnierek, Z.; Arko, A.J.; Koelling, D.D. (J. Less-Common Metals **122** [1986] 89/93).
[12] Möbius, S.; Hellwig, L.; Keller, C. (J. Less-Common Metals **121** [1986] 43/8).
[13] Seitchik, J.A.; Jaccarino, V.; Wernick, J.H. (Phys. Rev. [2] **138** [1965] A 148/A 152).

31.2 Ternary Alloys

Rh-U-Sb. Argon arc-melting was used on a water-cooled copper hearth to prepare RhUSb material for X-ray powder measurements; the material was homogenized in evacuated Vycor capsules prior to testing. The compound was found to have an MgAgAs-type structure with a = 6.531 Å and V/M = 69.64 (V/M is volume/formula weight) [1]. The electronic structure of URhSb has been investigated using synchrotron radiation photoemission spectroscopy; the maximum of the d-like valence band density of states was found to be shifted from the Fermi level to higher binding energies [8]. The compound $U_3Rh_3Sb_4$ was prepared by argon arc-melting on a water-cooled copper hearth under argon and the crystal structure investigated by X-ray diffraction; it was found to have a $Y_3Au_3Sb_4$-type structure with $a = 9.531 \pm 0.002$ Å [9].

Rh-U-Al. The compound RhUAl crystallizes in the hexagonal form with a ZrNiAl-type structure and a = 6.9647, c = 4.0192 Å [6]; the same compound was found to have an Fe_2T-type structure with $a = 6.947 \pm 0.0006$ Å and $c = 4.0192 \pm 0.0006$ Å [10]. It is reported to become ferromagnetic below 35 K [7].

Rh-U-Ga. The compound RhUGa has an Fe_2P-type structure with $a = 7.0063 \pm 0.0006$ Å and $c = 3.9449 \pm 0.0006$ Å [10].

Rh-U-Ge. The magnetic ordering of Rh_2UGe_2 has been investigated by neutron diffraction techniques; it was found to have a $ThCr_2Si_2$-type structure with $a = 4.155 \pm 0.002$ Å and $c = 9.771 \pm 0.005$ Å. Polycrystalline samples became paramagnetic at 4.2 K in a zero magnetic field; neither separate magnetic reflections nor contributions to nuclear peaks were detected [2]. Neutron diffraction studies on the compound URh_2Ge_2 were interpreted in terms of a $CaBe_2Ge_2$-type structure; magnetometric measurements indicated antiferromagnetic behaviour at temperatures below 8 K in contrast to the neutron diffraction data taken at 4.2 K. The magnetic susceptibility and $1/\chi$ curves are shown in **Fig. 122** [3].

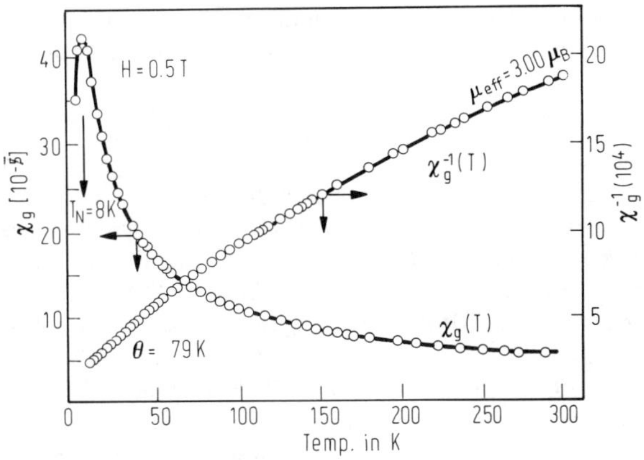

Fig. 122. Magnetic susceptibility and reciprocal magnetic susceptibility curves for URh_2Ge_2.

The same compound has been studied by resistivity measurements and the effect of pressure investigated, which was found to be negligible. A steep fall in ϱ(T) at 2.0 K suggests that magnetic ordering is taking place in the system (see **Fig. 123**) [4].

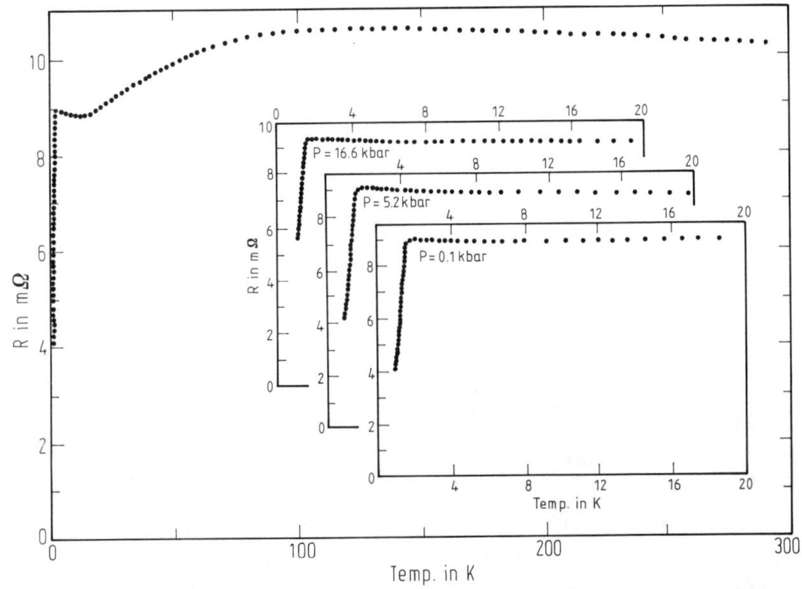

Fig. 123. Resistive behaviour of URh_2Ge_2. The precipitous fall of R(T) at 2.0 K signals magnetic ordering in the system. The inset shows resistivity below 20 K at 5.2 and 16.2 kbar.

The compound $U_3Rh_4Ge_{13}$ had X-ray powder patterns suggesting a derived structure of $Yb_3Rh_4Sn_{13}$-type; some diffraction lines corresponding to the Pm3n space group including the (222) one are split, and so a simple tetragonal distortion is excluded. So far the structure is unknown. It orders antiferromagnetically at 22 K. Susceptibility and magnetization curves are given [5].

Rh-U-Sn. The compound RhUSn was prepared in the same way as the RhUSb described above and was similarly examined to give a structure of the Fe_2P-type with a = 7.375, c = 3.996 Å, c/a = 0.54 and V/M = 62.75 [1].

References:

[1] Dwight, A.E. (J. Less-Common Metals **34** [1974] 279/84).
[2] Ptasiewicz-Bak, H.; Leciejewicz, J.; Zygmunt, A. (J. Phys. F **11** [1981] 1225/35).
[3] Ptasiewicz-Bak, H.; Leciejewicz, J.; Zygmunt, A. (Solid State Commun **55** [1985] 601/4).
[4] Thompson, J.D.; Fisk, Z.; Gupta, L.C. (Phys. Letters A **110** [1985] 470/2).
[5] Lloret, B.; Chevalier, B.; Gravereau, P.; Darriet, B.; Etourneau, J. (J. Phys. Colloq. [Paris] **49** [1988] C8-487/C8-488).
[6] Veenhuizen, P.A.; Klaasse, J.C.P.; de Boer, F.R.; Sechovsky, V.; Havela, L. (J. Appl. Phys. **63** [1988] 3064/6).
[7] Andreev, A.V.; Bartashevich, M.I. (Fiz. Met. Metalloved. **62** [1986] 266).
[8] Höchst, H.; Tan, K.; Buschow, K.H.J. (J. Mag. Magn. Mater. **54/57** [1986] 545/6).
[9] Dwight, A.E. (J. Nucl. Mater. **79** [1979] 417).
[10] Dwight, A.E.; Mueller, M.H.; Conner, R.A.; Downey, J.W.; Knott, H. (Trans. Am. Inst. Mining Metall. Petrol. Eng. **242** [1968] 2075/80).

32 Alloys with Manganese

32.1 The Rh-Mn System

Phase Diagram. The diagram shown in **Fig. 124** is produced from data from [1, 2].

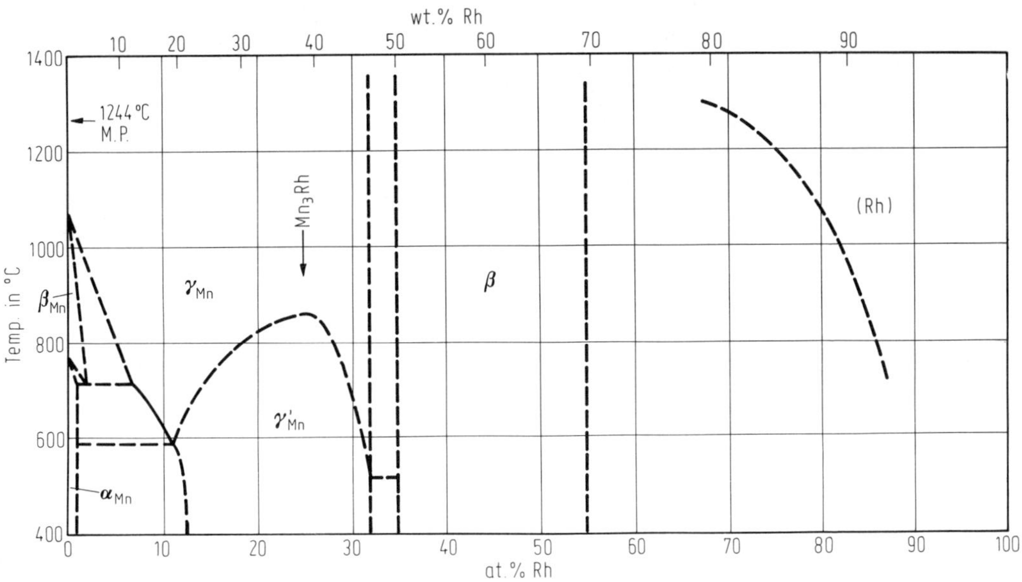

Fig. 124. Partial phase diagram Rh-Mn.

Little work has been done on the rhodium-rich end of the diagram. Detailed thermal analysis and microscopic examination of alloys containing up to ~6 at% rhodium show the effect of rhodium on the β, γ and δ forms of manganese; the liquidus and solidus fall to a peritectic where δ(4%)+liq. (5.5%)=γ(4.5%); the γ field extends and the β/γ boundary falls steeply, confirming the earlier results of [1]; **Fig. 125**, p. 192, shows the results [3]. On theoretical grounds, a limited range of solid solution of rhodium in manganese is predicted by [4].

Intermetallic compounds of the formulae $RhMn_3$ and RhMn are predicted by [5]. The main interest in this system has been in the low-temperature martensitic change from cubic to tetragonal (CuAu 1) of the β (CsCl) phase and the associated complex magnetic changes, which were not found by [1]; these transformation temperature lines which show hysteresis between heating and cooling need to be added to the diagram. There is fairly good agreement between authors on the compositional range of the β-phase at room temperature when quenched from above 500 °C.

Range of composition in at% Mn: 45 to 65 [1], 50.0±0.5 to ~58 [6], 50 to 60 [7]. Because of the speed of the martensitic transformation it is impossible to retain the high-temperature CsCl phase by quenching except in very small sections; the pure cubic phase was retained in a thin-walled quartz tube of ≦3 mm internal diameter, quenched in iced water [6].

The low-temperature face-centred tetragonal phase resulting from the martensitic transformation was found to be para- or antiferromagnetic, but at high manganese contents and low temperatures it is ferromagnetic [8]. The compositional dependence of the marten-

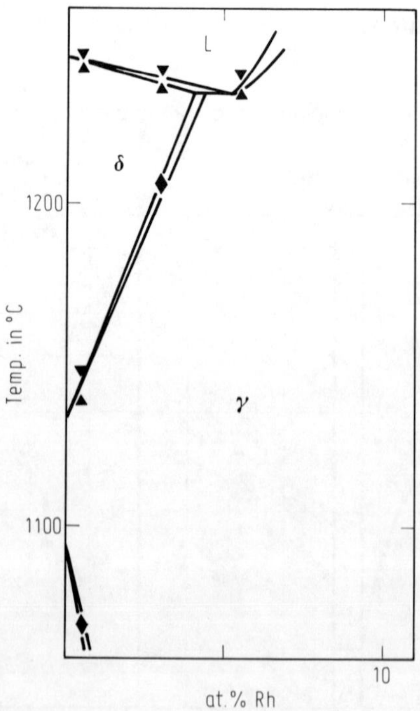

Fig. 125. Rh-Mn diagram between 0 and ~6 at% Rh.

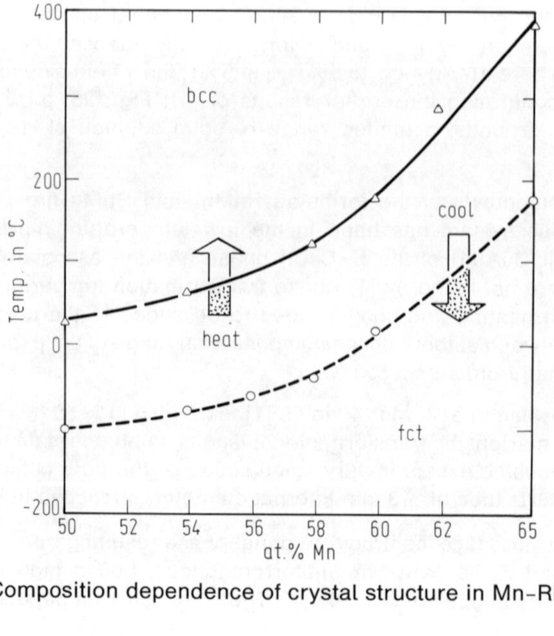

Fig. 126. Composition dependence of crystal structure in Mn-Rh system.

sitic transformation and its hysteresis has been studied by hot-stage microscopy, magnetic susceptibility, and dilatometry yielding a diagram showing the paramagnetic Curie temperature and transformation temperatures which is shown in a figure [8]. Similar work using differential thermal analysis, X-ray diffraction, and magnetic measurements produced the transformation diagram shown in **Fig. 126** [7].

Dilatometric measurements on a 62 at% Mn alloy show the hysteresis in the transformation, see **Fig. 127**; it was found in this work that the stresses induced by the volume changes resulted in the development of cracks in the specimen; the authors checked that this was not connected with oxygen absorption [8].

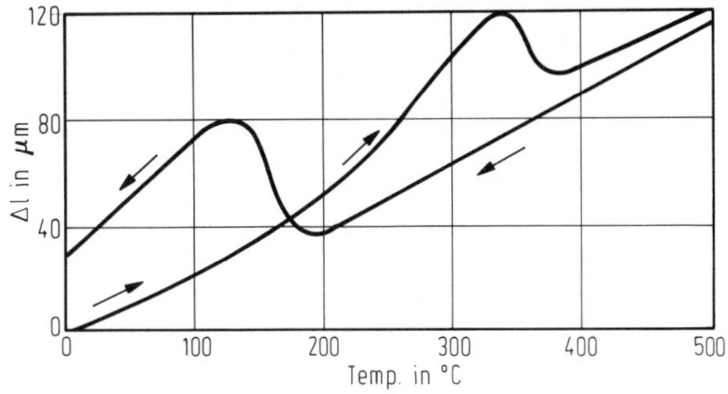

Fig. 127. Dilatometric behaviour of an Rh alloy with 62 at% Mn under the influence of temperature.

Crystallography. The table below gives details of the phases present after specified heat treatments, the crystal structure, and the lattice parameters reported by [1].

at% Mn	quenching temp. in °C	phase	a in kX	c/a
19.2	900	α (fcc)	3.807	
24.4	900	α	3.806	
30.0	1300	α	3.822	
	1100	α	3.812	
		β (bcc; CsCl)	3.040	
	1000	α	3.811	
	1000	β	3.03	
	850	α	3.801	
	850	β	3.03	
	650/550	α	3.798	
	650/550	β	3.03	
40.0	800	β	3.040	
	800	α	3.808	
50.0	900/550	β	3.045	
60.3	800	β	3.035	
70.3	1050	γ-Mn (fcc)	3.820	
	900	γ-Mn	3.81	
	530	γ-Mn (fcc)	3.804	
75.2	1050	γ-Mn (fcc)	3.804	
	800	γ-Mn (fcc, ordered)	3.804	
80.8	1050	γ-Mn (fcc)	3.801	
	800/550	γ-Mn (fcc, ordered)	3.801	

at% Mn	quenching temp. in °C	phase	a in kX	c/a
89.5	1200/1050/900	γ-Mn (fcc)	3.776	
	750	γ-Mn (fcc)	3.776	
	750	α-Mn	trace only	
	550	γ-Mn (fcc, ordered)	3.791	
	550	α-Mn	8.960	
94.5	1200/1050/900	γ-Mn (fct)	3.735	1.018
	750	γ-Mn (fct)	3.735	1.018
	750	α-Mn	8.964	
	550	α-Mn	8.964	
	550	γ-Mn (fct, ordered)	3.798	

bcc = body-centred cubic, fcc = face-centred cubic, fct = face-centred tetragonal

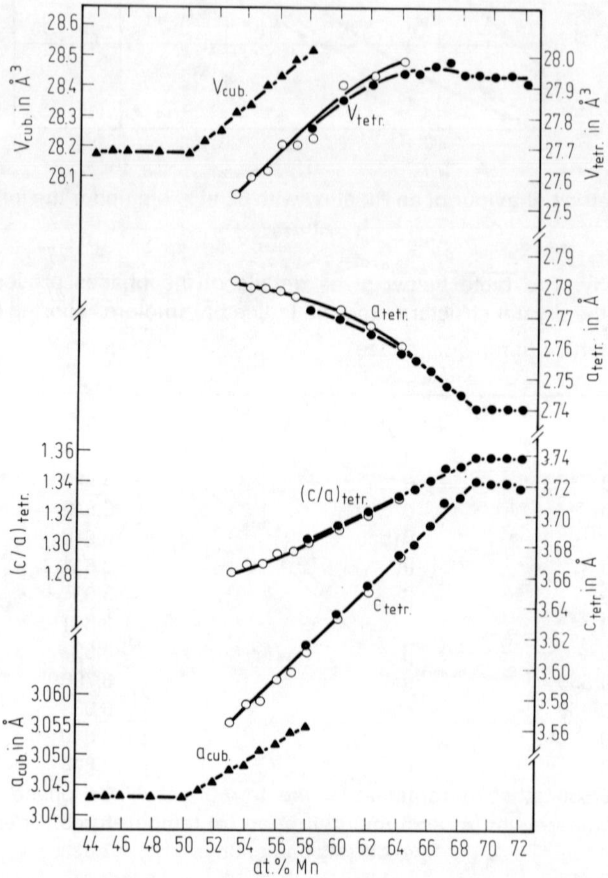

Fig. 128. Unit cell dimensions versus composition for the cubic and tetragonal RhMn phases at room temperature. Filled symbols represent samples quenched from ≧500 °C, open circles denote values obtained for samples cycled from room temperature through liquid nitrogen temperature and back to room temperature. The estimated inaccuracies of the parameters do not exceed the size of the symbols.

Lattice parameters for alloys in the range 44 to 72 at% Mn are given in **Fig. 128** and **Fig. 129**; the authors in this work were able to retain untransformed β-CsCl phase in small sections. Lattice parameters for both the cubic and tetragonal RhMn phases are plotted against manganese content at room temperature and at liquid helium and liquid nitrogen temperatures. Comparison of observed and calculated X-ray intensities confirmed both the CsCl- and CuAu(I)-type crystal structures. The same authors also carried out neutron diffraction work in connection with the determination of the magnetic structures, see table on p. 196.

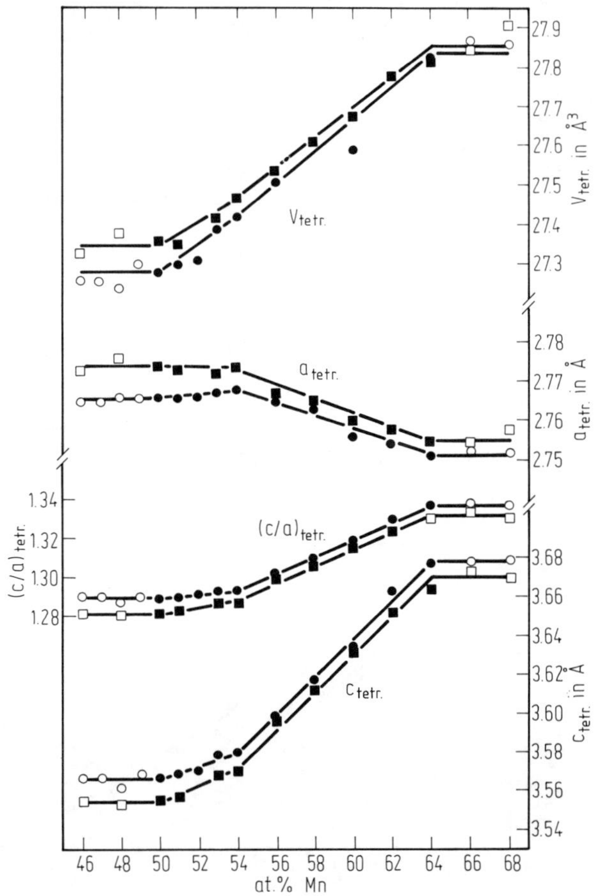

Fig. 129. Variation in lattice parameters of the tetragonal RhMn phase as a function of composition at liquid helium (●) and liquid nitrogen (■) temperatures. Open symbols represent two-phase alloys.

Density. The compositional dependences of the cubic and tetragonal RhMn phases are shown in **Fig. 130**, p. 196. They are compared with figures calculated from unit cell dimensions on the basis of substitutional solid solution [6].

Comparison of observed and calculated neutron diffraction data [6]:

	phase T in K	tetr. $Rh_{45}Mn_{55}$ [a)] 80		tetr. $Rh_{40}Mn_{60}$ 80		tetr. $Rh_{40}Mn_{60}$ 293		tetr. $Rh_{35}Mn_{65}$ 293	
hkl	type	jF_o^2	jF_c^2	jF_o^2	jF_c^2	jF_o^2	jF_c^2	jF_o^2	jF_c^2
100	magn	4.34	4.40	4.73	4.74	4.63	4.72	4.12	4.21
001	nucl	5.62	5.92	4.29	4.72	4.71	4.72	3.63	3.59
110	nucl	11.92	11.84	9.96	9.44	9.50	9.44	7.46	7.18
101	magn	8.87	8.72	9.91	9.92	10.16	9.89	8.64	8.77
111	nucl	0	0.64	0	0.08	0	0.08	0	0.08
200	nucl	0	0.32	0	0.04	0	0.04	0	0.04
002	nucl	0	0.16	0	0.02	0	0.02	0	0.02
210	magn	2.01	1.94	22.33	2.18	21.56	2.17	16.98	1.94
201	nucl	24.93	23.68		18.88		18.88		14.36
102	magn	2.31	3.10	3.85	3.34	3.60	3.33	3.08	2.97
211	magn	3.00	3.46	3.39	3.73	3.65	3.71	3.64	3.28
112	nucl	22.08	23.68	17.24	18.88	17.48	18.88	14.04	14.36
220	nucl			0	0.04	0	0.04	0	0.04
202	nucl			0	0.08	0	0.08	0	0.08
300	magn			19.40	0.33	19.63	0.33	13.83	0.28
221	nucl				18.88		18.88		14.36
	R	0.056		0.052		0.033		0.037	

a) cubic $Rh_{45}Mn_{55}$; T = 293 K; $h^2+k^2+l^2$, jF_o^2, jF_c^2; 1, 4.99, 4.48; 2, 0.28, 0.25; 3, 6.01, 5.97; 4, 0, 0.13; 5, 17.37, 17.92; R = 0.039

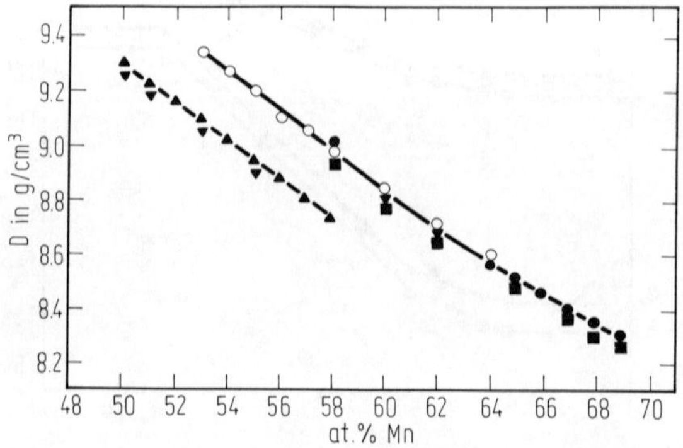

Fig. 130. Compositional dependences of the observed densities of the cubic (▼) and tetragonal (■) RhMn phases compared with those calculated from the unit cell dimensions on the basis of substitutional solid solution (▲, ●, ○).

Thermal Expansion. A linear thermal coefficient of expansion of about 1.3×10^{-6} was obtained for three different alloys of the cubic RhMn phase by [6]. The thermal dilatation curve for an alloy of 62 at% Mn on heating and cooling through the martensitic transformation is shown in Fig. 127, p. 193 [8].

Specific Heat. The low-temperature specific heat plotted against temperature of $Rh_{1-x}Mn_x$ where x=0.001 to 0.03 is shown in **Fig. 131**. The results together with susceptibility measurements suggest solute-solute interactions [10].

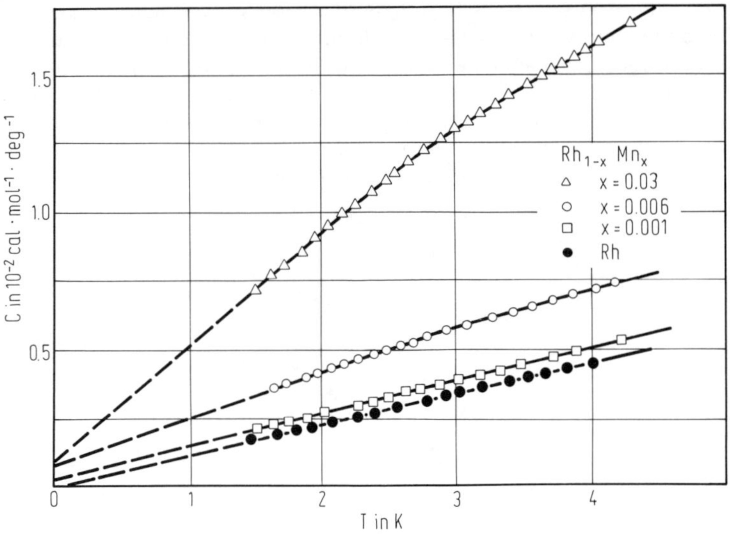

Fig. 131. Low-temperature specific heat.

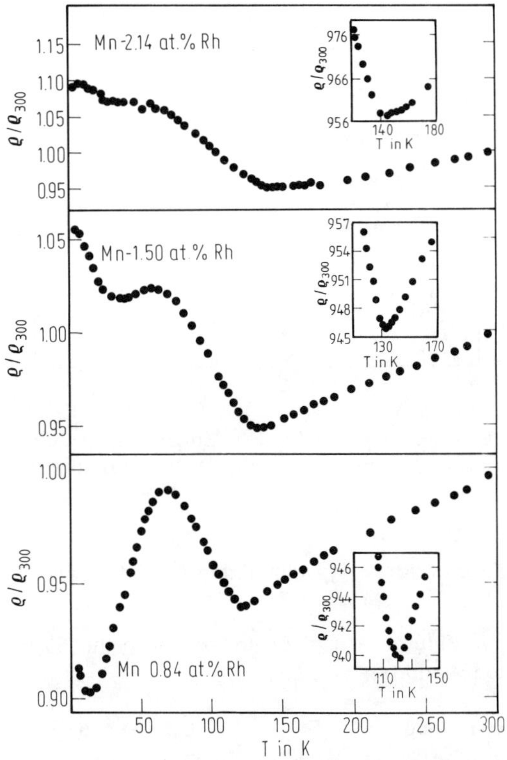

Fig. 132. Normalized resistivity for several Mn-Rh alloys.

Electric Resistance. The normalized resistivity of manganese containing 0.84, 1.50 and 2.14 at% Rh are shown in **Fig. 132**, p. 197, determined at temperatures between 0 and 300 K [9]. Measurements of ϱ have been made on alloys of 14.2, 9.4, 4.6 and 1.0 at% manganese at temperatures between 4 and 45 K; magnetoresistance was also measured in fields up to 18 kOe. The behaviour was found to be qualitatively similar to that of rhodium-iron alloys, except that their magnetoresistance decreased with decreasing temperature. **Fig. 133** shows the resistivity increment ($\varrho - \varrho_{Rh}$) plotted against temperature [11]. The electric resistivity was measured on $Mn_{1-x}Rh_x < 0.36 \leqq x \leqq 0.46$ between 77 and 700 K; the maximum in the ϱ/T curve was found to increase with increased ordering [15].

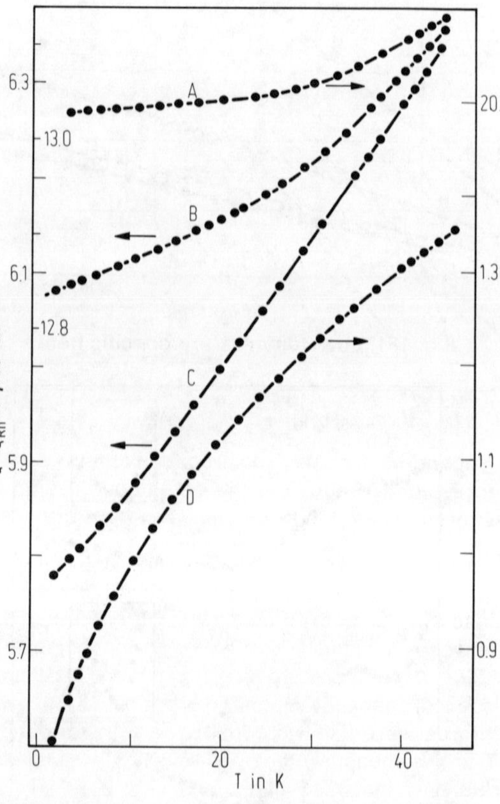

Fig. 133. Resistivity increments ($\varrho - \varrho_{Rh}$) for Rh-Mn alloys: A=14.2 at% Mn; B=Rh-9.4 at% Mn; C=Rh-4.6 at% Mn; D=Rh-1.0 at% Mn.

Néel Temperature. Figures for alloys containing 0.84, 1.50 and 2.14 at% Rh are given by [9].

Curie Temperature θ_P. Figures are given for alloys in the 50 to 68 at% Mn range in the following table due to [8] and [7].

at% Mn	50	54	56	58	60	65
θ_P in K	−93	150	293	325	403	497

Magnetic Susceptibility. Figures for χ^{-1} plotted against temperature are given by [6] and [7]. **Fig. 134** shows the figures by [6].

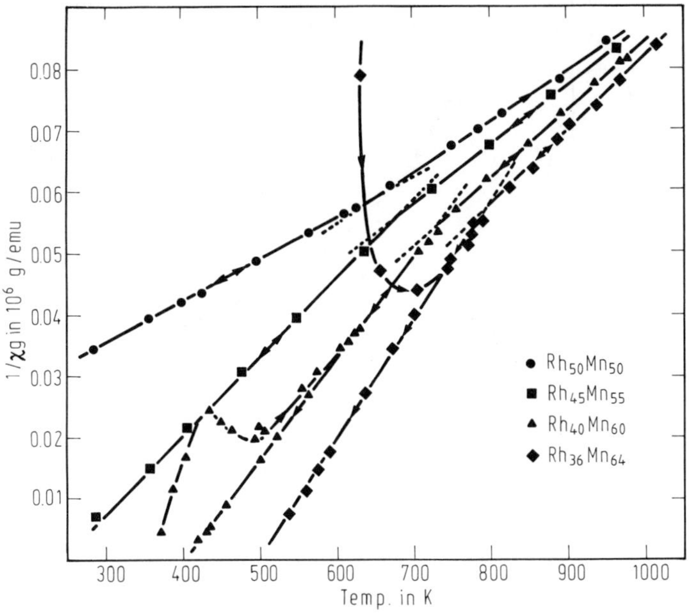

Fig. 134. Temperature dependence of the reciprocal magnetic susceptibility above room temperature for four representative alloys in the equiatomic region of the Rh-Mn system. The arrows indicate the direction of temperature change.

The results in this case are considered as curves with a change in slope at a transition temperature varying between 600 and 800 K with varying composition. Results showing a smooth curve in relation to temperature are given by [7]. The transition temperature is interpreted as being due to a change in magnetic order by [6]. Susceptibility measurements made on alloys in the β area were used to construct a transformation diagram (see figure in original paper). This shows measurements on an alloy with 62 at% Mn; in the heating cycle and at a temperature T_1 the $\beta_2+\beta_1$ becomes only β_1. At T_3 β_2 begins to form and the susceptibility rises to T_4 when only the β_2-phase exists. In the cooling curve the hysteresis in the transformation of the β_2-phase reaches T_2 before the transformation starts [8]. The magnetic susceptibility of several dilute transitional alloys of rhodium including two containing 0.2 and 0.6 at% manganese have been measured at temperatures between 4 and 800 K; the pronounced field dependence of $\Delta\chi$, coupled with specific heat anomalies suggested that the steep rise in $\Delta\chi$ was due to solute-solute interactions [10]. Measurements on alloys containing 9.4 and 14.2 at% manganese at temperatures between 4 and 120 K using fields up to 18 kOe are shown in **Fig. 135**, p. 200; the alloys were cooled in a field of 8 kOe [11].

Magnetization. The field and temperature dependence of the magnetization of $Mn_{61}Rh_{39}$ is shown in **Fig. 136**, p. 200. The measurements were made on polycrystalline samples in a field of 12 kOe. The various portions of the curve are explained as follows: section

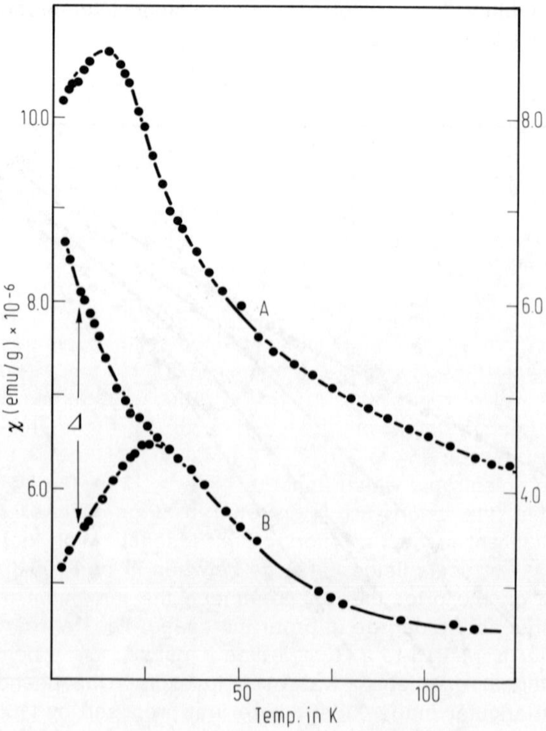

Fig. 135. Total susceptibility against temperature for two Rh-Mn alloys; A=Rh-9.4% Mn; B=Rh-14.2% Mn. The alloys were cooled in a field of 8 kOe.

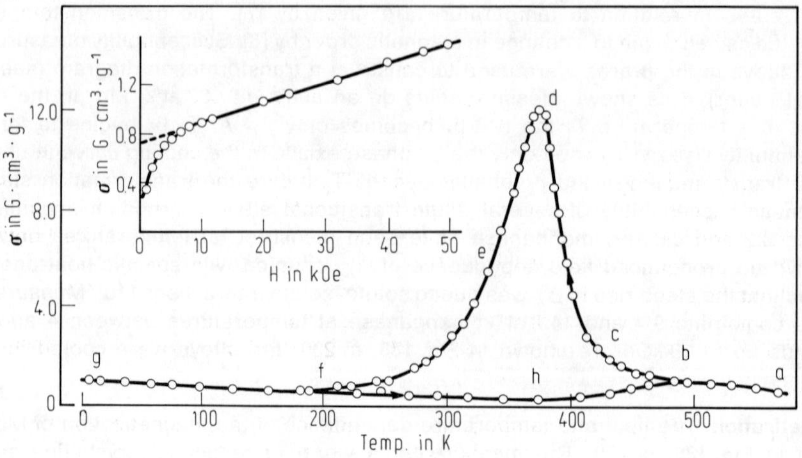

Fig. 136. Magnetization of $Mn_{61}Rh_{39}$.

ab corresponds to the paramagnetism of the β-phase; the rise in magnetization over cd is due to ordering of ferromagnetic clusters caused by non-stoichiometry from MnRh, and further cooling first converts the clusters from the superparamagnetic to the single-domain state and then causes the $\beta \rightarrow \gamma$ transition, the magnetization decreasing because the γ-phase is antiferromagnetic. The $\beta \rightarrow \gamma$ transition is extended along the temperature scale and ends only at f. In the section there is some increase in magnetization which is atypical of the antiferromagnetic state. The reheating curve was found to follow gfhba. The hypothesis that the rise in magnetization in the fg section might be due to the appearance of a noncollinear structure was confirmed by testing. However, textured samples were tested with the magnetic field along the easy and difficult magnetization axes [13].

The ferromagnetic properties of the β-phase are said to be caused by cells in which Rh atoms are replaced by Mn atoms, and this replacement is responsible for the noncollinear structure of the γ-phase; it is concluded that the weak ferromagnetism is due to the forces of exchange and is not of relativistic origin [13]. Alloys containing 36 to 46 at% rhodium were used to study the transition from the fct→bcc structure and antiferromagnetic→ferromagnetic transition between 254 to 376 K. The magnetic hysteresis loop exhibits an effect typical of exchange anisotropy, which appears not only in the $\gamma+\beta$ region but also where the β-phase is stable; the results are interpreted in terms of an exchange interaction between regions of different magnetic ordering [14]. Magnetization measurements on alloys of composition $Mn_{1-x}Rh_x$ using fields $\leqq$12 kOe between 77 and 700 K showed that the temperature depended on the prior thermal treatment, the maximum magnetization moving to lower temperatures as the degree of order increased. The degree of ordering is caused by an order-disorder transition to a cluster-type β-phase [15]. A density functional theory to describe noncollinear magnetism was used to explain the phenomenon in $RhMn_3$; it is shown that the triangular magnetic structure first proposed by [16] has the lowest total energy [17].

Magnetic Moment. The following table gives the effective magnetic moment μ_{eff} against Mn content [7].

at% Mn	50	54	56	58	60	65
μ_{eff} in μ_B	3.82	3.37	2.93	2.81	2.60	2.37

Neutron diffraction studies on γ-phase alloys containing 4 to 24 at% rhodium gave the magnetic moment per site between these composition limits as 1.6 to 2.0 μ_B [18].

References:

[1] Raub, E.; Mahler, W. (Z. Metallk. **46** [1955] 282/90).
[2] Hansen, M.; Anderko, K. (Constitution of Binary Alloys, 2nd Ed., McGraw-Hill, New York – Toronto – London 1958, p. 948).
[3] Hellawell, A. (J. Less-Common Metals **1** [1959] 343/7).
[4] Topchiashvili, L.I. (Tr. Inst. Prikl. Khim. Elektrokhim. Akad. Nauk Gruz. SSR **1** [1960] 51/61).
[5] Yao, Y.L. (Trans. Am. Inst. Mining Metall. Petrol. Eng. **224** [1962] 1146/53).
[6] Selte, K.; Bjerkelund, E.; Kjekshus, A.; Andresen, A.F.; Pearson, W.B.; Meisalo, V. (Acta Chem. Scand. **26** [1972] 719/32).
[7] Nakayama, Y.; Asanuma, M. (Japan. J. Appl. Phys. **4** [1965] 315/6).
[8] Kussmann, A.; Müller, K.; Wollenberger, H. (Z. Angew. Physik **20** [1965/66] 461/5).
[9] Williams, W.; Stanford, J.L. (J. Magn. Magn. Mater. **1** [1976] 271/85).
[10] Claus, H. (Phys. Rev. B [3] **5** [1972] 1134/43).

[11] Murani, A.P.; Coles, B.R. (J. Phys. F **2** [1972] 1137/44).
[12] Zavadskii, E.A.; Medvedeva, L.I. (Fiz. Tverd. Tela [Leningrad] **15** [1974] 2397/402; Soviet Phys. Solid State **15** [1974] 1595).
[13] Medvedeva, L.I.; Zavadskii, E.A. (Fiz. Tverd. Tela [Leningrad] **16** [1975] 3139/40; Soviet Phys. Solid State **16** [1975] 2033).
[14] Medvedeva, L.I. (Ukr. Fiz. Zh. **25** [1980] 1427/31).
[15] Medvedeva, L.I. (Ukr. Fiz. Zh. **26** [1981] 194/8).
[16] Kouvel, J.S.; Kasper, J.S. (J. Phys. Chem. Solids **24** [1963] 529/36).
[17] Kübler, J.; Höck, K.-H.; Sticht, J.; Williams, A.R. (J. Phys. F **18** [1988] 468/83).
[18] Cowlam, N.; Bacon, G.E.; Gillott, L.; Harmer, G.R.; Self, A.G. (J. Phys. F **9** [1979] 1387/407).

32.2 Ternary Alloys

Rh-Mn-As. Mössbauer spectroscopy, neutron and X-ray diffraction and magnetic measurements were used to examine the factors in the ordering of compounds of the type MM'As where M=3d transition metal and M'=4d transition metal (Rh). Four ordering factors; size factor, d-electron number, crystalline field, and s-d electron transfer are examined [27]. Magnetic measurements on material of composition RhMnAs at low temperatures show that between the Curie temperature T_c at 200 K and a transition temperature T_t of 158 K the alloy shows ferromagnetic behaviour with strongly field-dependent magnetization; at T_t it undergoes a first-order transition and is antiferromagnetic at low temperature. X-ray crystallography investigations failed to find any change in crystal structure to explain this. However, the lattice parameters show abrupt changes at T_t which are related to changes in electric resistivity [28]. The magnetic structures of RhMnAs are further discussed by [29].

Rh-Mn-Sb. These alloys were investigated by X-ray and magnetic analysis, in particular the compositions $Rh_{0.90}Mn_{1.2}Sb_{0.90}$ and $Rh_{1.10}Mn_{1.10}Sb_{0.80}$; these were found to have a $C1_b$-type structure and lattice parameters of 6.145 and 6.170 Å, respectively, at room temperature. The Curie temperatures were 325 and 350 K, and the saturation magnetizations 79.5 and 75.0 emu/g at absolute zero. Magnetic moments of 3.19 and 3.29 μ_B at absolute zero were reported; their reciprocal susceptibilities obeyed the Curie-Weiss law above the Curie temperature [22, 23]. The $C1_b$ alloy $Rh_{0.98}Mn_{1.04}Sb_{0.98}$ had a lattice parameter of a=6.142 Å, was ferromagnetic with a Curie temperature of 338 K and had a magnetic moment of 3.35 μ_B/Mn atom [19].

The crystal structures and magnetic properties of alloys of composition $Rh_1Mn_{1-x}Sb_x$ were examined in detail by [24, 25]. It was found that alloys in the composition range of $Rh_{1.00}Mn_{0.66}Sb_{0.34}$ to about $Rh_{1.00}Mn_{0.55}Sb_{0.45}$ are solid solutions of the fct $L1_0$ type with lattice parameters of 4.200 to 4.182 Å and axial ratios of 0.806 to 0.829. The typical $Rh_{1.00}Mn_{0.55}Sb_{0.45}$ alloy has a saturation magnetization of 52.3 emu/g at absolute zero, a magnetic moment of 3.20 μ_B per Mn atom, a Curie point of 330 K and a paramagnetic Curie point of 286 K. The observed relationship between reciprocal susceptibility and temperature indicates weak ferromagnetic behaviour [24, 25]. Rh_2MnSb was found to have a large tetragonal distortion of the Heusler structure with a=5.898, c=6.987 Å, a paramagnetic Curie temperature of 312 K and a ferromagnetic Curie temperature of 304 K [2]. The alloy $Rh_2Mn_{1.12}Sb_{0.88}$ was found to have the distorted $L2_1$, Heusler structure with a=5.85 Å and c/a=1.16 and a Curie temperature of 348 K; the magnetic moment was 3.42 μ_B [5]. The same alloy was examined by NMR and Mössbauer techniques and the hyperfine field at the tin impurity site found to be negative [26]. The same alloy has been examined crystallographically and by low field ac susceptibility measurements. The lattice parameters were

$a = 5.85$ Å, $c/a = 1.16$, the Curie temperature was 348 K, and the magnetic moment per formula unit 3.4 μ_B. The absence of the peak found on the susceptibility/temperature curves in other ferromagnetic Heusler alloys examined is discussed [6]. Mössbauer measurements on the hyperfine magnetic field at the Sb site in $Rh_2Mn_{1.12}Sb_{0.86}{}^{119}Sn_{0.02}$ are compared with results on similar Ge, Sn and Pb alloys [18].

Rh-Mn-Bi. The compound of composition $Rh(Mn_{0.8}Bi_{0.2})$ is reported as ferromagnetic with a Curie temperature of 185 K and a magnetic moment of 3.5 μ_B per formula unit [20]. A new class of magnetic Rh-Mn-Bi alloys, in particular $Rh_2Mn_5Bi_4$ with a face-centred cubic lattice have been examined; lattice parameter $a = 12.31$ Å, ferromagnetic Curie temperature −7 °C [21].

Rh-Mn-Al. A Heusler alloy of composition Rh_2MnAl with $L2_1$-type crystal structure is described by [19]. After homogenizing treatment at 900 °C for 100 h and cooling at 50 °C/h, the antiferromagnetic alloy was found to be single-phased, $a = 6.022$ Å, Néel temperature $T_N = 26$ K, magnetic moment $= 4.7$ μ_B [19].

The same alloy was found to have a B2-type structure with $2a = 6.005$ Å; the parametric Curie temperature was nonlinear. The Curie temperature was 85 to 105 K. The structure is a disordered Heusler one, with the Mn and Al atoms disordered. Magnetic measurements are given [2].

Rh-Mn-Ga. The alloy Rh_2MnGa was found to have a B2-type structure with $2a = 6.056$ Å, a paramagnetic Curie temperature of 60 K, and a Curie temperature of 75 to 85 K. The structure is a disordered Heusler one where the Mn and Ga atoms are disordered; magnetization and reciprocal susceptibility curves are given plotted against field strength and temperature, respectively [2].

Rh-Mn-In. The alloy Rh_2MnIn was found to have a B2-type structure with $2a = 6.287$ Å, a paramagnetic Curie temperature of 81 K, and a Curie temperature of 100 to 110 K. The structure is a disordered Heusler one where the Mn and In atoms are disordered; magnetization and reciprocal susceptibility curves are given plotted against field strength and temperature, respectively [2]. The same alloy has been examined to determine the hyperfine field at the In site using ^{119}Sn in $Rh_2MnIn_{0.98}Sn_{0.02}$ and Mössbauer techniques; the results at 77 K were 4.86(4), 19.7(2), 35.8(4) kOe and 0 at 293 K [18].

Rh-Mn-Tl. The alloy Rh_2MnTl was found to have a B2-type structure with $2a = 6.324$ Å, and a paramagnetic Curie temperature of 90 K. The structure is a disordered Heusler one where the Mn and Tl atoms are disordered; magnetization and reciprocal susceptibility curves are given plotted against field strength and temperature, respectively [2].

Rh-Mn-Ge. The compound Rh_2MnGe was found to have the $L2_1$ Heusler-type structure with $a = 6.030$ Å, a paramagnetic Curie temperature of 312 K, and a Curie temperature of 450 K. The effective moment per formula unit was 4.64 μ_B; the magnetization $\sigma = 77.5$ emu/g at 5 K and 6 kOe, and 64.2 at 300 K and 19 kOe [2]. TDPAC technique was used to examine the hyperfine fields in Rh_2MnGe at the nonmagnetic impurities Ru and Cd; measurement of the field at the Ru site used the 20.7 ns 90 keV state in ^{99}Ru populated in the decay of 16 d ^{99}Rh, see table below [15].

Hmf in kG	T in K	site	impurity
195 ± 11	77	Rh	Ru
126 ± 4	293	Rh	Ru
49 ± 4	355	Rh	Ru

The same $L2_1$ Heusler alloy with a Curie temperature of 450 K which X-ray diffraction measurements showed to be single-phased was examined to determine its hyperfine magnetic field. The technique used involved the substitution of 2 at% Ge by In and using the reaction $In^{113}(p, 3n)Sn^{111} \rightarrow In^{111} \rightarrow Cd^{111}$ induced in a cyclotron producing 30 MeV protons. The hyperfine magnetic fields were determined by the TDPAC technique utilizing the 247 keV excited state of Cd^{111}. For Rh_2MnGe, $B = 188 \pm 6$ and 160 ± 3 kG at 77 and 293 K [4].

Similar TDPAC work using Cd as an impurity at the Ge site gave a value for the hyperfine field at 0 K as -203 ± 3 kOe [7]. Mössbauer and TDPAC techniques on the same alloy doped with Sn were used for the former method and In for the latter. The lattice parameter was determined as a = 6.030(5) Å, and the hyperfine fields at 0 K at the Cd and Sn impurity sites −198(3) and −47(2) kOe, respectively [8].

Tin hyperfine fields were measured using Mössbauer spectroscopy in the Heusler alloy Rh_2MnGe; 2 at% Sn was substituted for germanium. The results are given in table below. The most probable Sn hyperfine field and the isomer shift δ are shown. All Sn isomer shifts are relative to $CaSnO_3$ [16].

T in K	H in kOe	δ in mm/s
77	-48 ± 1	1.43 ± 0.02
295	-42 ± 1	1.41 ± 0.02

The crystallographic and magnetic properties of the compound RhMnGe were determined by [30]. The structure was found to be C22-type with a = 6.55, c = 3.56 Å; the Curie temperature was 636 K and the paramagnetic Curie temperature 675 K. The effective Bohr magneton number was 4.4 μ_B/Mn atom [30].

The magnetic properties of the same compound were investigated using magnetometric measurements; it was found to be ferromagnetic at low temperatures with a Curie temperature of 622 K and a magnetic moment of 3.56 μ_B [17].

Rh-Mn-Sn. The structure and magnetic properties of Rh_2MnSn have been determined and are given below [2, 9]; the figures in parentheses refer to [2]. Structure at 20 °C: Heusler structure $L2_1$, a = 6.242 (6.252), c = 6.242 Å, c/a = 1;

$\theta = 413$ (412) K, $T_c = 412$ K, $\mu = 3.10\ \mu_B$ [1], $\mu_1 = 3.30\ \mu_B$ [2]

[1] moment per formula unit measured at 4.2 K and 66 kOe; [2] moment per formula unit calculated from $2\sqrt{(\mu/2)(1+\mu)}$

Curves for the magnetization plotted against applied field and the inverse susceptibility against temperature (70 to 670 K) are shown for Rh_2MnX alloys where X = Sn, In, Ga, Al, Pb and Tl [2].

Rh_2MnSn has a Heusler-type structure with a = 6.252 Å, a Curie temperature of 410 K [10]. Mössbauer measurements on Rh_2MnSn on the Sn site hyperfine field gave the following figure, +21 kOe at 77 K; a relationship is indicated between the Heusler alloys and several body-centred cubic binary alloys of iron [11]. According to hyperfine field measurements on the same compound, made using the spin-echo method, the field at the Sn site is negative (−26 kOe) [12]. Other work on hyperfine fields using ruthenium as a dilute impurity, yielded figures of 190.4 ± 6.0 kG at 77 K and 136.7 ± 4.3 kG at 293 K at the rhodium site [13]. Mössbauer studies using a single-line $Ca^{119}SnO_3$ source gave a hyperfine field of $\pm 17.4 \pm 0.5$ kOe at 300 K; the Curie temperature was found to be 420 K [14]. The hyperfine magnetic fields for Rh_2MnSn determined by TDPAC (Time Differential Perturbed Angular Correlation) techniques are given below [15].

Hmf using ^{111}Cd at Sn site [9].

B in kG	189±5	113±5	33±3
T in K	77	293	373

Using ^{99}R at Rh site, B = 137 ± 2 kG at 293 K, 37 ± 6 at 355 K [15].

The hyperfine field at the tin site measured by Mössbauer spectroscopy in the same compound was found to be 31 ± 1 kOe at 77 K and 18 ± 1 kOe at 295 K [16]. Mössbauer and TDPAC measurements of the hyperfine fields in a variety of different Heusler alloys gave a figure for the field at the tin site in Rh_2MnSn of 21 kOe at ~0 K [7].

Measurements of the low field ac susceptibility of a number of ferromagnetic Heusler alloys including Rh_2MnSn showed peaks when plotted against temperatures in the range 100 to 500 K; these could be taken to imply spin glass behaviour, but the authors considered that it probably was a consequence of a rapid build-up of an anisotropy field below the ordering temperature [5, 6].

Rh-Mn-Pb. The compound Rh_2MnPb was found to have the fully ordered $L2_1$-type structure of the Heusler alloy with lattice parameter a = 6.332 Å and a Curie temperature of 338 K. This alloy was investigated with several others of the general formula Rh_2MnX, where X = Sn, Tl, In, Ga, Al. Magnetization and susceptibility plotted against field strength and temperature are shown for these alloys in **Fig. 137** and **Fig. 138**, p. 206. Differences in crystal

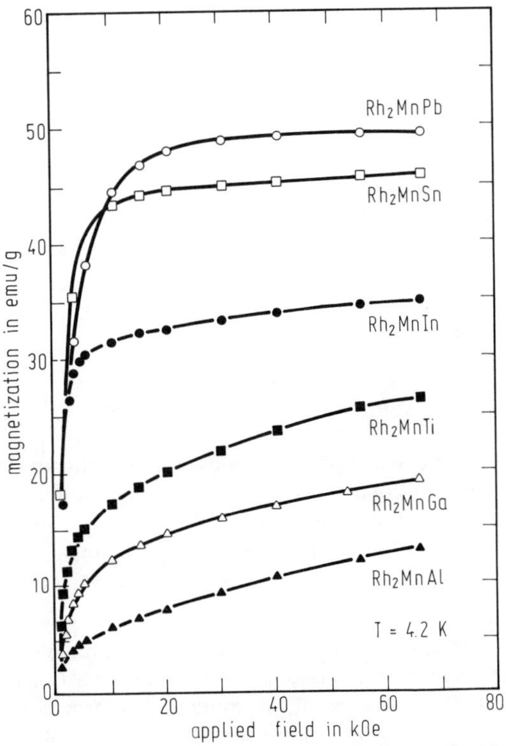

Fig. 137. Magnetization vs. applied magnetic field for Rh_2MnX compounds.

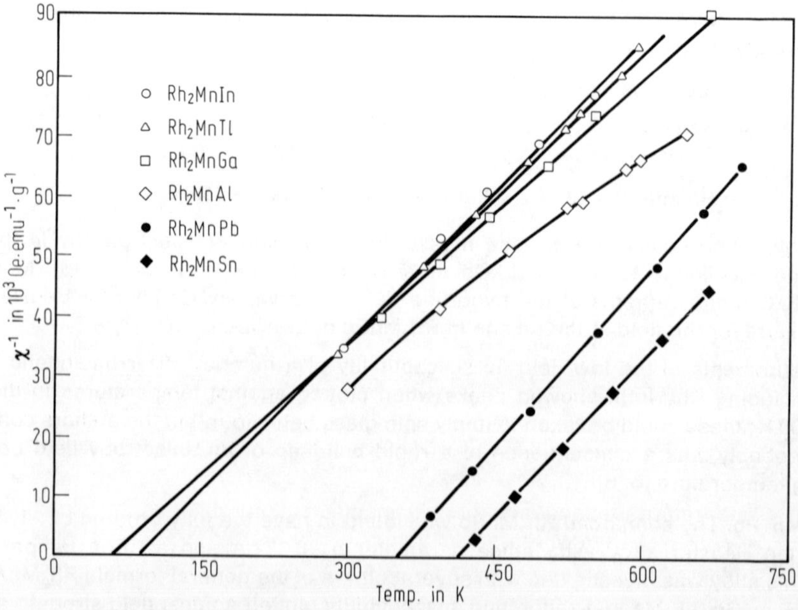

Fig. 138. Reciprocal susceptibility as a function of temperature for Rh_2MnX compounds.

structure were attributed to electron concentration, and differences in magnetic properties to the presence of AF (antiferromagnetic) Mn-Mn interactions caused by variations in atomic order [2].

Mössbauer and NMR spin-echo spectra were used to study the hyperfine field of tin as a dilute component in Rh_2MnPb and at all three nuclear sites. A Curie temperature of 335 K was found and this hyperfine field data: H(Rh) 278 kOe, 205 kOe (Pb site), H(Mn) 394 kOe, H(Sn) 25 kOe, H(Pb) 53 kOe. The signs of HF at Pb and Rh could not be found [3]. Measurements on the hyperfine magnetic field of Rh_2MnPb have been made after substituting 2 at% Pb with In and bombarding with 30 MeV protons; hyperfine fields were measured by the TDPAC technique utilizing the 247 keV excited state of ^{111}Cd. The compound is a Heusler alloy of crystal type $L2_1$ [4].

B in kG	143 ± 2	114 ± 2	84 ± 2	53.1 ± 1
T in K	77	200	273	293

The low field ac magnetic susceptibility measured on Heusler alloys Rh_2MnPb and Rh_2MnSn showed peaks which could be interpreted as indications of spin build-up of an anisotropy field below the ordering temperature. The coercive field H_c for Rh_2MnPb was 70 Oe and 260 Oe at 300 and 77 K, respectively; the magnetic moment per formula unit was 2.9 μ_B [5, 6]. Further Mössbauer and TDPAC work has been carried out on the hyperfine fields in Rh_2MnPb by [7, 8].

Rh-Mn-Cr. Alloys of composition $Mn_{58}Rh_{42-x}Cr_x$ $(0 \leqq x < 8)$ have been studied in an investigation of the magnetic change in equiatomic alloys from the CuAu γ-phase to the CsCl β-phase. The temperature dependence of the magnetization in alloys where $x = 0$ to 6 is shown in **Fig. 139**. The doping strongly increases the maximum magnetization and lowers the transition temperature of the AF$\rightarrow$F magnetic change. The field dependence of the

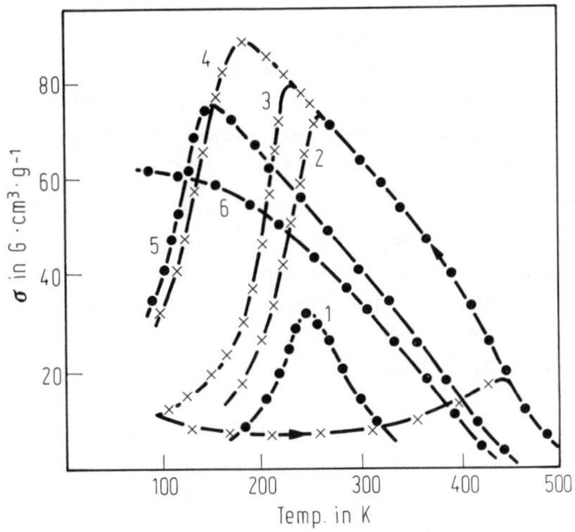

Fig. 139. Temperature dependence of the magnetization obtained for samples of the $Mn_{58}Rh_{42-x}Cr_x$ system with different values of x : 1) 0; 2) 1; 3) 2; 4) 4; 5) 6; 6) 8.

magnetization suggests that at a certain temperature below the Curie point superparamagnetic behaviour is shown, changing to ferromagnetism at lower temperatures; this is interpreted as showing the existence of superparamagnetic clusters. It is considered that the addition of chromium increased the density of the valence electrons, stabilizing the body-centred cubic lattice [1].

References:

[1] Medvedeva, L.I. (Fiz. Tverd. Tela [Leningrad] **17** [1975] 3446/8; Soviet Phys. Solid State **17** [1975] 2257/8).
[2] Suits, J.C. (Phys. Rev. B [3] **14** [1976] 4131/5).
[3] Pillay, R.G.; Grover, A.K.; Tandron, P.N.; Khoi, L.D.; Veillet, P. (Proc. Nucl. Phys. Solid State Phys. Symp. C **21** [1978] 622/4).
[4] Jha, S.; Black, R.D.; Julian, G.M.; Blue, J.W.; Liu, D.C. (J. Appl. Phys. **50** [1979] 7507/9).
[5] Dhar, S.K.; Grover, A.K.; Malik, S.K.; Vijayaraghavan, R. (Proc. Nucl. Phys. Solid State Phys. Symp. C **21** [1978] 549/51).
[6] Dhar, S.K.; Grover, A.K.; Malik, S.K.; Vijayaraghavan, R. (Solid State Commun. **33** [1980] 545/7).
[7] Jha, S.; Seyoum, H.M.; Demarco, M.; Julian, G.M.; Stubbs, D.A.; Blue, J.W.; Silva, M.T.X.; Vasquez, A. (Hyperfine Interact. **16** [1983] 685/8).
[8] Jha, S.; Seyoum, H.M.; Julian, G.M.; Dunlap, R.A.; Vasquez, A.; da Cunha, J.G.M.; Ramos, S.M.M. (Phys. Rev. B [3] **32** [1985] 3279/81).
[9] Suits, J.C. (Solid State Commun. **18** [1976] 423/5).
[10] Uhl, E. (J. Magn. Magn. Mater. **49** [1985] 101/5).

[11] Campbell, C.C.M.; Birchall, T.; Suits, J.C. (J. Phys. F **7** [1977] 727/43).
[12] Itoh, J.; Shimitzu, K.; Mizutani, H.; Grover, A.K.; Gupta, L.C.; Vijayaraghavan, R. (J. Phys. Soc. Japan **42** [1977] 1777/8).
[13] Pillay, R.G.; Tandon, P.N.; Devare, H.G. (Solid State Commun. **23** [1977] 439/41).
[14] Pillay, R.G.; Tandon, P.N. (Phys. Status Solidi A **45** [1978] K109/K112).
[15] Jha, S.; Saleh, N.; Ward, T.E.; Blue, J.W.; Liu, D.C. (J. Appl. Phys. **50** [1979] 2069/71).

[16] Dunlap, R.A.; Jha, S.; Julian, G.M. (Can. J. Phys. **62** [1984] 396/9).
[17] Bazela, W. (J. Less-Common Metals **100** [1984] 341/6).
[18] Jha, S.; Seyoum, H.M.; Yehia, S.; Mitros, C.; Julian, G.M.; Dunlap, R.A.; Vasquez, A.; da Cuna, J.G.M.; Ramos, S.M.M. (Hyperfine Interact. **28** [1986] 491/4).
[19] Masumoto, H.; Watanabe, K. (J. Phys. Soc. Japan **32** [1972] 281).
[20] Suits, J.C. (IBM J. Res. Develop. **19** No. 4 [1975] 422/3).

[21] Street, G.B.; Suits, J.C.; Lee, K. (Solid State Commun. **14** [1974] 33/6).
[22] Masumoto, H.; Watanabe, K. (Nippon Kinzoku Gakkaishi **36** [1972] 680/5).
[23] Masumoto, H.; Watanabe, K. (Trans. Japan Inst. Metals **14** [1973] 177/82).
[24] Masumoto, H.; Watanabe, K. (Trans. Japan Inst. Metals **17** [1976] 303/7).
[25] Masumoto, H.; Watanabe, K. (Nippon Kinzoku Gakkaishi **39** [1975] 1065/9).
[26] Pillay, R.G.; Grover, A.K.; Tandon, P.N.; Khoi, Le, D.; Veillet, P. (J. Magn. Magn. Mater. **15/18** [1980] 647/8).
[27] Sénateur, J.P.; Fruchart, D.; Boursier, D.; Rouault, A.; Montreuil, J.R.; Deyris, B. (J. Phys. Colloq. [Paris] **38** [1977] C7-61/C7-66).
[28] Chenevier, B.; Fruchart, D.; Bacmann, M.; Sénateur, J.P.; Chaudouet, P.; Lundgren, L. (Phys. Status Solidi A **84** [1984] 199/206).
[29] Chenevier, B.; Bacmann, M.; Fruchart, D.; Sénateur, J.P.; Fruchart, R. (Phys. Status Solidi A **90** [1985] 331/41).
[30] Masumoto, H.; Watanabe, K.; Mitera, M. (J. Phys. Soc. Japan **34** [1973] 1414).

33 Alloys with Nickel

33.1 The Rh-Ni System

Phase Diagram

The diagram showing uninterrupted solid solution is due to [1]. The liquidus and solidus points are based on measurements by [2, 3, 4]. The dotted miscibility gap boundary shown in **Fig. 140** is deduced from thermodynamic and magnetic data from [5 to 10].

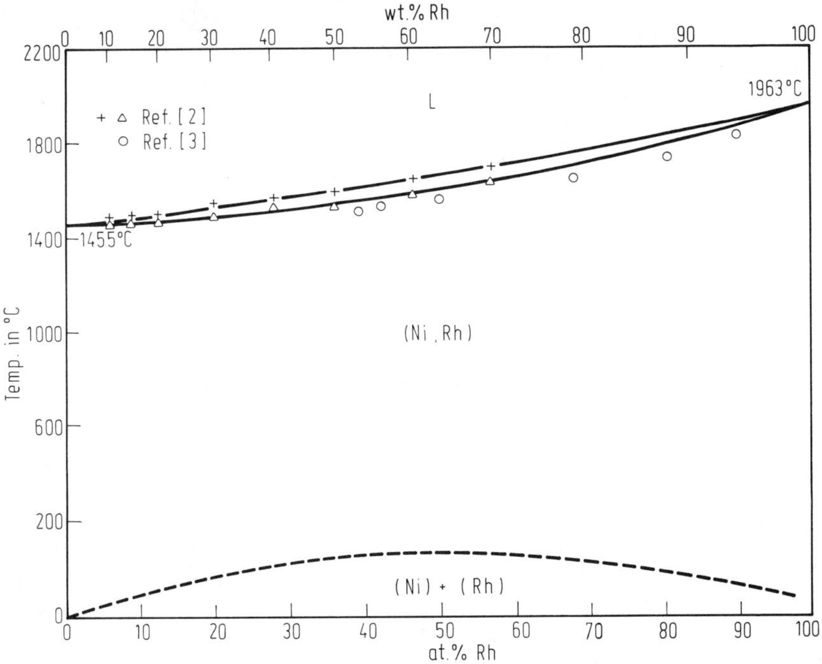

Fig. 140. Phase diagram Rh-Ni.

References:

[1] Nash, A.; Nash, P. (Bull. Alloy Phase Diagrams **5** [1984] 403/5).
[2] Kornilov, I.I.; Myasnikova, K.P. (Izv. Akad. Nauk Met. **1965** No. 2, pp. 175/9; Russ. Metall. **1965** No. 2, pp. 107/9).
[3] Mit'ko, M.M.; Dubinin, E.L.; Timofeev, A.I.; Chegodaev, A.I. (Izv. Vysshikh Uchebn. Zaved. Tsvetn. Metall. **1978** No. 3, pp. 84/8; Soviet Non-Ferrous Metals. Res. **3** [1978] 115/7).
[4] Melting Points of the Elements (Bull. Alloy Phase Diagrams **2** [1981] 145/6).
[5] Pratt, J.N.; Bird, J.M.; Martosudirdjo, S. (AD-786609-8GA [1974] 22/9; C.A. **83** [1975] No. 14569).
[6] Pratt, J.N. (CALPHAD VII **2** [1978] 203/4).
[7] Miedema, A.R.; Boom, R.; deBoer, F.R. (J. Less-Common Metals **41** [1975] 283/98).
[8] Bucher, E.; Brinkman, W.F.; Maita, J.P.; Williams, H.J. (Phys. Rev. Letters **18** [1967] 1125/9).
[9] Hahn, A.; Wohlfarth, E.P. (Helv. Phys. Acta **41** [1968] 857/68).
[10] Triplett, B.B.; Phillips, N.E. (Phys. Letters A**37** [1971] 443/4).

Crystallographic Properties

The lattice parameter/composition diagram given in **Fig. 141** is by [1]. The measurements were made by [2 to 5].

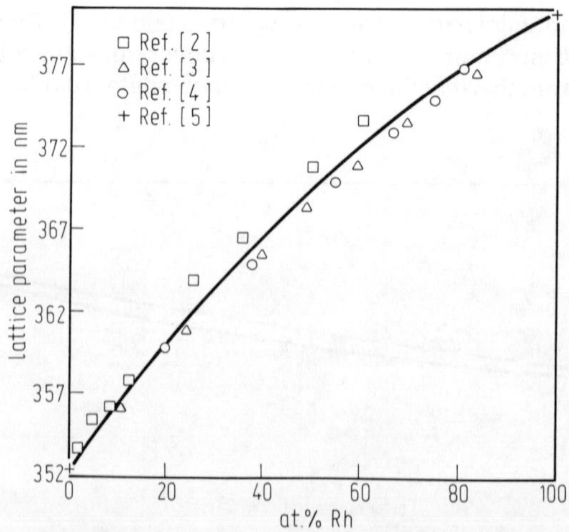

Fig. 141. Lattice parameter of Rh-Ni alloys.

Measurements of diffuse X-ray scattering established that short-range order existed in these alloys; agreement between calculated and experimental values of the SRO parameters was obtained [6]. Similar measurements found the short-range order (SRO) parameters α_1 to have anomalously large negative values; theoretically calculated values were found to be in agreement [7, 8]. The theoretical pseudopotential and experimental diffuse X-ray scattering methods were used to study the short-range order in 18, 33 and 48 at% Rh alloys, its existence was established in all three. It was found that the experimental and calculated values of α_1 agreed in sign [9]. Alloys of rhodium and nickel are considered not to be random solid solutions, but to have varying degrees of clustering or short-range order depending on the heat treatment given; by systematically changing the degree of clustering, information can be obtained on the magnetic phase diagram [10].

References:

[1] Nash, A.; Nash, P. (Bull. Alloy Phase Diagrams **5** [1984] 403/5).
[2] Kornilov, I.I.; Myasnikova, K.P. (Izv. Akad. Nauk SSSR Met. **1965** No. 2, pp. 175/9; Russ. Metall. **1965** No. 2, pp. 107/9).
[3] Hofer, F. (Solid State Commun. **45** [1982] 303/8).
[4] Luo, H.L.; Duwez, P. (J. Less-Common Metals **6** [1964] 248/9).
[5] Phillips, W.L. (Trans. Am. Inst. Mining. Metall. Petrol. Eng. **230** [1964] 526/9).
[6] Katsnel'son, A.A.; Silonov, V.M.; Khawaja, F.A.; Mekhrabov, A.O.O.; Prozorov, A.N. (Acta Crystallogr. A **34** Suppl. [1978] S 325).
[7] Abbas, T.; Khawaja, F.A. (IC-83/79 [1983] 17 pp.).
[8] Abbas, T.; Khawaja, F.A. (Solid State Commun. **49** [1984] 641/4).
[9] Katsnel'son, A.A.; Prozorov, A.N.; Silonov, V.M. (Fiz. Met. Metalloved. **57** No. 5 [1984] 985/92; Phys. Metals Metallogr. [USSR] **57** No. 5 [1984] 135/41).
[10] Crane, S.; Carnegie, D.W.; Claus, H. (J. Appl. Phys. **53** [1982] 2179/84).

Mechanical Properties

Density, Surface Tension (Melting Point). The density D and surface tension γ at the indicated melting point mp is related to composition in the table below [1].

at% Ni	γ in mJ/m^2	D in g/cm^3	mp in °C
9.9	1900	10.5	1830
19.1	1890	10.25	1740
31.7	1852	9.45	1650
49.9	1815	9.52	1565
57.7	1805	9.24	1530
60.8	1800	9.15	1512
79.8	1762	8.67	1490

Hardness. The hardness of an alloy containing 1 wt% nickel in the as-cast state is 134 kg/mm^2 according to [2]. The hardness of alloys containing between 2.3 and 61.08 at% rhodium was studied at temperatures between 20 and 1000 °C by [3].

Tensile Properties. Alloys containing 0.57, 1.7 and 5.95 at% rhodium were tested at room and elevated temperatures. The room-temperature yield stress is plotted against composition in **Fig. 142a**, and the UTS (σ_B) and elongation in **Fig. 142b**.

The yield stress, UTS and elongation % of the same alloys tested at 500, 800 and 1000 °C are shown in **Fig. 143**, p. 212 [4].

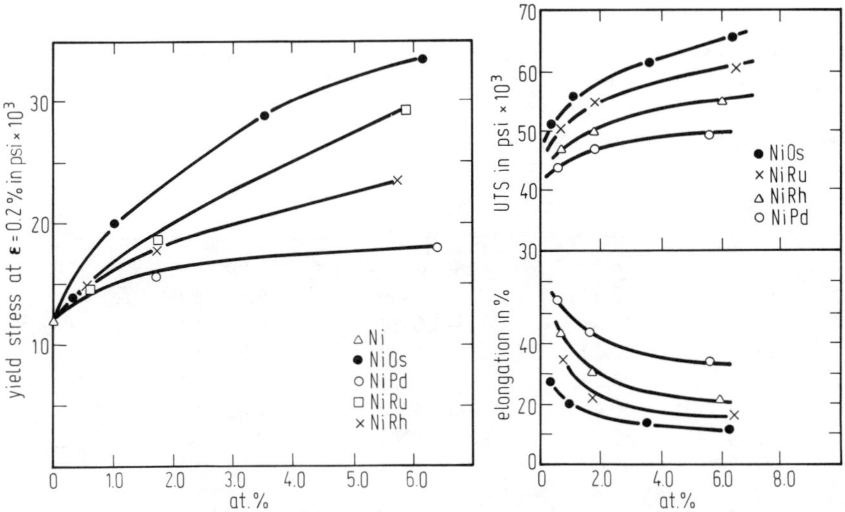

Fig. 142. a) Room-temperature yield strength of nickel and the nickel alloys listed as a function of atomic percent platinum group element. b) Room-temperature ultimate tensile strength and elongation of nickel and the nickel alloys listed as a function of atomic percent platinum group element.

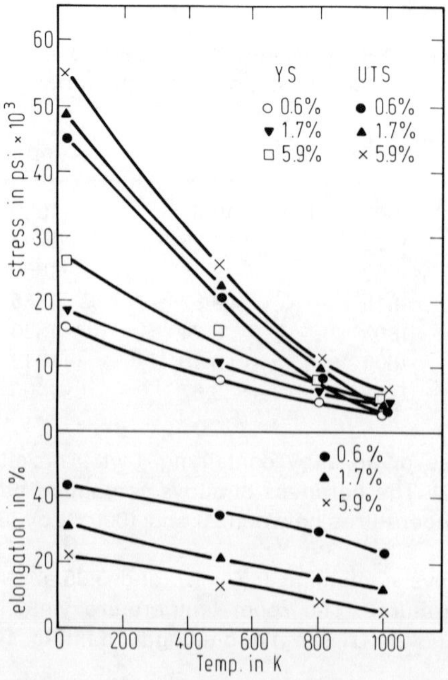

Fig. 143. Yield strength, ultimate tensile strength, and elongation of Ni-Rh alloys as a function of temperature.

Creep Properties. Some estimate of creep rates has been made by [3] by measuring the lengths of hardness indentation diagonals whilst heating at 800 °C in vacuum for 1, 5, 10 and 20 min; the results are shown in the table below. The creep rate is V_d; for the derivation of the creep rate see original paper [3].

No.	at% Rh	ln V_d	V_d
2	4.85	+1.0611	2.89
3	8.90	−0.0889	9.15×10^{-1}
4	12.05	−0.6866	5.03×10^{-1}
5	17.60	−4.1389	1.59×10^{-2}
6	26.30	−6.2289	1.97×10^{-3}
7	36.40	−16.8762	4.68×10^{-8}
8	55.50	−7.1278	8.02×10^{-4}
9	61.08	−6.6533	1.29×10^{-3}

Stress-rupture tests were made on the same alloys used for the tensile tests; the tests were carried out in air at 650 and 800 °C, the stress level being 10000 psi (7.031 kg/mm²) at the lower temperature and 5000 psi (3.516 kg/mm²) at the higher one. The time to rupture plotted against composition is shown in **Fig. 144** [4].

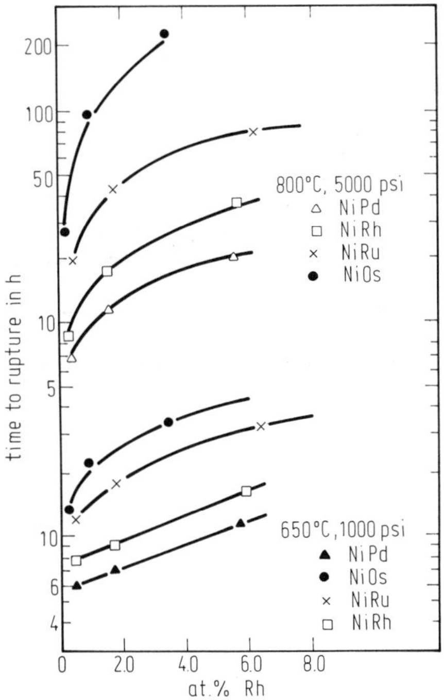

Fig. 144. Rupture time at the stress and temperature indicated of nickel alloys as a function of atomic percent platinum group element.

References:

[1] Mit'ko, M.M.; Dubinin, E.L.; Timofeev, A.I.; Chegodaev, A.I. (Izv. Vysshikh Uchebn. Zaved. Tsvetn. Metall. **1978** No. 3, pp. 84/8; Soviet Non-Ferrous Metals Res. **3** [1978] 115/7).

[2] Handley, J.R. (Platinum Metals Rev. **33** [1989] 64/72).

[3] Kornilov, I.I.; Myasnikova, K.P. (Izv. Akad. Nauk SSSR Metall. Gorn. Delo **1963** No. 6, pp. 146/51; Russ. Metall. Mining **1963** No. 6, pp. 75/83).

[4] Phillips, W.L. (Trans. Am. Inst. Mining Metall. Petrol. Eng. **230** [1964] 526/9).

Thermal Properties

Melting Point. Composition/melting-point figures are given by [1], see table on p. 211.

Specific Heat. Most measurements have been made on alloys containing sufficient nickel to be ferromagnetic, and most have been made at low temperature to probe the electronic structure. Such measurements were made to confirm a theoretical prediction that interaction between low-energy spin electrons and conduction electrons would result in increased specific heat in a 63 at% nickel-rhodium alloy [2]. Similar tests on a wider range of alloys were carried out by [3]. Both sets of figures are shown in **Fig. 145**, p. 214, **Fig. 146**, p. 215. A theoretical analysis of existing data suggested an alternative explanation for the observed anomaly proposing the formation of clusters of nickel-rich particles which behave superparamagnetically [4].

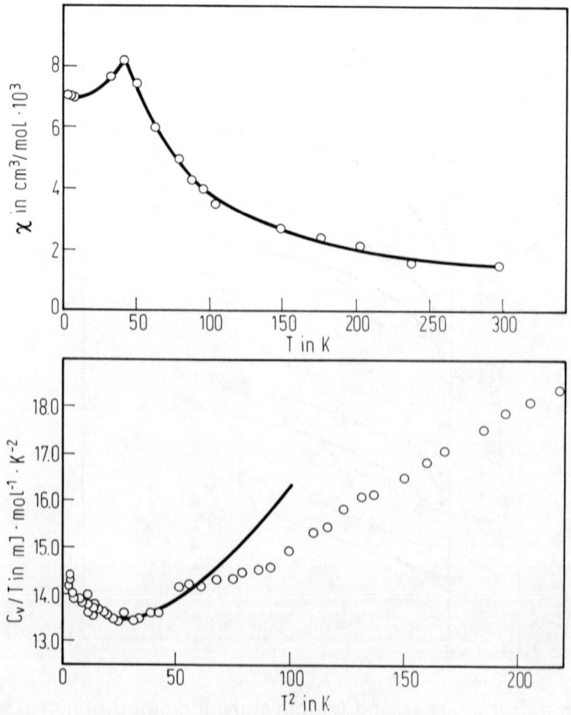

Fig. 145. The magnetic susceptibility and specific heat of $Ni_{0.63}Rh_{0.37}$, the sample with the largest enhancement effects.

Further investigation of a 62% nickel alloy showed that the anomalous contribution to the heat capacity was strongly dependent on the magnetic field and this was interpreted as strong evidence for the cluster theory [5]. Two theoretical papers calculate the specific heat coefficient γ, the first from the relationship $\gamma = \varrho_{\uparrow}(E'_F) + \varrho_{\downarrow}(E'_F)$ where (E'_F) is the Fermi energy in the alloy at concentration c, and $\varrho_{\uparrow}(E'_F)$ the density of states in the majority-spin d-band; $\varrho_{\downarrow}(E'_F)$ is the density of states in the minority-spin band in the neighbourhood of E_F. In the second paper the magnetization and specific heat coefficients were investigated as a function of concentration. Plots of $\gamma(c)$, $\varrho_{\uparrow}(E'_F)$ and $\varrho_{\downarrow}(E'_F)$ against concentration are shown in a figure [6, 7].

References:

[1] Mit'ko, M.M.; Dubinin, E.L.; Timofeev, A.I.; Chegodaev, A.I. (Izv. Vysshikh Uchebn. Zaved. Tsvetn. Metall. **1978** No. 3, pp. 84/8; Soviet Non-Ferrous Metals Res. **3** [1978] 115/7).

[2] Bucher, E.; Brinkman, W.F.; Maita, J.P.; Williams, H.J. (Phys. Rev. Letters **18** [1967] 1125/9).

[3] Brinkman, W.F.; Bucher, E.; Williams, H.J.; Maita, J.P. (J. Appl. Phys. **39** [1968] 547/8).

[4] Hahn, A.; Wohlfarth, E.P. (Helv. Phys. Acta **41** [1968] 857/68).

[5] Triplett, B.B.; Phillips, N.E. (Phys. Letters A **37** [1971] 443/4).

[6] Jacobs, R.L.; Takahashi, Y. (J. Phys. F **12** [1982] 529/35).

[7] Takahashi, Y.; Jacobs, R.L. (J. Magn. Magn. Mater. **31/34** [1983] 49/50).

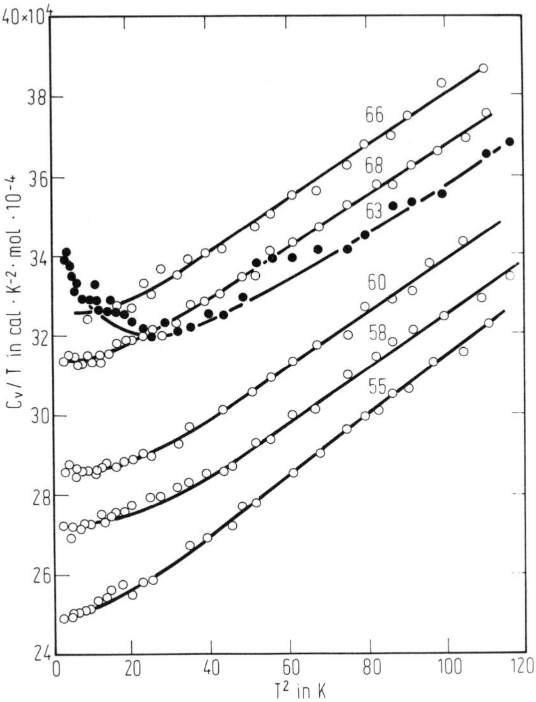

Fig. 146. The specific heat C_v/T plotted versus T^2. The numbers near each curve give the Ni concentration. The filled-in circles are $Ni_{0.63}Rh_{0.37}$. The lines are smooth curves drawn through the data to facilitate viewing.

Electrical and Magnetic Properties

Electronic Density of States. X-ray isochromat spectroscopy has been used to investigate the electronic density of states in the Rh-Ni alloy system; the measured isochromats could be described by a superposition model based on the density of states of the component metals [1].

Miscellaneous Magnetic Properties. NMR studies on $Ni_{1-x}Rh_x$ alloys and measurements of the Knight shift K, spin-lattice relaxation time T_1 and inhomogeneous linewidth Δ have been carried out by [2]. The results are shown in the table below.

x	K in %	$T_1 \cdot T$ in s·K	Δ in %
0.80	$+0.0 \pm 0.04$	7.1 ± 0.7	0.8 ± 0.1
0.60	-1.3 ± 0.1	5.0 ± 0.5	2.3 ± 0.2
0.50	-2.7 ± 0.3	3.5 ± 0.3	5 ± 1
0.45	-5.1 ± 0.6	2.3 ± 0.2	6 ± 1
0.42	-6.6 ± 0.9	2.4 ± 0.2	9 ± 2
0.40	-8.1 ± 1.0	1.2 ± 0.3	12 ± 2
0.38	-11.6 ± 0.6	1.8 ± 0.3	6 ± 1

Curie Temperature. Measurements have been made by [3, 4]. The results were plotted against rhodium content; except at the lowest rhodium level there is excellent agreement. Experiments on the effect of pressure on T_c have been made by [5] and [6].

Magnetization. This was measured in a field of H = 14240 G for alloys of varying composition between 0 and 300 K; **Fig. 147** gives the results [7].

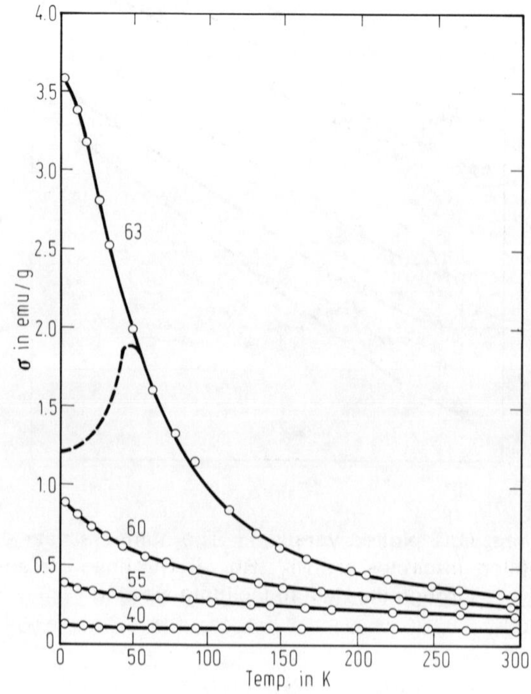

Fig. 147. The magnetization σ at H = 14240 G as a function of temperature. The numbers give the nickel concentration. The dotted line is the magnetization of $Ni_{0.63}Rh_{0.37}$ with the remanence subtracted off.

The spontaneous magnetization of some Rh-Ni alloys was measured by [3]. The results are given in the table below [3].

at% Rh	1.59	3.46	6.63	10.09	19.69	29.82
σ in cgs units/g . . .	59.5	59.0	56.3	53.6	39.1	18.3

Detailed measurements of the magnetization of Rh-Ni alloys have been made as a function of temperature and magnetic field around the critical concentration for the onset of ferromagnetism. It is shown that the magnetization of these disordered alloys can be described by a model based on two pieces of evidence, the presence of superparamagnetic clusters and the power of the Landau theory in explaining the properties near the magnetic phase transition [15]. A theoretical treatment of the magnetization of Rh-Ni alloys as a function of impurity concentration is given by [16]. Experimental values of the magnetization σ plotted against magnetic field were determined for alloys containing between 51 and 65 at% nickel; these are shown in **Fig. 148** [17].

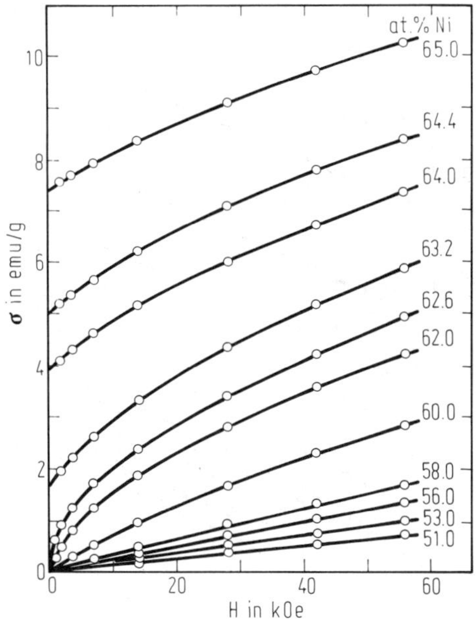

Fig. 148. Magnetization vs. field at 4.2 K for Ni-Rh alloys (compositions in at% Ni).

Bohr Magneton. The number of magnetons p related to rhodium content is given by [2, 7]. The results are shown in the following tables by [3] and [8].

at% Rh	1.59	3.46	6.63	10.09	19.69	29.82	–	–
p	0.633	0.636	0.621	0.606	0.472	0.235	–	–

Rh %	2.0	4.0	7.0	10.0	20.0	24.0	28.0	32.0
p	0.624	0.628	0.620	0.604	0.484	0.398	0.280	0.171

Magnetic Moment. The magnetic moment distribution in ferromagnetic Ni-Rh alloys was determined from the diffuse scattering of polarized and unpolarized neutrons; the average Ni moment remains near 0.6 μ_B up to 12 at% Rh and then decreases towards zero at the critical concentration of 37 at%. There is an initial rapid decrease in the rhodium moment that follows a P_{12} dependence which corresponds to a moment of 2 μ_B for isolated Rh atoms. The data indicate moment fluctuations at both the Ni and Rh sites that are associated with local environment [9, 10]. A detailed theoretical analysis of the distribution of magnetic moment in these alloys is given by [11]. The magnetocrystalline anisotropy of the alloys has been studied by [12, 13]. Oscillations of the Mössbauer line shift near the Curie temperature have been examined; it was concluded that it is probably connected with the distribution of the atomic magnetic moments [14].

Magnetic Susceptibility. Theoretical calculations of susceptibility have been made by [18, 19, 20]; the volume paramagnetostriction has been calculated by [21]. The magnetic properties of alloys containing 51 to 65 at% rhodium are given by [17]. **Fig. 149**, p. 218, shows the inverse initial susceptibility-temperature curves [17].

Inverse susceptibility-temperature curves for alloys containing between 2 and 32 at% Rh are shown in a figure by [8]. The susceptibilities of alloys near the critical composition

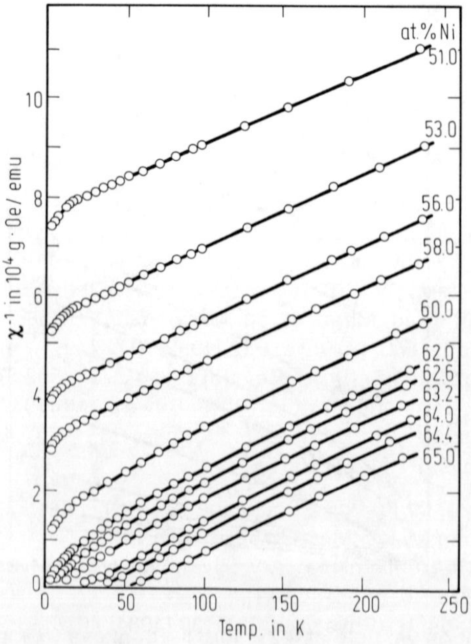

Fig. 149. Inverse initial susceptibility vs temperature for Ni-Rh alloys (compositions in at% Ni).

for ferromagnetism (63 at%) containing 58, 60 and 62 at% Ni were measured and each resolved into a Curie-Weiss component and a mildly temperature-dependent component; the approach to ferromagnetism arises entirely from the Curie-Weiss component which is attributed to a dilute concentration of superparamagnetic polarization clouds with giant moments $\sim 25\ \mu_B$ per cloud [22]. The inverse susceptibility plotted against temperature for alloys containing between 1.7 and 16.3 wt.% rhodium is shown in a figure by [23]. Low-field AC susceptibility measurements have been determined in relation to "giant" effects of metallurgical factors on the magnetic ordering temperature which were thought to be associated with the degree of atomic short-range order. By careful control of the latter, the existence of a spin glass state was demonstrated [24].

Hall Constant. The values of R_0 and R_1, the paramagnetic and ferromagnetic Hall constants were measured by [25]. The figures are given in the table below.

Hall constants in 10^{-5} cm³/A · s:

at% Rh	2	4	6	10	25	50	100
R_0	−5.5	−5.9	−6.0	−6.2	−6.1	−2.0	+4.8
R_1	−258	−374	−610	−990	–	–	–

References:

[1] Kleber, R. (Z. Physik **264** [1973] 309/20).
[2] Narath, A.; Weaver, H.T. (J. Appl. Phys. **41** [1970] 1077/8).
[3] Crangle, J.; Parsons, D. (Proc. Roy. Soc. [London] A **255** [1960] 509/19).
[4] Vetter, R.; Vuik, J. (Phys. Status Solidi A **63** [1981] 637/43).
[5] Kadomatsu, H.; Fujiwara, H. (Solid State Commun. **29** [1979] 255/8).

[6] Fujiwara, H.; Kadomatsu, H.; Oishi, K.; Yamamoto, Y. (J. Phys. Soc. Japan **40** [1976] 1010/6).
[7] Brinkman, W.F.; Bucher, E.; Williams, H.J.; Maita, J.P. (J. Appl. Phys. **39** [1968] 547/8).
[8] Bölling, F. (Physik Kondensierten Materie **7** [1968] 162/84).
[9] Cable, J.W.; Wollan, E.O. (Physica B+C **86/88** [1977] 745/6).
[10] Cable, J.W. (Phys. Rev. [3] B**15** [1977] 3477/84).

[11] Hicks, T.J. (J. Phys. F**10** [1980] 879/92).
[12] Kadamatsu, H.; Tokunaga, T.; Fujiwara, H. (J. Phys. Soc. Japan **50** [1981] 1409/10).
[13] Oishi, K.; Kasai, A.; Fujiwara, H. (J. Phys. Soc. Japan **51** [1982] 3504/7).
[14] Delyagin, N.N.; Nesterov, V.I.; Semenov, S.I. (Phys. Status Solidi A**117** [1983] K27/K29).
[15] Acker, F.; Huguenin, R. (J. Magn. Magn. Mater. **12** [1979] 58/76).
[16] Takahashi, Y.; Jacobs, R.L. (J. Phys. F**12** [1982] 517/28).
[17] Muellner, W.C.; Kouvel, J.S. (Phys. Rev. [3] B**11** [1975] 4552/9).
[18] Leviv, K.; Bass, R.; Bennemann, K.H. (Phys. Rev. [3] B**6** [1972] 1865/79).
[19] van der Rest, J.; Gautier, F.; Brouers, F. (J. Phys. F**5** [1975] 995/1013).
[20] van der Rest (J. Phys. F**7** [1977] 1051/68).

[21] Hirooka, S.; Shimizu, M. (J. Phys. Soc. Japan **43** [1977] 477/82).
[22] Köster, W.; Gmöhling, W. (Z. Metallk. **52** [1961] 713/20).
[23] Vasil'yeva, R.P.; Cheremushkina, A.V.; Ivanova, N.N.; Myasnikova, K.P. (Izv. Akad. Nauk SSSR Met. **1976** No. 5, pp. 169/71).
[24] Carnegie, D.W.; Claus, H. (Phys. Rev. [3] B**30** [1984] 407/9).
[25] Köster, W.; Romer, O. (Z. Metallk. **55** [1964] 805/10).

Electrical Properties

Resistivity. The electric resistivity and the temperature coefficient of nickel-rhodium alloys were measured by [1]. The results are shown below.

at% Ni	0	50	75	90	94	96	98	100
ϱ in $\mu\Omega\cdot$cm	5.5	40.9	42.6	28.8	21.9	17.1	12.3	7.06
$\Delta\varrho/\Delta T$ in 10 $\mu\Omega\cdot$cm/°C	2.25	2.01	2.50	4.72	4.75	4.61	4.36	4.15

Resistivity measurements made between 2 and 700 K on Rh-Ni alloys revealed a magnetic scattering behaviour strikingly different from that found in other giant moment alloys such as Ni-Cu [2]. The resistivity plotted against temperature for alloys containing between 1.7 and 16.3 wt% rhodium is shown in **Fig. 150** [3].

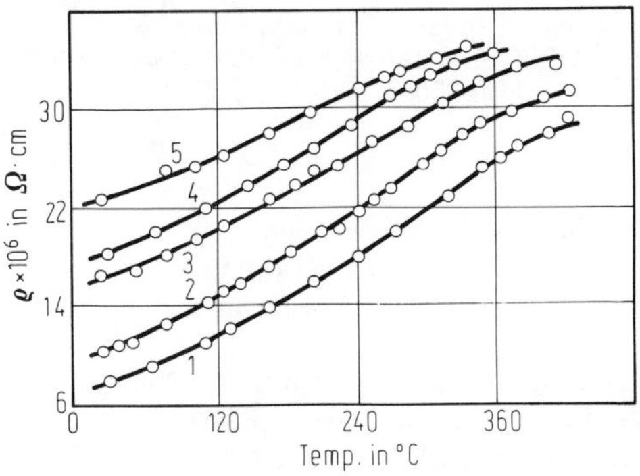

curve 1 = 1.7 at% Rh
curve 2 = 5.15 at% Rh
curve 3 = 8.47 at% Rh
curve 4 = 9.10 at% Rh
curve 5 = 16.3 at% Rh

Fig. 150. Resistivity of Rh-Ni alloys as a function of temperature.

Experimental results for the resistivity anisotropy of nickel alloys containing transitional metal impurities (including rhodium) have been determined; the saturation resistivity anisotropy $\Delta\varrho/\varrho$ was found to be $+0.05\%$ [4].

Work on residual resistivities of alloys containing between 0.5 and 5 at% rhodium showed the occurrence of an s-d virtual bound state at the Fermi level [5]. Residual resistivity measurements on dilute alloys of rhodium in nickel carried out at 4.77 and 297 K confirmed the existence of virtual bound states [6]. A residual resistivity increase of 1.80 $\mu\Omega \cdot cm \cdot at\%^{-1}$ rhodium is given by [7]. Resistivities of 3.1, 9.2, and 12.2 at% rhodium alloys were measured at 0.5 K temperature intervals in the Curie point region, the results being plotted in a figure. The origin of the horizontal scale has been taken at the Curie point of the respective alloys [8].

Thermoelectric Power. The absolute thermoelectric power ε was measured on alloys containing between 2 and 50 at% rhodium; the results are presented in the table below [1].

at% Rh	2	4	6	10	25	50	100
ε in 10^{-6} V/°C	−15.0	−10.1	−5.2	+3.14	+15.33	+12.2	+1.2

The thermoelectric power of alloys containing between 56 and 64 at% nickel was measured as a function of temperature by [9]. The thermal emf of alloys containing 1.7, 5.15, 8.47, 9.10, and 16.3 wt% rhodium are shown in **Fig. 151** for temperatures up to 400 °C [3].

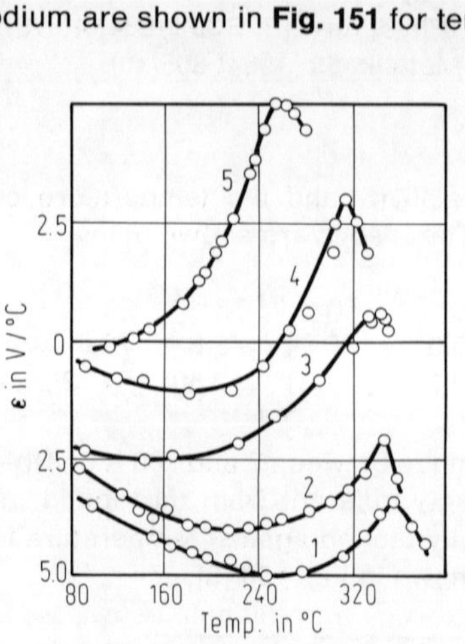

Fig. 151. Thermoelectric power of Rh-Ni alloys. Figures at the curves denote same Rh contents as in Fig. 150, p. 219.

References:

[1] Köster, W.; Gmöhling, W. (Z. Metallk. **52** [1961] 713).
[2] Houghton, R.W.; Sarachik, M.P.; Kouvel, J.S. (Solid State Commun. **10** [1972] 369/71).
[3] Vasil'yeva, R.P.; Cheremushkina, A.V.; Ivanova, N.N.; Myasnikova, K.P. (Izv. Akad. Nauk SSSR Met. **5** [1976] 169/71).
[4] Jaoul, O.; Campbell, I.A.; Fert, A. (J. Magn. Magn. Mater. **5** [1977] 23/34).
[5] Cadeville, M.C.; Durand, J. (Solid State Commun. **6** [1968] 399/401).
[6] Durand, J.; Gautier, F. (J. Phys. Chem. Solids **31** [1970] 2773/87).
[7] Dorelyn, J.F.W. (Philips Res. Rept. **31** [1976] 287).

[8] Vetter, R.; Vuik, J. (Phys. Status Solidi A **63** [1981] 637/43).
[9] Touger, J.S.; Sarachik, M.P. (Solid State Commun. **20** [1976] 1/4).

Ternary Alloys

Rh-Ni-As. The structure of RhNiAs is tetragonal, Cu_2Sb-type with a=3.565, c=6.161 Å [1]. The order of compounds of the RhNiAs-type is discussed by [2].

Rh-Ni-Bi. The quasi-binary $NiBi_3$-$RhBi_3$ section of the Ni-Bi-Rh system has been studied by thermal and X-ray analysis [3]. The structural properties of $Ni_{1-x}Rh_xBi_3$ were examined by powder X-ray, neutron diffraction, differential thermal analysis and differential scanning calorimetry measurements; complete solid miscibility was found. The crystal structure is of the $NiBi_3$-type and the positional parameters were found to vary insignificantly with x. The variation of the decomposition temperature T_d with composition is shown in **Fig. 152** [4].

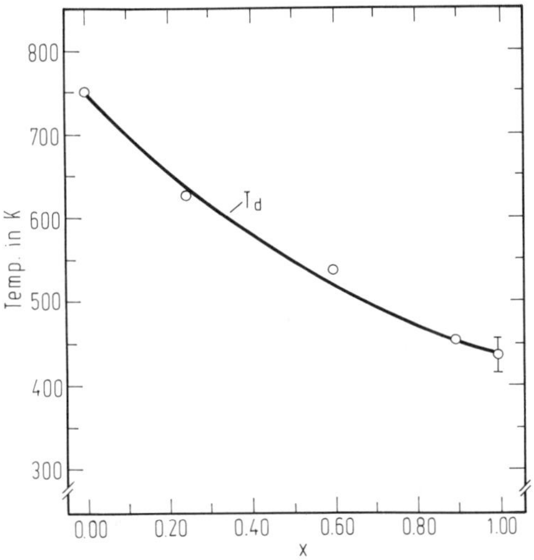

Fig. 152. Decomposition temperature T_d of $Ni_{1-x}Rh_xBi_3$ alloys.

Rh-Ni-In. Quenched randomly disordered samples of $Ni_{0.97}Rh_{0.03}In^{111}$ have been studied in the critical region by perturbed angular correlation; values of β_{eff} were insufficiently asymptotic to permit a comparison with theory [10].

Rh-Ni-Hf. A Ti_2Ni-type phase was examined by [6]. The lattice parameter was found to be 12.202±0.001 Å. For the approximate boundaries of the Ti_2Ni-type phase at 1100 °C see figure in original paper.

Rh-Ni-Sn. The crystal structure and magnetic properties of the Heusler compound Rh_2NiSn were measured. The crystal structure at room temperature was the fully ordered $L2_1$ Heusler structure. The structural and magnetic properties are given below.

Structure at 20 °C: $L2_1$; a=6.136, c=6.136 Å, c/a=1, unit cell volume 231 Å^3; moment per formula unit measured at 4.2 K and 66 kOe: $\mu=0.6\ \mu_B$ [7]. The hyperfine field has been examined in the same alloy by Mössbauer techniques; the hyperfine field at tin was found to be 17.4±0.5 kOe at 300 K [9].

Rh-Ni-Ta. The phase equilibrium at 1100 °C was studied by X-ray, microstructural and hardness measurements. The isothermal section so produced is shown in **Fig. 153** [5].

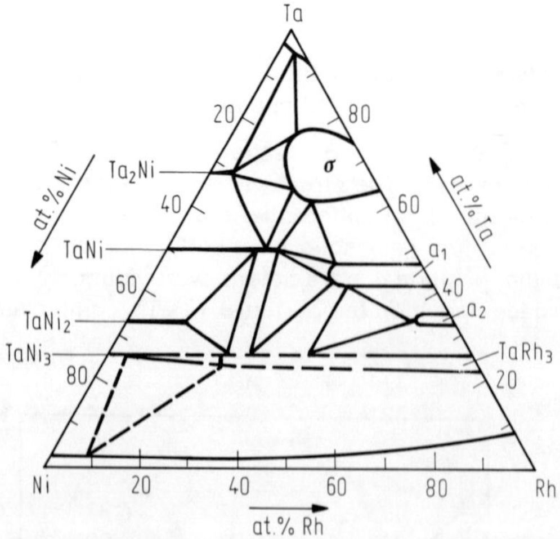

Fig. 153. Isothermal section of Rh-Ni-Ta phase diagram at 1100 °C.

Rh-Ni-Mn. Alloys of composition $(Rh_xNi_{1-x})Mn$ have been examined for local environment effect by NMR (nuclear magnetic resonance); the nuclear transverse relaxation time T_2 of ^{55}Mn was measured in the ferromagnetic alloy. The dependence of T_2 on the local environment was considered to be due mainly to the variation of the effective exchange coupling between an Mn moment and its neighbours [8].

References:

[1] Deyris, M.B.; Roy-Montreuil, J.; Rouault, M.A.; Krumbügel-Nylund, A.; Sénateur, J.-P.; Fruchart, R.; Michel, A. (Compt. Rend. C **278** [1974] 237/9).

[2] Sénateur, J.P.; Fruchart, D.; Boursier, D.; Rouault, A. (J. Phys. [Paris] **38** [1977] 61/6).

[3] Kuz'min, R.N.; Nikitina, S.V. (Izv. Akad. Nauk SSSR Met. **1969** 215/7; Russ. Metall. **1969** No. 4, pp. 139/41).

[4] Fjellåg, H.; Furuseth, S. (J. Less-Common Metals **128** [1987] 177/83).

[5] Khokhlova, L.M.; Kuprina, V.V. (Vestn. Mosk. Univ. Khim. **36** [1981] 404/5; Moscow Univ. Chem. Bull. **36** No. 4 [1981] 84/5).

[6] Nevitt, M.V.; Downey, J.W.; Morris, R.A. (Trans. Am. Inst. Mining Metall. Petrol. Eng. **218** [1960] 1019/23).

[7] Suits, J.C. (Solid State Commun. **18** [1976] 423/5).

[8] Ueno, K.; Asayama, K. (J. Phys. Soc. Japan **47** [1979] 1094/101).

[9] Pillay, R.G.; Tandon, P.N. (Phys. Status Solidi A **45** [1978] K109/K112).

[10] Chowdhury, A.R.; Allard, C.; Suter, R.M.; Collins, G.S.; Hohenemser, C. (Hyperfine Interact. **10** [1981] 893/9).

Quaternary Alloys

Rh-Ni-Mn-Sn. A new series of Heusler alloys of composition $(Rh_{1-x}Ni_x)_2MnSn$ is reported. They proved to be single-phased having the cubic $L2_1$ structure. The table on p. 223 gives the magnetic properties and lattice parameters of some of these alloys.

Ferromagnetic saturation moment μ_{00} in μ_B, Curie temperature θ_F in K, paramagnetic Curie temperature θ_P in K, paramagnetic moment μ_{eff} in μ_B, ratio of magnetic carriers in the para- and ferromagnetic state, lattice parameter a in Å of $(Rh_{1-x}Ni_x)_2MnSn$.

alloy	μ_{00}	θ_F	θ_P	μ_{eff}	$\frac{\mu_P}{\mu_{00}}$	a
Rh_2MnSn	4.14	410	412	4.83	0.95	6.252
$Rh_{1.9}Ni_{0.1}MnSn$	4.03	392	401	4.86	0.98	6.245
$Rh_{1.8}Ni_{0.2}MnSn$	4.00	382	392	4.94	1.01	6.233
$Rh_{1.6}Ni_{0.4}MnSn$	4.06	364	380	4.91	0.99	6.209
$Rh_{1.4}Ni_{0.6}MnSn$	4.05	346	364	4.88	0.98	6.187
$Rh_{1.2}Ni_{0.8}MnSn$	4.05	337	352	4.87	0.98	6.170
$Rh_{1.0}Ni_{1.0}MnSn$	4.04	327	345	4.90	0.99	6.149
$Rh_{0.8}Ni_{1.2}MnSn$	4.05	324	340	4.89	0.99	6.132
$Rh_{0.6}Ni_{1.4}MnSn$	4.03	321	336	4.94	1.00	6.117
$Rh_{0.4}Ni_{1.6}MnSn$	4.02	325	340	4.91	1.00	6.095
$Rh_{0.2}Ni_{1.8}MnSn$	4.13	331	348	4.93	0.98	6.075
$Rh_{0.1}Ni_{1.9}MnSn$	4.07	335	355	4.91	0.99	6.068
Ni_2MnSn	3.98	342	363	4.90	1.00	6.052

Uhl, E. (J. Magn. Magn. Mater. **49** [1985] 101/5).

34 Alloys with Cobalt

34.1 The Rh-Co System

Phase Diagram. The diagram shown in **Fig. 154** is based on lattice parameter, magnetic, dilatometric, microscopic and hardness measurements. The system is considered to be one of continuous solid solution and is shown with a minimum, since alloys containing 57.2 and 75.2 at% rhodium showed incipient melting at 1400 °C. Rhodium raises the transformation temperature of the hexagonal to face-centred cubic structure change, and the hysteresis in this diffusionless transformation increases with rhodium content [1].

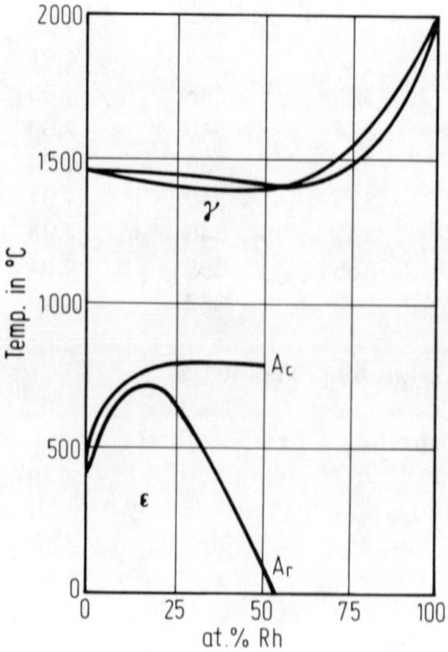

Fig. 154. Phase diagram Rh-Co. A_c = transformation $\varepsilon\to\gamma$ (heating); A_r = transformation $\gamma\to\varepsilon$ (cooling).

Preparation. Samples were annealed at 1200 °C for from 2 to 10 d, the longer times being given to the alloys with higher rhodium content.

Crystallography. The lattice parameters are shown in **Fig. 155**; the phases present are indicated [1]. High-temperature surface segregation has been studied on alloys containing 19.3, 47.8, and 84.7 at% rhodium using Auger electron spectroscopy; cobalt segregation occurred on the 84.7 at% rhodium alloy with a heat of segregation of -3.7 ± 7 kJ/mol, while rhodium enrichment was found on the 19.3% alloy with a heat of segregation of 1 ± 0.1 kJ/mol. No variation from bulk composition was found on the surface of the 47.8% alloy [2].

Density, Surface Tension. The density D and the surface tension σ at the melting point mp for various compositions were determined by [3], the figures are given below.

at% Co	10	20.7	34	49.9	59.9	66.9	79.8
σ in mJ/m^2 . . .	1905	1915	1892	1930	1923	1875	1860
D in g/cm^3 . . .	10.55	10.15	10.0	9.65	9.2	8.95	8.55
mp in °C	1860	1760	1660	1540	1502	1490	1460

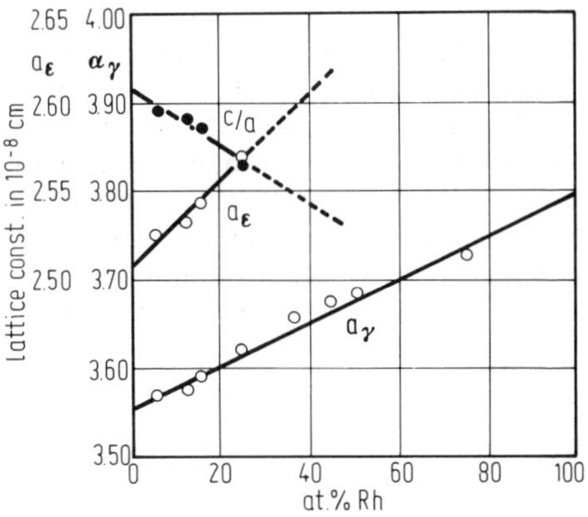

Fig. 155. Phases and lattice parameters in the Rh-Co system.

Hardness. The Vickers hardness H_v related to rhodium content is shown in a figure [1]. According to [4], the addition of 1 wt% of cobalt resulted in an as-cast hardness of 130 kg/mm^2.

Creep Properties. Rhodium containing 1 wt% of cobalt when tested at a stress of 345 bar at temperatures of 1200 and 1400 °C had failure times of 46.2 and 4.9 h, respectively [4].

Electrical and Magnetic Properties. Considerable interest has been shown in the magnetic and electrical properties of these alloys, since both constituents are transitional; many measurements have been made and theories tailored to fit the findings into one of the electronic theories. Alloys containing 2.76, 5.86, 7.41, and 10.80 at% rhodium were studied by [5]. The work was confined to the fcc phase obtained by heating the alloys for three days at 1100 °C. The spontaneous magnetization at absolute zero, the Curie point and the Bohr magneton number were determined, the first property being obtained by extrapolation from higher temperatures; the spontaneous magnetization σ_T is defined as the value of the intrinsic or domain magnetization at zero applied field. The results are shown in the following table; several simple models based on d-band exchange were applied to explain the results, none being satisfactory.

Rh in at%	2.76	5.86	7.41	10.80
σ_0 in cgs units/g	157.8	150.2	145.3	—
Curie point θ in K	1344	1300	1277	1206
Bohr magneton number p	1.70_0	1.65_5	1.61_8	—

The NMR signals of rhodium as an impurity (<2 at%) were used by [6] to determine the internal fields at the nuclei of rhodium in cobalt. NMR shifts at the first three shells of Rh^{103} nuclei which neighbour a cobalt impurity site were obtained by [7] using SEDOR technique (spin-echo double resonance); the SEDOR spectrum of Rh^{103} in $Co_{0.0025}Rh_{0.9975}$ is shown in the original paper. The electric resistivity of dilute alloys of rhodium in cobalt was measured at 4.77 and 297 K and the results suggested the existence of virtual bound states; the impurity concentration dependence of the deviations from Matthiessen's rule were used to derive the resistivities of both s↑ and s↓ bands [8]. Spin-echo measurements at low and high temperatures on cobalt containing small amounts of rhodium showed satellite lines which arise from the change in stability and extension of d shells between impurity and host atoms; a new interpretation of the satellite line is suggested for alloys with virtual bound states, attributing it to the neighbours of a pair of impurity atoms [9]. Alloys containing 28 to 44 at% cobalt were studied in the temperature range 1.6 to 300 K by [10]; the magnetization/temperature curves obtained are shown in **Fig. 156**. The magnetic phase diagram for temperatures between 0 and 20 K is shown in **Fig. 157**; this is due to [11, 12] and supporting evidence is given in [10] for the existence, in the alloys containing local moments, of two stages of order, with alloys containing up to 42 at% cobalt having the characteristics of spin glasses. The author indicates that alloys containing in excess of 36 at% cobalt are ferromagnetic at low temperatures. The magnetic properties change at 44 at% cobalt, possibly due to changes in crystal structure. Electric resistance measurements made by [13] tend to confirm some of these conclusions.

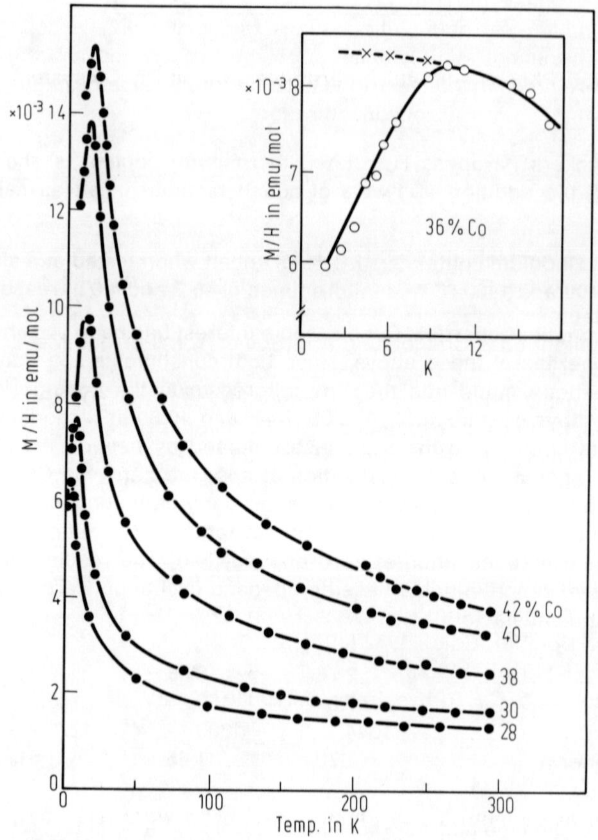

Fig. 156. Magnetization against temperature for RhCo alloys; alloys were cooled in zero field. Measurements were made in a field of 500 Oe. Inset: the enlarged low temperature magnetization results for $Rh_{64}Co_{36}$; points (×) show the results obtained if the sample is cooled below T_m in an applied field (500 Oe).

X-ray Photoelectron Spectroscopy. The spectra of the 2p levels of Co in a rhodium-20 at% cobalt alloy have been used to study its core state; the features of the spectrum are discussed in terms of the possible electronic arrangements [14].

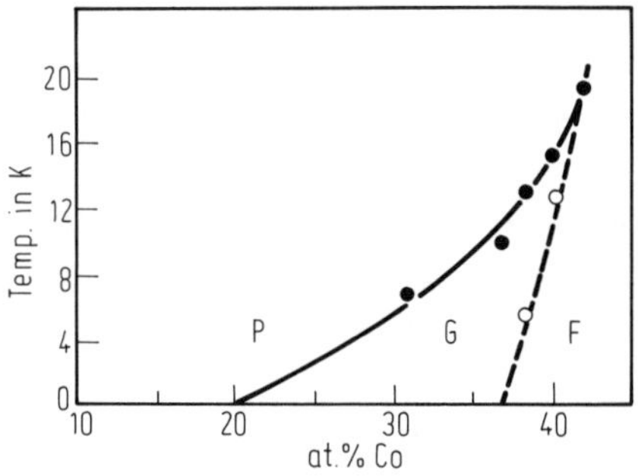

Fig. 157. The magnetic phase diagram; P, G and F designate the paramagnetic region, the magnetic glass and the ferromagnetic regions, respectively. The solid line is a plot of T_m as a function of concentration. Points o, on the broken line are the temperatures at which anomalies were noted in the electric resistance measurements, plotted as a function of concentration.

References:

[1] Köster, W.; Horn, E. (Z. Metallk. **43** [1952] 444/9).
[2] Ellison, K.A.; Underhill, P.R.; Smeltzer, W.W. (Surf. Sci. **182** [1987] 69/84).
[3] Handley, J.R. (Platinum Metals Rev. **33** [1989] 64/72).
[4] Mit'ko, M.M.; Dubinin, E.L.; Chegodaev, A.I. (Soviet Non-Ferrous Metals Res. **6** [1978] 115/7).
[5] Crangle, J.; Parsons, D. (Proc. Roy. Soc. [London] A **255** [1960] 509/19).
[6] Kotani, M.; Itoh, J. (J. Phys. Soc. Japan **22** [1967] 345/6).
[7] Walstedt, R.E.; Wernick, J.H. (Phys. Rev. Letters **20** [1968] 856/9).
[8] Durand, J.; Gautier, F. (J. Phys. Chem. Solids **31** [1970] 2773/87).
[9] Durand, J. (J. Phys. [Paris] **32** [1971] 899/901).
[10] Jamieson, H.C. (J. Phys. F **5** [1975] 1021/30).
[11] Tari, A. (Diss. Imperial College, London 1972).
[12] Coles, B.R.; Tari, A.; Jamieson, H.C. (Low Temp. Phys.-LT 13 Proc. 13th Intern. Conf. Low Temp. Phys., Boulder, Colo., 1972 [1974]).
[13] Tari, A. (J. Phys. F **6** [1976] 1313/23).
[14] Andrews, P.T.; Johnson, C.E. (Phys. Letters A **70** [1979] 140/2).

34.2 Ternary Alloys

Rh-Co-As. Alloys of the type MM′As where M = 3 d transition metal and M′ = 4 d transition metal were studied, CoRhAs being amongst them. The ordering factors were examined by Mössbauer spectroscopy, neutron and X-ray diffraction and by magnetic measurements [13]. These alloys were also investigated by X-ray diffraction, and CoRhAs was found to have a deformed hexagonal structure [12].

Rh-Co-Sb. The lattice constants, magnetic susceptibility, magnetic moment and Curie temperature were determined for the ferromagnetic, tetragonally distorted Heusler alloys Rh_2SbCo and Rh_2SbFe. The Co alloy measurements of susceptibility at temperatures between 100 and 500 K showed the cobalt alloy to behave similarly to spin-glass alloys obtained by rapid freezing. The lattice parameters of the two alloys are given on p. 228.

alloy	a in Å	c/a	μ in μ_B	T_c in K
Rh_2FeSb	5.75	1.21	2.77	510
Rh_2CoSb	5.71	1.24	1.42	450

T_c is the Curie temperature, μ_B is the magnetic moment [14].

Rh-Co-Ce. The pseudo-binary system $Ce(Co_{1-x}Rh_x)_2$ has been examined by X-ray diffraction and magnetic methods and the following lattice spacings obtained for the series of solid solutions; x = mole fraction of $CeRh_2$ [10].

x	0	0.166	0.333	0.500	0.666	0.833	1.0
a in kX . . .	7.1462	7.222	7.302	7.339	7.417	7.4755	7.5259
	($CeCo_2$)	±0.002	±0.002	±0.002	±0.002		($CeRh_2$)

Rh-Co-Ho. A detailed study has been made of the magnetic properties of the Laves phases $Ho(Co_{1-c}Rh_c)_2$ where c = 0.02, 0.05, 0.08, 0.12, 0.16, and 0.25, with particular attention to the nature of the magnetic transitions. The temperature dependence of the susceptibility in fields of 100 to 500 Oe is shown in **Fig. 158** and the change in Curie temperature with composition in a figure in the original paper [11]. Dependence of the Curie temperature on the Rh contents C.

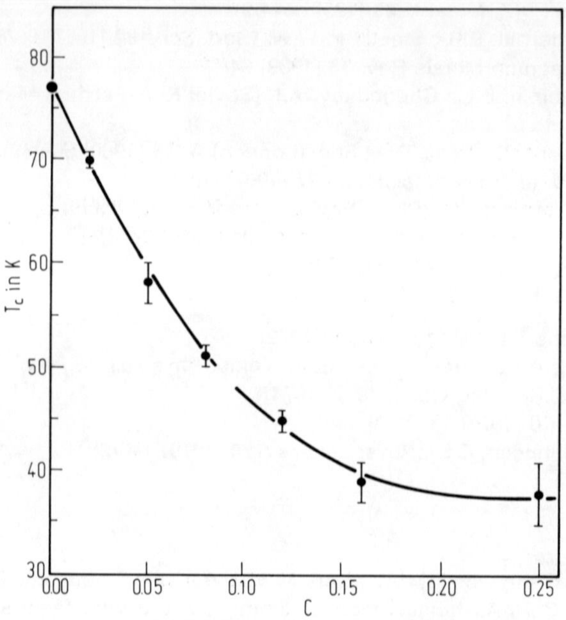

Fig. 158. Change of T_c with Rh contents C.

Rh-Co-Sn. The crystal structure and magnetic properties were determined for a number of alloys of the form Rh_2TSn where T = Mn, Ni, Cu, V, Cr, Fe, Co. The figures obtained for T = Co were:

Structure at 20 °C tetragonal, a = 4.114, c = 6.906 Å, c/a = 1.19; Curie temperature 444 K, magnetic moment per formula unit measured at 4.2 K and 66 kOe: 2.44 μ_B [3].

The a.c. low-field susceptibility was studied on Rh_2CoSn between 100 and 600 K; a peak found at ~420 K was attributed to rapid build-up of an anisotropy field below the ordering temperature, rather than to spin glass behaviour [4]. Mössbauer studies on the hyperfine field at tin sites in the same alloy have been carried out by [5 to 9]. Some results are given below.

T in K	hyperfine field in kOe	Ref.
77	−51	[5]
77	52	[6]
78	−54.9	[7]
88	55.4±1.5	[9]
293	42.4±1.5	[9]

The critical exponent β of the equation relating the hyperfine field of ^{119}Sn in Rh_2CoSn with temperature has been determined. The suggested equation was $H(T)=A(1-T/T_c)^3=H(0)\ Dt^{\beta}$. D is a constant, H(0) is the hyperfine field at T=0 K, $t=1-T/T_c$, T_c being the Curie temperature. β was found to be 0.388, A=55.8 kOe, and T=444.12 K [8].

Rh-Co-V. The lattice spacings of some ternary compositions having the Cr_3O-type structure are given below.

composition	$V_3Co_{0.9}Rh_{0.1}$	$V_3Co_{0.7}Rh_{0.3}$	$V_3Co_{0.5}Rh_{0.5}$	$V_3Co_{0.3}Rh_{0.7}$	$V_3Co_{0.1}Rh_{0.9}$
a in Å . . .	4.6903±0.0001	4.7147±0.0001	4.7364±0.0001	4.7557±0.0001	4.7752±0.0001

V_3Rh is mutually soluble with V_3Co at 1000 °C [2].

Rh-Co-Nb. The ternary alloys of the general composition $Nb_3Rh_{1-x}B_x$ having the Cr_3Si-type crystal structure were examined for superconducting properties down to 1.7 K; in the general formula, B=Co, Ru, Pd, Os, Ir, Pt, Au. The results on cobalt-containing alloys are given in the following table which gives the lattice parameters and critical temperatures [1].

composition	a in Å	T_c in K
$Nb_3Rh_{0.98}Co_{0.02}$	5.132	2.28
$Nb_3Rh_{0.95}Co_{0.05}$	5.135	1.96
$Nb_3Rh_{0.90}Co_{0.10}$	5.1347	1.90

References:

[1] Zegler, S.T. (Phys. Rev. 5 [2] **137** [1965] A1438/A1440).
[2] Zegler, S.T.; Downey, J.W. (Trans. Metall. Soc. AIME **227** [1963] 1407/11).
[3] Suits, J.C. (Solid State Commun. **18** [1976] 423/5).
[4] Dhar, S.K.; Grover, A.K.; Malik, S.K.; Vijayaraghavan, R. (Solid State Commun. **33** [1980] 545/7).
[5] Campbell, C.C.M.; Birchall, T.; Suits, J.C. (J. Phys. F **7** [1977] 727/43).
[6] Itoh, J.; Shimizu, K.; Mizutani, H.; Grover, A.K.; Gupta, L.C.; Vijayaraghavan, R. (J. Phys. Soc. Japan **42** [1977] 1777/8).
[7] Nikolaev, I.N.; Potapov, V.P.; Bezotosnyi, I.Yu.; Mar'in, V.P. (Soviet Phys. Solid State **20** [1978] 2135/6).
[8] Delyagin, N.N.; Zonnenberg, Yu.D.; Kornienko, E.N.; Krylov, V.I.; Nesterov, V.I. (Soviet Phys. Solid State **20** [1978] 148/9).

[9] Grover, A.K.; Pillay, R.G.; Nagarajan, V.; Tandon, P.N. (Phys. Status Solidi B **98** [1980] 495/505).
[10] Mansey, R.C.; Raynor, G.V.; Harris, I.R. (J. Less-Common Metals **14** [1968] 337/47).
[11] Tari, A. (J. Magn. Magn. Mater. **69** [1987] 247/52).
[12] Deyris, B.; Roy-Montreuil, J.; Rouault, A.; Krumbügel-Nylund, A.; Sénateur, J.P.; Fruchart, R.; Michel, A. (Compt. Rend. C **278** [1974] 237/9).
[13] Sénateur, J.P.; Fruchart, D.; Boursier, D.; Rouault, A.; Roy-Montreuil, J.; Deyris, B. (J. Phys. Colloq. [Paris] **38** [1977] C7-61/C7-66).
[14] Dhar, S.K.; Grover, A.K.; Malik, S.K.; Vitayaraguavan, R. (Proc. Nucl. Phys. Solid State Phys. Symp. C **21** [1978] 549/51).

34.3 Quaternary Alloys

Rh-Co-Cr-V. In an X-ray study of solid solutions of Cr_3Si-type intermetallic compounds, the alloy of composition Cr_3V_3RhCo was examined by [1]. The results are given below.

treatment	phases	a in Å
untempered	γ	2.94
tempered 550 °C	γ	2.94
tempered 850 °C	γ, β	2.94, 4.69
tempered 1050 °C	β, γ	4.69, 2.94

Rh-Co-Mn-Sn. Heusler alloys of composition $(Rh_{1-x}Co)_2MnSn$ have been examined. These proved to be single-phased, having the $L2_1$ structure. Magnetic measurements were made in the ferromagnetic and paramagnetic regions and the dependency of the Curie temperatures on composition established; the ferromagnetic behaviour is discussed on the basis of d-d interaction. **Fig. 159** shows the lattice parameter and Curie temperature related

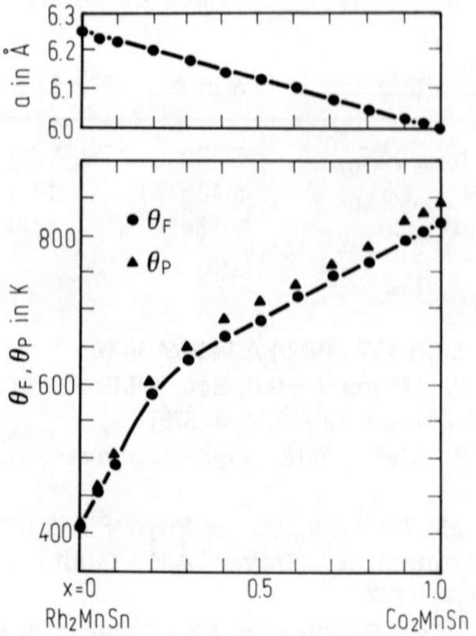

Fig. 159. Lattice parameter α(Å), Curie temperature θ_F(K) and paramagnetic Curie temperature θ_P(K) of the alloy series $(Rh_{1-x}Co_x)_2MnSn$. The straight line connects the lattice parameters of the ternaries.

to composition. The following table gives the magnetic properties and lattice parameters of the various alloys studied [2].

Experimental ferromagnetic saturation moment μ_{00}, calculated saturation moment of Mn $\mu_{00\,Mn,\,calc}$, calculated overall saturation moment $\mu_{00\,calc}$, Curie temperature θ_F, paramagnetic Curie temperature θ_P, experimental paramagnetic moment μ_{eff}, calculated paramagnetic moment $\mu_{eff,\,calc}$, lattice parameter a of $(Rh_{1-x}Co_x)_2MnSn$:

alloy	$\mu_{00\,exp}$ in μ_B	$\mu_{00\,Mn,\,calc}$ in μ_B	$\mu_{00\,calc}$ in μ_B	θ_F in K	θ_P in K	$\mu_{eff,\,exp}$ in μ_B	$\mu_{eff,\,calc}$ in μ_B	a in Å
Rh_2MnSn	4.14	4.10	4.10	410	412	4.83	4.90	6.252
$Rh_{1.9}Co_{0.1}MnSn$	4.20	4.08	4.16	455	457	4.90	4.92	6.236
$Rh_{1.8}Co_{0.2}MnSn$	4.31	4.05	4.20	492	507	4.79	4.94	6.225
$Rh_{1.6}Co_{0.4}MnSn$	4.49	4.00	4.30	590	606	4.92	4.98	6.204
$Rh_{1.4}Co_{0.6}MnSn$	4.50	3.95	4.40	635	651	5.06	5.02	6.176
$Rh_{1.2}Co_{0.8}MnSn$	4.59	3.90	4.50	666	691	5.09	5.06	6.148
$Rh_{1.0}Co_{1.0}MnSn$	4.72	3.85	4.60	687	715	5.12	5.10	6.130
$Rh_{0.8}Co_{1.2}MnSn$	4.79	3.80	4.70	720	738	4.98	5.14	6.106
$Rh_{0.6}Co_{1.4}MnSn$	4.85	3.75	4.80	749	767	5.03	5.18	6.074
$Rh_{0.4}Co_{1.6}MnSn$	4.94	3.70	4.90	770	792	5.06	5.22	6.047
$Rh_{0.2}Co_{1.8}MnSn$	5.00	3.65	5.00	800	827	5.27	5.26	6.028
$Rh_{0.1}Co_{1.9}MnSn$	5.02	3.63	5.06	811	838	5.24	5.28	6.013
Co_2MnSn	5.02	3.60	5.10	825	856	5.29	5.30	5.999

References:

[1] von Philipsborn, H. (Z. Kristallogr. **131** [1970] 73/87).
[2] Uhl, E. (J. Magn. Magn. Mater. **49** [1985] 101/5).

35 Alloys with Iron

35.1 The Rh–Fe System

Phase Diagram

Little has been published on the rhodium–iron phase diagram. Thermal analysis investigations have established the liquidus, solidus and γ/δ phase transformation temperatures for alloys containing up to 21 at% rhodium and this work showed that the system was of the expanded γ-field type, in which the temperature of the A_4 transformation was raised by the addition of rhodium [1]. Earlier work carried out by magnetic measurements had shown that the γ-phase underwent an austenitic → martensitic transformation at lower temperatures. Changes in the magnetic susceptibility and the Curie temperature suggested other changes at compositions of 20 and 50 at% rhodium [2]. The former was later shown to be associated with a crystallographic change of the bcc α-phase to an ordered CsCl-type structure. A later investigation showed an unexpected and sudden rise in the magnetization curve of alloys containing ~50 at% rhodium at a temperature of ~100 °C [3]. In an X-ray investigation which followed, the phenomenon was wrongly associated with a crystalline phase change between an ordered CsCl and a fcc phase [4]. Subsequent X-ray work showed that the ordered CsCl structure was stable above and below the magnetic transformation temperature, and that it remained stable up to temperatures of ~1300 °C; a small change in lattice parameter was found associated with the magnetic transformation. Above ~1300 °C, the CsCl structure changed into the fcc γ-phase [5]. This conclusion was supported by Mössbauer work on FeRh [6]. Neutron diffraction studies showed evidence of an antiferromagnetic ordering of the type found in Heusler alloys, Fm3m [7]. A considerable volume of research on alloys near the FeRh composition has followed; this will be considered separately.

The phase diagram shown in **Fig. 160** depends heavily on the work already outlined and in addition makes use of thermodynamic calculations and extrapolation to produce

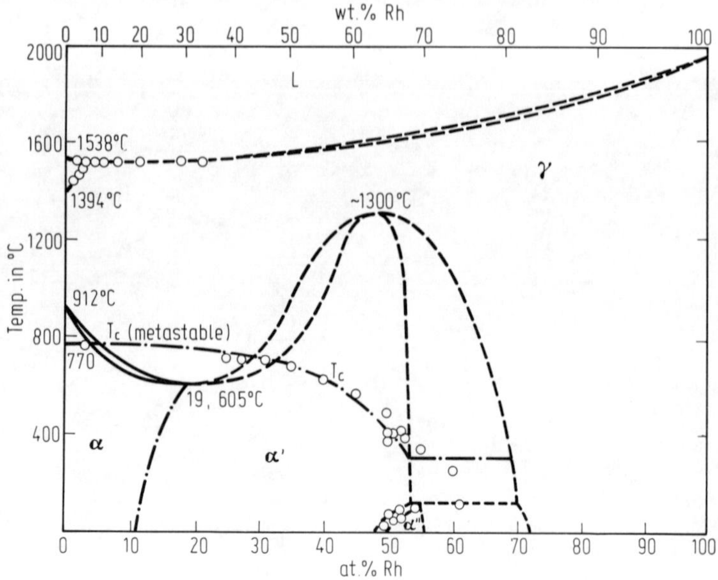

Fig. 160. Phase diagram Rh–Fe.

a possible diagram [8]. Also, the following papers were consulted; most of these associated the Curie temperature and the antiferromagnetic ⇌ ferromagnetic changes near the FeRh composition [9 to 18].

References:

[1] Gibson, W.S.; Hume-Rothery, W. (J. Iron Steel Inst. [London] **189** [1958] 243/50).
[2] Fallot, M. (Compt. Rend. **205** [1937] 558/60).
[3] Fallot, M. (Ann. Phys. [Paris] [11] **10** [1938] 291/332).
[4] Fallot, M.; Hocart, R. (Rev. Sci. **8** [1939] 498/9).
[5] de Bergevin, F.; Muldawer, L. (Compt. Rend. **252** [1961] 1347/9).
[6] Chao, C.C.; Duwez, P.; Tsuei, C.C. (J. Appl. Phys. **42** [1971] 4282/4).
[7] Muldawer, L.; de Bergevin, F. (J. Chem. Phys. **35** [1961] 1904/5).
[8] Swartzendruber, L.J. (Bull. Alloy Phase Diagrams **5** [1984] 456/62).
[9] Kouvel, J.S.; Hartelius, C.C. (J. Appl. Phys. Suppl. **33** [1962] 1343/4).
[10] Shirane, G.; Chen, C.W.; Flinn, P.A.; Nathans, R. (J. Appl. Phys. **34** [1963] 1044/5).

[11] Zakharov, A.I.; Kadomtseva, A.M.; Levetin, R.Z.; Ponyatovskii, E.G. (Zh. Eksperim. Teor. Fiz. **46** [1964] 2003/10; Soviet Phys.-JETP **19** [1964] 1348/53).
[12] Kren, E.; Pal, L.; Szabo, P. (Phys. Letters **9** [1964] 297/8).
[13] Estrin, E.I. (Izv. Akad. Nauk SSSR Met. **1966** No. 3, pp. 97/100; Russ. Met. **1966** No. 3, pp. 146/9).
[14] Ponyatovskii, E.G.; Kut-sar, A.R.; Dubovka, G.T. (Kristallografiya **12** [1967] 79/83; Soviet Phys.-Crystallogr. **12** [1967] 63/6).
[15] Léger, J.M.; Susse, C.; Vodar, B. (Compt. Rend. C **265** [1967] 892/5).
[16] Wayne, R.C. (Phys. Rev. [2] **170** [1968] 523/7).
[17] Tu, P.; Heeger, A.J.; Kouvel, J.S.; Comly, J.B. (J. Appl. Phys. **40** [1969] 1368/9).
[18] McKinnon, J.B.; Melville, D.; Lee, E.W. (J. Phys. C Suppl. **1** [1970] S 46/S 58).

Crystallographic Properties

Crystal Structure Data

phase	approx. range in at%	crystal structure	Ref.
δ	0 to 3	bcc	[1]
α	0 to 19	bcc	[2]
γ	0 to 100	fcc	[3]
α′	11 to 55	CsCl	[4 to 7]
α″	48 to 55	CsCl	[4 to 7]
α′/γ	50 to 64	fcc, CsCl	[8, 9]

Lattice Parameters in Å

at% Rh	structure	parameter	heat treatment	Ref.
2	bcc	2.874	100 h, 700 °C; fce, cool	[10]
2	bcc	2.8736	q. 1000 °C; ann. 1 h, 700 to 800 °C	[11]
4	bcc	2.881	100 h, 700 °C; fce, cool	[10]
4	bcc	2.8808	q. 1000 °C; ann. 1 h, 700 to 800 °C	[11]
6	bcc	2.8885	100 h, 700 °C; fce, cool	[10]

Lattice Parameters in Å (continued)

at% Rh	structure	parameter	heat treatment	Ref.
6	bcc	2.8880	q. 1000 °C; ann. 1 h, 700 to 800 °C	[11]
10	bcc	2.899	sl. cool from 1000 °C	[2]
20	bcc	2.929	12 d at 510 °C	[2]
25	bcc	2.943	6 d at 570 °C	[2]
25	fcc	3.671	q. from 1000 °C	[2]
30	bcc	2.955	3 d at 500 °C	[2]
30	bcc	2.942 + 2.970	3 d at 700 °C	[2]
35	bcc	2.965	6 d at 600 °C	[2]
35	bcc	2.958 + 2.978	3 d at 800 °C	[2]
40	bcc	2.981	10 d at 1000 °C	[2]
40	bcc	2.991	144 h, 530 °C	[12]
	fcc	3.73	900 h, 440 °C	
48	bcc	2.989	fce, cool from 1000 °C	[2]
49	bcc	2.993	fce, cool from 1000 °C	[2]
50	bcc	2.99	90 h 450 °C	[12]
50	bcc	2.991	900 h 440 °C	[12]
50	fcc	3.74	q. from melt, 2 h, liq. O_2 cool RT	[12]
50	bcc	2.986	fce, cool from 1000 °C	[2]
50	bcc	2.983		[15]
50	bcc	2.983		[15]
50	bcc	2.986		[15]
50	bcc	2.986		[15]
52	bcc	2.986	fce, cool from 1000 °C	[2]
	fcc	3.764		
52	bcc	2.99	1 d 950 °C; sl. cool	[16]
	fcc	3.74		
52.5	bcc	2.99	tested 20 °C	[12]
	fcc	3.77	predominant phase	[12]
	bcc	2.99	predominant phase	[12]
	fcc	3.77	tested 120 °C	[12]
53	bcc 15 °C	2.987	35 h 100 °C	[17]
	bcc 65 °C	2.997		
	bcc 20 °C	2.987		
	bcc 90 °C	2.998		
53.1	bcc/fcc < 5%	2.985	24 h 1000 °C; fce cool	[14]
55	bcc 66.7%	2.986	24/50 h 1000 °C vacuum	[9]
	fcc	3.766		
57	bcc 53.3%	2.986	24/50 h 1000 °C vacuum	[9]
	fcc	3.766		
59	bcc 40%	2.986	24/50 h 1000 °C vacuum	[9]
	fcc	3.766		
64	fcc only	3.764	24/50 h 1000 °C vacuum	[9]
60	bcc	2.99	144 h 530 °C	[12]
	fcc	3.780		[12]
60	bcc	2.986	fce cool from 1000 °C	[2]
	fcc	3.763		
60	fcc	3.7594	q. 1200 °C, H_2	[12]
	bcc	2.9859		[12]

Lattice Parameters in Å

at% Rh	structure	parameter	heat treatment	Ref.
61.2	bcc 5 to 10% fcc	2.986	24 h 1000 °C; fce cool	[14]
64	—	3.785	144 h 530 °C	[12]
70	fcc	3.7717	q. 1200 °C H_2	[13]
80	fcc	3.780	fce cool from 1000 °C	[2]
80	fcc	3.7820	q. 1200 °C H_2	[13]
90	fcc	3.7929	q. 1200 °C H_2	[13]

Measurements on alloys between 27.5 and 90 at% rhodium quenched from the liquid state were carried out by [3]; the results are shown in **Fig. 161**. Abbreviations used above: fce = furnace, sl. = slow, q. = quenched, ann. = annealed.

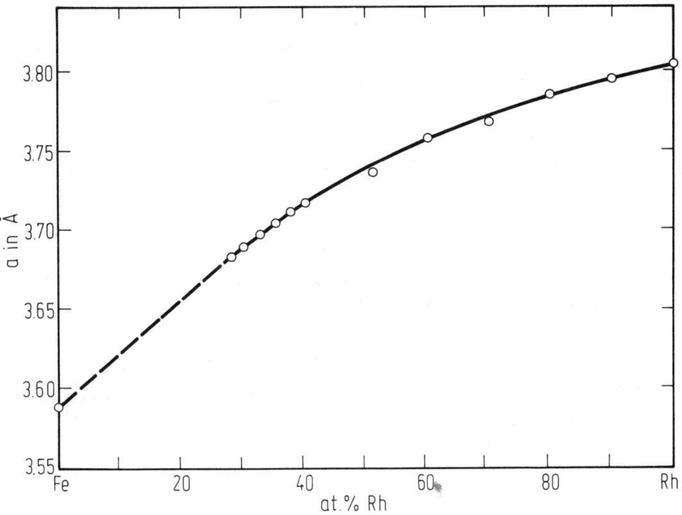

Fig. 161. Lattice parameters of fcc Rh-Fe alloys.

All alloys except those containing up to and including 10 at% rhodium with bcc structure are of the ordered CsCl type. The figures given under [10] are approximate only, since they are taken from a figure in the paper. The figures in [14] have an estimated error of ±0.002. The four figures under [15] were measured from (111) reflections and at temperatures of 20, 90, 95 and 105 °C. The lattice parameter is a linear function of the Rh concentration; it can be calculated from the formula $a = a_{Fe} + (36 \pm 2) \times 10^{-4} c_{Rh}$, where c_{Rh} is the rhodium concentration in at% [11].

References:

[1] Gibson, W.S.; Hume-Rothery, W. (J. Iron Steel Inst. [London] **189** [1958] 1243/50).

[2] Shirane, G.; Chen, C.W.; Flinn, P.A. (Phys. Rev. [2] **131** [1963] 183/90).

[3] Chao, C.C.; Duwez, P.; Tsuei, C.C. (J. Appl. Phys. **42** [1971] 4282/4).

[4] de Bergevin, F.; Muldawer, L. (Compt. Rend. **252** [1961] 1347/9).

[5] Bertaut, E.F.; de Bergevin, F.; Roult, G. (Compt. Rend. **256** [1963] 1688/91).
[6] Shirane, G.; Nathans, R.; Chen, C.W. (Phys. Rev. [2] **134** [1964] A1547/A1553).
[7] Fallot, M. (Ann. Phys. [Paris] [11] **10** [1938] 291/332).
[8] Muldawer, L.; de Bergevin, F. (J. Chem. Phys. **35** [1961] 1904/5).
[9] Hofer, E.M.; Cucka, P. (J. Phys. Chem. Solids **27** [1966] 1552/5).
[10] Abrahamson, E.P.; Lopata, S.L. (Trans. Metall. Soc. AIME **236** [1966] 76/87).

[11] Zwell, L.; Speich, G.R.; Leslie, W.C. (Metall. Trans. A **4** [1973] 1990/2).
[12] Fallot, M.; Hocart, R. (Rev. Sci. **8** [1939] 498/500).
[13] Schwerdtfeger, K.; Zwell, L. (Trans. Metall. Soc. AIME **242** [1968] 631/3).
[14] McKinnon, J.B.; Melville, D.; Lee, E.W. (J. Phys. C Suppl. **1** [1970] S46/S58).
[15] Kren, E.; Pal, L.; Szabo, P. (Phys. Letters **9** [1964] 297/8).
[16] Kouvel, J.S.; Hartelius, C.C. (J. Appl. Phys. **33** [1962] 1343/4).
[17] de Bergevin, F.; Muldawer, L. (Compt. Rend. **252** [1961] 1347/9).

Mössbauer Effect

This was used to study alloys between 0 and 52 at% Rh, measuring hyperfine fields and the isomer shifts of ^{57}Fe. It was found that at rhodium contents exceeding 20 at%, chemical ordering of the CsCl type occurred and two distinct hyperfine fields were found, labelled as FeI and FeII. FeI meant Fe on Fe sites and FeII Fe on Rh sites. Neutron diffraction studies of the 48 at% rhodium alloy gave magnetic moments of $\mu_{FeI} = 3.2\ \mu_B$ but no direct measurements were possible for μ_{FeII}. The alloy containing 50 at% rhodium transformed on cooling from ferromagnetic to antiferromagnetic at 65 °C and at this phase transition, the hyperfine field showed a discontinuous decrease of 14 kOe. The Mössbauer spectra of the ordered and disordered 25 at% rhodium alloy are shown; the peaks due to FeI and FeII are shown [1]. Mössbauer measurements were also used by [2] to examine the antiferromagnetic-ferromagnetic change in 50 and 54 at% rhodium alloys using ^{57}Co as a source; they supplement the data of [1]. Work on metastable alloys in the range of 27.5 to 90 at% rhodium quenched from the liquid state and measured at room temperature was done by [3]; it was shown that by rapid quenching from the liquid state, the γ-phase can be retained, except for compositions from 40 to 60 at% rhodium. It is suggested that difficulty in retaining the γ-phase in this composition range was due to the high temperature of the immiscibility gap. A tentative equilibrium diagram is shown. Mössbauer work suggested the following phases present in alloys at 5 to 50 at% rhodium after quenching and annealing [4].

at% Rh	phases quenched	phases annealed
5	$\gamma_P + \alpha_F$	α_F
25	$\gamma_P + \alpha_F$	$\alpha_F + \gamma_P$
33	$\gamma_P + \alpha_F$	$\gamma_P + \alpha_F$
50	$\gamma_P + \alpha_{+\alpha}$	$\gamma_P + \alpha_F + \alpha_{AF}$

A Mössbauer study of a face-centred cubic sample of FeRh containing 49 at% iron was made at temperatures between 5 and 295 K. The temperature dependence of the hyperfine field and the Néel temperature are given [5].

References:

[1] Shirane, G.; Chen, C.W.; Flinn, P.A.; Nathans, R. (Phys. Rev. [2] **131** [1963] 183/90).
[2] Cser, L.; Dezsi, I.; Keszthelyi, L. (KFKI Kozl. **12** [1964] 119/24).

[3] Chao, C.C.; Duwez, P.; Tsuei, C.C. (J. Appl. Phys. **42** [1971] 4282/4).
[4] Aquino, J.M. (Diss. Moscow Univ. 1976 referred to by Fuentes, J.; Rivero, A. (Phys. Status Solidi B**135** [1986] K125/K128).
[5] Sumiyama, K.; Shiga, M.; Nakamura, Y. (Phys. Status Solidi A**13** [1972] K75/K77).

Mechanical Properties

Density. The densities at the melting point of alloys containing between 10.9 and 79.7% iron are given by [1]:

% Fe	10.9	19.9	32.8	49.9	59.1	65.5	79.7
density at melting point	10.38	10.15	9.67	8.91	8.5	8.25	7.74

Elastic Moduli. The Young's modulus, shear modulus and the effect of temperatures up to 200 °C, for alloys containing up to 8.17 at% rhodium were determined by [2]. The same two moduli and Poisson's ratio were measured using velocities of ultrasonic waves in the longitudinal and transverse directions; the effect of temperature up to 360 K was also determined for an alloy of 49.5 at% rhodium [3].

Tensile Strength. The room temperature tensile properties of a 9.5 wt% rhodium alloy were found to be 0.2% offset strength 75.0 ksi, ultimate tensile strength 89.9 ksi, elongation 31%, reduction of area 77% [4].

Hardness. Some alloys were studied at temperatures between 77 and 411 K. The results are shown in the table below according to [5].

at% Rh	H_v in kg/mm²			
	at 77 K	188 K	300 K	411 K
1.31	228	112	90	85
2.01	202	129	103	98
4.18	209	135	123	113
8.06	237	183	159	149

The strain-hardening exponent, n, is given as 0.159 for a 9.5 wt% rhodium alloy by [4].

Impact Strength. The effect of an addition of 9.5 wt% rhodium to iron on the impact energy at temperatures between −200 °C and room temperature was investigated by [4]; see figure in original paper.

Surface Tension. The surface tensions σ measured at the melting point of various alloys were determined by [1], the results are given below.

at% Rh	10.9	19.9	32.8	49.9	59.1	65.5	79.7
σ in mJ/m²	1915	1912	1912	1832	1835	1832	1825

References:

[1] Mit'ko, M.M.; Timofeef, A.I.; Chegodaev, A.I. (Soviet Non-Ferrous Metals Res. **6** [1978] 115/7).
[2] Speich, G.R.; Schwoeble, A.J.; Leslie, W.C. (Metall. Trans. **3** [1972] 2031/8).
[3] Palmer, S.B.; Dentschuk, P.; Melville, D. (Phys. Status Solidi A**32** [1975] 503/8).
[4] Floreen, S.; Hayden, H.W. (Trans. Metall. Soc. AIME **239** [1967] 1405/7).
[5] Stephens, J.R.; Witzke, W.R. (J. Less-Common Metals **48** [1976] 285/308).

Thermal Properties

Thermal Expansion. The temperature dependence of a 50 at% rhodium alloy was determined by [1], and the results are given in **Fig. 162**. Similar determinations made to a lower temperature are shown by [2]. In this case the two alloys used had compositions of 53.1 and 61.2 at% rhodium [2]. The effect of rhodium additions to iron-nickel low expansion alloys has been studied; additions of 1 to 19% were found to increase the coefficient of thermal expansion [3].

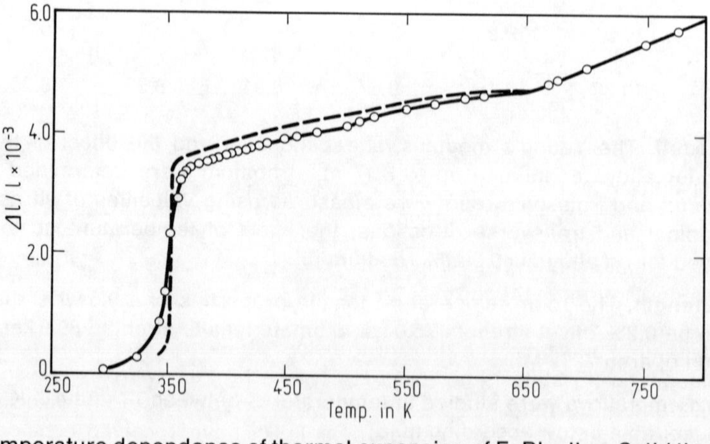

Fig. 162. Temperature dependence of thermal expansion of FeRh alloy. Solid line, experimental data taken on heating of the alloy; dashed line, calculation by formula.

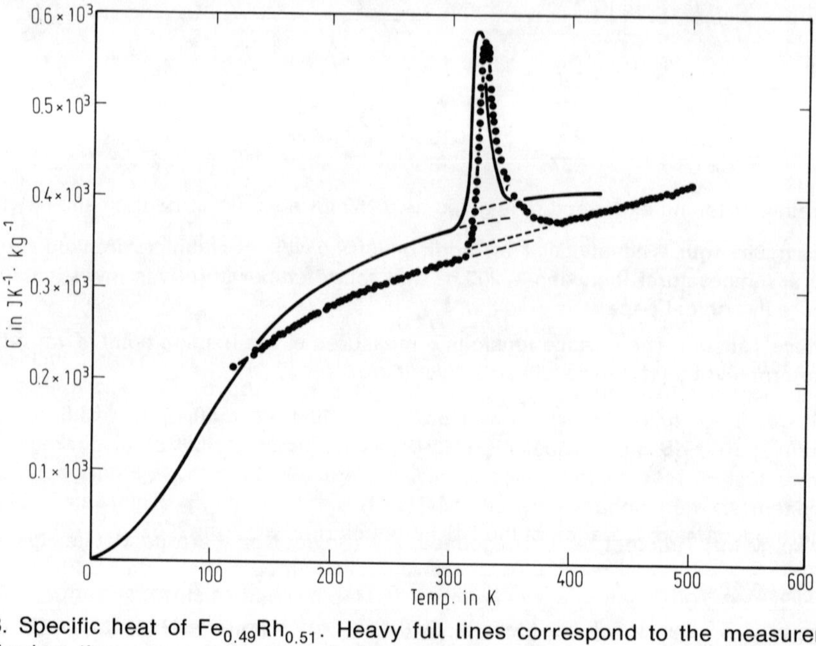

Fig. 163. Specific heat of $Fe_{0.49}Rh_{0.51}$. Heavy full lines correspond to the measurements. Heavy broken lines are extrapolated Debye-like curves fitted to the data. Thin lines are the theoretical predictions.

Specific Heat C. This was determined for a 48 at% rhodium alloy in the temperature range 0.015 to 1 K by [4]. The specific heat of a 51 at% rhodium alloy was determined at 100 to 500 K by [5]; the results are shown in **Fig. 163**.

Electronic Specific Heat γ. Some results are given in the table below. Under phase heading, AF = antiferromagnetic, F = ferromagnetic, T_c = critical temperature in K.

at% Rh	phase	T_c	γ in $\mu J \cdot g^{-1} \cdot K^{-2}$	Ref.
48	F	—	66	[6]
50.9	AF	328	10	[6]
47	F	—	59	[7]
49	F	—	60	[7]
51	AF	310	16	[7]
48.2	F	—	62.5	[8]
50.9	AF	520	10.5	[8]

γ for a 50 at% rhodium alloy was calculated by [9] and reasonable agreement obtained with experimental figures of [4] and [7].

Thermodynamic Functions. The heat of transformation from the antiferromagnetic (AF) to the ferromagnetic (F) state is given as $H = T_k \Delta S = 1.6 \pm 0.2$ cal/g by [1], compared with figures of 0.92 and ~1.0 cal/g by [10] and [11]. The entropy changes associated with the AF→F transition in alloys of composition near FeRh have received considerable study. Some results for ΔS are given below.

T_c in K	ΔS in $J \cdot kg^{-1} \cdot K^{-1}$	Ref.
~350	19.7 ± 1.7	[1]
328	14.0	[12]
310	13.6*)	[7]
—	13.5	[14]
—	18.0	[2]
—	17.23	[9]

*) This figure was estimated from measurements of the electronic heat coefficient in the F and AF states from the equation $\Delta S = (\gamma_F - \gamma_{AF}) T_c$, where T_c is the critical temperature [7].

The total entropy change in the transformation of 14 $J \cdot kg^{-1} \cdot K^{-1}$ was considered to be composed of a lattice entropy ΔS_{lat} of 4.5 $J \cdot kg^{-1} \cdot K^{-1}$ and a magnetic entropy $\Delta S_{mag} = \Delta S - \Delta S_{lat}$; these entropies are shown plotted against the mean T_c (see figure in original paper) [12]. The equilibrium between Rh-Fe alloys and CO_2-CO atmospheres was determined at 1200 °C and the data used to derive the activities in these binary solid solutions; the system showed negative deviations from Raoult's law [15]. This data was used for the thermodynamic evaluation of the Rh-Fe equilibrium diagram [16].

References:

[1] Zakharov, A.I.; Kadomtseva, A.M.; Levetin, R.Z.; Ponyatovskii, E.G. (Zh. Eksperim. Teor. Fiz. **46** [1964] 2003/10; Soviet Phys.-JETP **19** [1964] 1348/53).

[2] McKinnon, J.B.; Melville, D.; Lee, E.W. (J. Phys. C Suppl. **1** [1970] S 46/S 58).

[3] Solov'eva, N.A.; Yudevich, M.I.; Pasternak, I.I.; Pogosov, V.Z. (Metalloved. Term. Obrab. Met. **1968** No. 4, pp. 45/6).
[4] Dreyfus, B.; Stetsenko, P.; Thoulouze, D. (Phys. Letters A **24** [1967] 454/5).
[5] Richardson, M.J.; Melville, D.; Ricodeau, J.A. (Phys. Letters A **46** [1973] 153/4).
[6] Fogarassy, B.; Kemény, T.; Pál, L.; Tóth, J. (Phys. Rev. Letters **29** [1972] 288/91).
[7] Tu, P.; Heeger, A.J.; Kouvel, J.S.; Comly, J.B. (J. Appl. Phys. **40** [1969] 1368/9).
[8] Ivarsson, J.; Pickett, G.R.; Tóth, J. (Phys. Letters A **35** [1971] 167/8).
[9] Koenig, C. (J. Phys. F **12** [1982] 1123/37).
[10] Flippen, R.B.; Darnell, F.J. (J. Appl. Phys. **34** [1963] 1094).

[11] Kouvel, J.S.; Hartelius, C.C.; Lawrence, P.E. (Bull. Am. Phys. Soc. [2] **8** [1963] 54).
[12] Kouvel, J.S. (J. Appl. Phys. **37** [1966] 1257/8).
[13] Kouvel, J.S.; Hartelius, C.C. (J. Appl. Phys. Suppl. **33** [1962] 1343/4).
[14] Lommel, J.M. (J. Appl. Phys. **40** [1969] 3880/1).
[15] Schwerdtfeger, K.; Zwell, L. (Trans. Metall. Soc. AIME **242** [1968] 631/3).
[16] Swartzendruber, L.J. (Bull. Alloy Phase Diagrams **5** [1984] 456/62).

Magnetic Properties

The interesting magnetic changes associated with the composition near 50 at% rhodium were first discovered by Fallot [1, 2]. These have been the reason for the large number of papers involving magnetic measurements of various kinds aimed at unravelling the mechanism of the AF → F change, since it was established that no crystallographic change was involved in this transformation, although there was a small change in lattice parameter [3]. Studies on the 53 at% alloy showed that in the F region rhodium had a small magnetic moment and that the iron in the alloy had a much greater magnetic moment (~3 μ_B) than that of α-iron (2.2 μ_B) [4]. The magnetization in a field of 5 kOe of a 52 at% rhodium alloy was measured by [5] between 77 and 770 K and a sharp transition at 350 K was found; above this temperature the alloy behaved ferromagnetically with a Curie temperature of 675 K. An average magnetic moment of 1.86 μ_B was found or 3.85 μ_B per atom if the rhodium moment is assumed to be zero; if the iron moment is taken to be 2.2 μ_B, the rhodium moment has to be 1.5 μ_B. This work suggested that the rhodium atoms contribute to the ferromagnetically aligned moment [5], a deduction confirmed by [6, 7] using the Mössbauer effect and neutron diffraction; it was also shown that the rhodium atoms carry a reduced moment in the AF state, a fact theoretically arrived at by [9] using the Landau theory [10], and experimental data on pressure effects from [11]. Magnetization curves for alloys of 51 to 63 at% rhodium were measured between room temperature and 450 °C. Composition, magnetic moment, transition temperature and paramagnetic Curie point are given in the table on p. 241.

The decrease of magnetic moment with increasing rhodium content is attributed to an increasing proportion of paramagnetic γ-phase [12]. The authors in [5] showed that the antiferromagnetic → ferromagnetic (AF → F) transition was sensitive to external fields and it was suggested that this behaviour could be explained as an exchange inversion of the type described by [13]; this was considered to be so by [14]. However, poor numerical agreement with experimental results has been found [11], and experiments on pressure sensitivity by [15] seem to disprove an exchange inversion explanation. Work by [16] and [17] further reinforces this view. The following table gives a summary of the magnetic moments and transition temperatures obtained by various investigators [18].

Magnetic properties of Fe-Rh alloys.

% Rh	magnetic moment per alloy atom in μ_B *)	average transition temperature in °C	paramagnetic Curie point in °C
51	1.73	26	367
52	1.67	30	368
54	1.63	33	370
55	1.52	41	364
56	1.27	55	368
57	1.26	66	366
58	1.10	67	365
59	1.03	68	367
60	0.69	74	365
61	0.48	77	364
62	0.33	78	345
63	0.18	78	343

*) extrapolated to 27 °C

Sample composition, temperature for abrupt transition (T_{crit}, alloys still ferromagnetic at low temperatures designated by FM), electronic-heat coefficient (γ), and entropy change (ΔS).

composition	T_{crit} in K	γ in $\mu J/g \cdot K^2$	ΔS in $\mu J/g \cdot K$
$Fe_{53}Rh_{47}$	FM	59	—
$Fe_{51}Rh_{49}$	FM	60	—
$Fe_{49}Rh_{51}$	310	16	13.6

Later work on the temperature dependence of the critical field for the AF → F transition in FeRh using static magnetic fields, instead of the pulsed ones used by [17], showed it to have a nearly linear relationship with temperature; this suggests that besides any electronic contribution, there must be a considerable lattice one to the entropy change observed [19]. A model for the AF → F transition based on a thermodynamical approach gave a value of 14 $J \cdot kg^{-1} \cdot K^{-1}$ which agreed with measured values. The changes in physical properties at the transition were attributed to a large change in the electronic density of states at the Fermi level due to a halving of the magnetic Brillouin zone on passing from the F → AF state [20]. However, a study of the transition by the positron annihilation method showed that there was no significant change in the electronic structure during this transition [21, 22]. Ellipsometric studies on the transition using alloys containing 48, 50, 52, and 54 at% iron have been carried out between 25 and 125 °C in the 1 to 3 eV range. Results showed that the band structure of FeRh is not drastically effected as the AF → F change occurs, and is in fact less changed than calculations would suggest. Hence the large change found in the linear term of the electronic specific heat between the AF and F phases must be attributed to a magnetic contribution rather than to a change in the density of states at the Fermi surface [23].

Dilute Alloys. A considerable number of magnetic measurements has been made on rhodium containing small quantities of iron and iron containing small amounts of rhodium.

Rhodium. The magnetic hyperfine field of rhodium containing ~1 at% iron was found by Mössbauer measurements to be negative [24]. Isomer shifts in dilute alloys were measured by Mössbauer techniques and a correlation found with the standard electrode potential [25]. The magnetic susceptibilities of alloys containing <1 at% iron measured at temperatures between 0.37 and 293 K in magnetic fields up to 11000 G indicated that the system does not have normal local moments [26]. Mössbauer technique was again used to study alloys containing <0.1 at% iron using applied fields up to 62 Oe; a temperature range of 0.5 to 300 K was used. A localized magnetic moment was found correlated with the average concentration of electrons outside the last full shell [27]. NMR have been made on 0.06, 0.12 and 0.15% iron samples; the difficulties in using rhodium as a matrix element in NMR experiments are pointed out. Cold-working produced a strong increase in line width with decreasing temperature. The large value of the exchange parameter J found suggested a modified form of s-d mixing exchange, similar to that found in dilute cobalt alloys [28]. Mössbauer studies made on rhodium containing ~0.28 at% ^{57}Fe as an impurity were made in the temperature range 4 to 1000 K, and the recoilless fraction and second-order Doppler shift of the Mössbauer line measured; the iron-host force constant was found to be less than the equivalent host-host constant so that resonant band modes for the iron atoms would explain the result [29]. Similar experiments have been used to investigate deviations from dilute iron alloys below 100 K; the hyperfine field was plotted against temperature and the applied field. The findings are compared with the theoretical predictions of [30]. Mössbauer methods have been used to investigate the spin-glass alloy of rhodium-5 at% iron in the temperature range of 12.5 to 16 K where the hyperfine structure begins to appear [52]. The unusual linear relationship of electric resistivity with temperature near 1 K has been investigated by de Haas-van Alphen measurements of scattering anisotropy in a rhodium-0.1% iron alloy at 1.2 K; various models are discussed [31]. The same method was used with rhodium containing 500 and 700 ppm iron; wave slope analysis gave values of the exchange field showing no significant field dependence. No appreciable spin dependent anisotropy was observed [53].

Iron. The spatial distribution of magnetic moment disturbance round ~1.5 at% rhodium (and 14 other elements) in iron was determined using a neutron scattering technique, and rhodium was found to produce a small net increase in moment on the iron atoms round the impurity site; the implications are discussed [32]. Neutron magnetic resonance was used by [33] to study the internal field at nuclei in iron with <2 at% rhodium (and many other elements); some conclusions on the physical interpretation are given. The temperature dependence of the hyperfine field at iron atoms near 4d and 5d impurities was studied using Mössbauer technique at temperatures between that of liquid nitrogen and the Curie point [34]. Spin-echo nuclear resonance was used to study hyperfine field distributions in 0.5 to 10 at% alloys of rhodium, Ir and Mo; the findings are discussed in detail [35].

Effect of Pressure. This was first studied by [11] who measured the shift in transition temperature T_K with pressure on heating and cooling; a temperature hysteresis was found. The Landau theory of second-order phase transitions is used by [9] to deduce a pressure/temperature diagram, using data from [11]. The effect of pressure on the transition temperature of a 53 at% rhodium alloy was studied by [36]. Alloys containing 53 at% rhodium and an alloy with 50 at% iron, 46 at% rhodium and 4 at% iridium were used with pressures up to 60 kbar by [37]; a triple point was not found for the binary system, but the ternary system showed one. A pressure/temperature diagram was produced. The 50 at% alloy was similarly studied using pressures up to 23 kbar between 290 and 700 K; the possible existence of a triple point was discussed [38]. Two binary alloys containing 51 and 52.8 at% rhodium were examined at pressures up to 25 kbar, and the transition and Curie temperatures determined; it was considered that a triple point was probable [15]. This view is not supported by work on three alloys containing 49.5, 50.0 and 51 at% rhodium done by [39]. The effect of hydrostatic pressure on the AF → F transition and the question of the triple point have been further considered theoretically and experimentally [38, 40].

Hyperfine fields have been measured by an NMR method as a function of pressure in a number of impurities in iron, these including rhodium; the results are given below for ^{103}Rh.

T in K	H_1 in kOe	$(\delta H1/\delta P)_T$
195	−531.9	3.2(5)
273	−524.65	3.5(8)

H_1 is the hyperfine field, $(\delta H1/\delta P)_T$ is in $kbar^{-1} \times 10^{-4}$ [41].

Electronic Structure. Calculations of the band structure and the partial density of states for the F, AF, and P(paramagnetic) ordered FeRh alloy have been calculated by [42]; the results are compared with other theoretical and experimental data. The band structure of the antiferromagnetic $Fe_{0.5}Rh_{0.5}$ alloy has been calculated and the effect of the substitution of an iron atom as an impurity on a rhodium site examined [43]. A simplified tight-binding calculation of the density of states in non-magnetic and magnetic FeRh has been made and the predictions made from it have been found to agree with existing experimental results [44]. The phase transitions of FeRh into the AF(antiferromagnetic) and F(ferromagnetic) states are discussed in the light of self-consistent band structure calculations; a Peierls-type instability is given as the reason for the AF transition [45]. The ground state magnetic properties of ordered iron-rhodium alloys of general composition Fe_xRh_{1-x} are analysed using a realistic canonical d-band model within the Hatree-Fock and coherent particle approximations. The band energy calculations indicate that in iron-rich alloys, excess iron atoms substituted at rhodium sites, having a larger magnetic moment than rhodium atoms, play an important role in the AF → F transition at low temperatures [46]. A theoretical calculation of the optical absorption in the three magnetic phases of RhFe has shown that each has its own absorption peak. The real and imaginary dielectric constants are shown plotted against the photon frequency [47]. The X-ray spectra of these phases have been calculated as well as that of FeAl. The particular features of these spectra have been laid out to distinguish the three phases. Whilst no experimental confirmation has yet been made, the calculations made on FeAl have shown good agreement with experiment [48]. Experimental determinations made by an ellipsometric method of the dielectric functions of the three magnetic phases have been described; the results suggest that the band structure is not effected by the AF → F transition [49].

Magnetostriction Effects. These effects associated with the AF → F change in FeRh were measured by [11] and [50]. The temperature dependence of the longitudinal magnetostriction (γ) and the relative change of Young's modulus in a field of 1770 Oe is shown in the original paper [11]. The measurements on FeRh made by [50] were carried out in a field up to 150 kOe between 290 and 400 K; figures for the field and temperature dependence and the variation in the hysteresis with varying temperature are given. The magnitude of the magnetostriction could reach 3 to 3.6×10^{-3}. Measurements of the parallel and perpendicular magnetostriction of a polycrystalline sample of FeRh, in the AF state and at the AF → F transition induced by the field, have been made in a pulsed magnetic field of up to 15 T. The perpendicular AF magnetostriction was found to be of the order of 3×10^{-4} at 10 T and was temperature-dependent. The parallel magnetostriction at 10 T was $\sim 5 \times 10^{-5}$ and temperature-independent; the lattice parameter change at the transition agreed well with thermal expansion measurements [51].

References:

[1] Fallot, M. (Ann. Phys. [Paris] [11] **10** [1938] 291/332).
[2] Fallot, M.; Hocart, R. (Rev. Sci. **8** [1939] 498/9).
[3] de Bergevin, F.; Muldawer, L. (Compt. Rend. **252** [1961] 1347/9).
[4] Bertaut, E.F.; Delapalme, A.; Forrat, F.; Roult, G. (J. Appl. Phys. Suppl. **33** [1962] 1123/4).
[5] Kouvel, J.S.; Hartelius, C.C. (J. Appl. Phys. Suppl. **33** [1962] 1343/4).
[6] Shirane, G.; Chen, C.W.; Flinn, P.A.; Nathans, R. (Phys. Rev. [2] **131** [1963] 183/90).
[7] Shirane, G.; Nathans, R.; Chen, C.W. (Phys. Rev. [2] **134** [1964] A1547/A1553).
[8] Cser, L.; Dezsi, I.; Keszthelyi, L. (KFKI Kozl. **12** [1964] 119/24).
[9] Hagitai, C. (Phys. Letters **17** [1965] 178/9).
[10] Landau, L.D.; Lifshitz, E.M. (Statistical Physics, Pergamon, New York 1958).

[11] Zakharov, A.I.; Kadomtseva, A.M.; Levitin, R.Z.; Ponyatovskii, E.G. (Zh. Eksperim. Teor. Fiz. **46** [1964] 1348/53).
[12] Hofer, E.M.; Cucka, P. (J. Phys. Chem. Solids **27** [1966] 1552/5).
[13] Kittel, C. (Phys. Rev. [2] **170** [1960] 335/42).
[14] Pal, L. (Proc. Intern. Conf. Magnetism, Nottingham 1964 [1965], pp. 158/60).
[15] Wayne, R.C. (Phys. Rev. [2] **170** [1968] 523/7).
[16] Kouvel, J.S. (J. Appl. Phys. **37** [1966] 1257/8).
[17] McKinnon, J.B.; Melville, D.; Lee, E.W. (J. Phys. C Suppl. **1** [1970] S46/S58).
[18] Schwartzendruber, L.J. (Bull. Alloy Phase Diagrams **5** [1984] 456/62).
[19] Pál, L.; Zimmer, G.; Picoch, J.C.; Tarnóczi, T. (Acta Phys. Acad. Sci. Hung. **32** [1972] 135/40).
[20] Ricordeau, J.A.; Melville, D. (J. Phys. F **2** [1972] 337/50).

[21] Ádám, A.; Cser, L.; Kajcsos, Z.; Zimmer, G. (Acta Phys. Acad. Sci. Hung. **32** [1972] 299/304).
[22] Ádám, A.; Cser, L.; Kajcsos, Z.; Zimmer, G. (Phys. Status Solidi B **49** [1972] K79/K82).
[23] Chen, L.-Y.; Lynch, D.W. (Phys. Rev. [3] B **37** [1988] 10503/9).
[24] Blum, N.; Grodzins, L. (Phys. Rev. [2] **136** [1964] A 133/A 137).
[25] Bara, J.; Hrynkiewicz, A.Z. (Phys. Status Solidi **15** [1966] 205/9).
[26] Knapp, G.S. (J. Appl. Phys. **38** [1967] 1267/8).
[27] Kitchens, T.A.; Steyert, W.A.; Taylor, R.D. (Phys. Rev. [2] **138** [1965] A467/A483).
[28] Descouts, P.; Page, J.L. (J. Magn. Resonance **6** [1972] 488/92).
[29] O'Connor, D.A.; Reeks, M.W.; Skyrme, G. (J. Phys. F **2** [1972] 1179/88).
[30] Clark, P.E.; Herbert, I.R. (Solid State Commun. **12** [1973] 469/71).

[31] Higgins, R.J.; Cheng, L.S.; Graebner, J.E.; Rubin, J.J. (Proc. 14th Intern. Conf. Low Temp. Phys., Otaniemi, Finland, 1975, Vol. 3, pp. 146/9).
[32] Collins, M.F.; Low, G.G. (Proc. Phys. Soc. [London] **86** [1965] 535/48).
[33] Kontani, M.; Itoh, J. (J. Phys. Soc. Japan **22** [1967] 345/6).
[34] Vincze, I. (Solid State Commun. **10** [1972] 431/45).
[35] Dean, R.H.; Jackson, G.A. (J. Phys. F **8** [1978] 1563/78).
[36] Bloch, D. (Ann. Phys. [Paris] [14] **1** [1966] 93/125).
[37] Léger, J.-M.; Susse, C.; Vodar, B. (Compt. Rend. C **265** [1967] 892/5).
[38] Ponyatovskii, E.G.; Kut-sar, A.R.; Dubovka, G.T. (Kristallografiya **12** No. 1 [1967] 79/83; Soviet Phys. Crystallogr. **12** [1967] 63/6).
[39] Vinokurova, L.I.; Vlasov, A.V.; Pardavi-Horváth, M. (Phys. Status Solidi B **78** [1976] 353/7).
[40] Grazhdankina, N.P.; Mirsayev, I.F.; Novikov, M.A.; Taluts, G.G. (Fiz. Met. Metalloved. **51** [1981] 547/55; Phys. Metals Metallog. [USSR] **51** No. 3 [1981] 81/8).

[41] Kasamatsu, Y.; Hihara, T.; Kojima, K.; Kamigaichi, T. (J. Magn. Magn. Mater. **54/57** [1986] 1107/8).
[42] Koenig, C. (J. Phys. F **12** [1982] 1123/37).
[43] Khwaja, Y.; Nauciel-Bloch, M. (Solid State Commun. **21** [1977] 529/32).
[44] Khan, M.A. (J. Phys. F **9** [1979] 457/72).
[45] Kulikov, N.I.; Kulatov, E.T.; Vinokurova, L.I.; Pardavi-Horváth, M. (J. Phys. F **12** [1982] L91/L96).
[46] Hasegawa, H. (J. Magn. Magn. Mater. **66** [1987] 175/86).
[47] Khan, M.A.; Koenig, C.; Riedinger, R. (J. Phys. F **13** [1983] L159/L164).
[48] Alouani, M.; Khan, M.A. (J. Phys. F **17** [1987] 519/41).
[49] Chen, L.-Y.; Lynch, D.W. (Phys. Rev. [3] B**37** [1988] 10503/9).
[50] Levitin, R.Z.; Ponomarev, B.K. (Zh. Eksperim. Teor. Fiz **50** [1966] 1478/80).

[51] Ricordeau, J.A.; Melville, D. (J. Phys. [Paris] **35** [1974] 149/52).
[52] Meyer, C.; Hartmann-Boutron, F.; Gros, Y. (J. Phys. [Paris] **47** [1986] 1395/404).
[53] Cheng, L.S.; Higgins, R.J.; Graebner, J.E. (Phys. Rev. [3] B**19** [1979] 3722/42).

Electrical Properties

Electric Resistance. The available information may be conveniently divided into two groups, namely determinations on alloys near the equiatomic composition, and those richer in rhodium.

Group I. The electric resistivity for the ordered FeRh alloy was determined between 0 and 800 K and the effect of increasing and decreasing temperature was measured [1]. The effect of magnetic field strength on the resistivity of 49.5 to 52.0 at% iron alloys has been measured, and is shown in a figure [2]. Alloys of similar composition were measured at zero field as a function of temperature; the authors noted that minor deviations in composition had a large effect on resistivity [3]. The effect of pressures up to 100 kbar on 49.5, 50, and 51 at% rhodium alloys was determined between 400 and 700 K [4, 5, 6]. Repeated measurements on an $Fe_{49}Rh_{51}$ alloy measured at temperatures between −6 and 250 °C through the AF → F transformation showed a decrease in scatter of the values due to some type of homogenization during the phase changes [7]. Measurements of AC susceptibility, magnetic moment and resistivity on FeRh alloys in the intermediate composition range between the AF and F phases, i.e. 50 and 52 at% iron, showed that a minimum occurred in values of the residual resistivity at 4.2 K (2.3 $\mu\Omega \cdot cm$) and the effective Bohr magneton number; this was associated with clustering and spin-glass behaviour [8].

Group II. Resistivity measurements on 3, 5, 11, and 15 at% iron alloys were carried out at temperatures between 2 and 18 K by [9]. An unusual resistance effect of rhodium alloys containing <1 at% iron was reported in 1964 by [10]; at very low temperatures the presence of iron gave rise to a strongly temperature-dependent resistivity with positive temperature coefficient [9]. The use of this effect to reliably measure very low temperatures is described by [11 to 17]. Explanations of the Coles effect based on disturbances by localized spin fluctuations have been offered by [18, 19, 20].

Thermoelectric Power. Measurements on the low-temperature thermoelectric power of dilute iron alloys have been made by [20, 21]. The results were in good agreement, suggesting a peak at ~8 K. The results are also in good agreement with the peak found in earlier measurements on rhodium which contained iron as an impurity [22].

References:

[1] Kouvel, J.S.; Hartelius, C.C. (J. Appl. Phys. **33** [1962] 1343/4).
[2] Schinkel, C.J.; Hartog, R. (AIP Conf. Proc. No. 10 [1972/73] 1365/7).
[3] Schinkel, C.J.; Hartog, R.; Hochstenbach, F.H.A.M. (J. Phys. F **4** [1974] 1412/22).
[4] Vinokurova, L.I.; Vlasov, A.V.; Pardavi-Horáth, M. (KFKI-75-51 [1975] 9 pp.; C.A. **83** [1975] No. 171735).
[5] Vinokurova, L.I.; Vlasov, A.V.; Pardavi-Horáth, M. (Phys. Status Solidi B **78** [1976] 353/7).
[6] Vinokurova, L.I.; Vlasov, A.V.; Kulikov, N.I. (J. Magn. Magnetic Mater. **25** [1981] 201/6).
[7] Myalikgulyev, G.M.; Tyurin, A.L.; Myasnikov, O.A.; Annaorazov, M.P. (Izv. Akad. Nauk Turkm. SSR Fiz. Tekh. Khim. Geol. **1983** No. 1, pp. 90/1).
[8] Vinokurova, L.I.; Vlasov, A.V.; Ivanov, V.; Pardavi-Horáth, M. (Acta Phys. Pol. A **73** [1988] 63/6).
[9] Murani, A.P.; Coles, B.R. (Metal. Phys. **3** No. 2 [1970] S159/S168).
[10] Coles, B.R. (Phys. Letters **8** [1964] 243/5).

[11] Rusby, R.L. (Platinum Metals Rev. **25** [1981] 57/61).
[12] Besley, L.M. (J. Phys. E **15** [1982] 824/6).
[13] Besley, M. (J. Phys. B **18** [1985] 201/5).
[14] Guy, D.R.P.; Friend, R.H. (J. Phys. E **19** [1986] 430/3).
[15] Logvinenko, S.P.; Mikhina, G.F. (Cryogenics **26** [1986] 484/5).
[16] Tamura, O.; Sakurai, H. (Japan. J. Appl. Phys. **26** [1987] L947/L950).
[17] Ricketson, B.W.A. (Platinum Metals Rev. **33** [1989] 55/7).
[18] Kaiser, A.B.; Doniach, S. (Intern. J. Magn. **1** [1971] 11/22).
[19] Rivier, N.; Zlatac, V. (J. Phys. F **2** [1972] L99/L104).
[20] Rusby, R.L. (J. Phys. F **4** [1974] 1265/74).

[21] Graebner, J.E.; Rubin, J.J.; Schutz, R.J.; Hsu, F.S.L.; Reed, W.A.; Higgins, R.J. (Am. Inst. Phys. **24** [1975] 445/7).
[22] Wassilieff, C.; Kaiser, A.B.; Trodahl, H.J. (J. Phys. F **10** [1980] 2761/8).
[23] Huntley, D.J. (Can. J. Phys. **49** [1971] 2610/2).

Chemical Reactions

Gases. The hydrogen solubility of iron-rhodium alloys containing 0 to 8.5 at% rhodium has been determined by [1].

Corrosion. Rhodium ions were found to accelerate corrosive attack of Armco iron in 0.2 M citric acid and 0.6 M HCl [2]. Contact with rhodium in a couple with iron in boiling 2 M HCl also resulted in increased rates of corrosion [3]. The stress-corrosion cracking of Fe-Cr-Ni based alloys in boiling $MgCl_2$ solutions was studied by [4]; alloying additions of all the platinum metals greatly increased the rate of stress-corrosion cracking. Work on a 0.005 wt% addition of rhodium to an iron-28.5% Cr-45% Mo alloy tested in boiling 10% H_2SO_4 reduced the rate of attack from 52180 mil/year to 14; the pitting corrosion in $KMnO_4$/NaOH solution was unaffected, but this was worse than the blank in $FeCl_3$ solution [5, 6].

References:

[1] Bagshaw, T.; Mitchell, A. (J. Iron Steel Inst. [London] **205** [1967] 769/71).
[2] Buck, W.R. (Corrosion [Houston] **14** [1958] 22/6).
[3] Buck, W.R. (Nature **181** [1958] 1681/2).
[4] Staele, R.W.; Royuela, J.J.; Raredon, T.L.; Serrate, E.; Morin, C.R.; Farrar, R.V. (Corrosion [Houston] **26** [1970] 451/86).

[5] Streicher, M.A. (Platinum Metals Rev. **21** [1977] 51/5).
[6] Streicher, M.A. (Corrosion [Houston] **30** [1974] 77).

35.2 Ternary Alloys

Rh-Fe-As. The structure of the compound RhFeAs was shown to be hexagonal of the type Fe_2P with a = 6.323, c = 3.618 Å [20].

Rh-Fe-Sb. The compound Rh_2FeSb was found to have a tetragonally distorted cubic structure with a = 5.75, c = 6.96 Å. The Curie temperature was 510 K, and the magnetic moment 2.77 μ_B [7].

Rh-Fe-Y. The hyperfine interactions of the compound $Y(Fe_xRh_{1-x})_2$ for $x \leqq 0.3$ were studied at 300, 80 and 5 K by Mössbauer ^{57}Fe measurements; a strong dependence of the magnetic hyperfine interaction on the local environment was shown [14]. Magnetization results as well as ^{57}Fe Mössbauer results on the Laves phase $Y(Fe_xRh_{1-x})_2$ for $x \leqq 0.3$; at high temperatures Curie-Weiss behaviour was found for $x \geqq 0.1$. The field dependence of the magnetization suggested the presence of non-interacting magnetic clusters. Micromagnetism is suggested for these compounds [15]. The same Laves phase (amongst others) has been studied using magnetization, Mössbauer and neutron depolarization measurements; Pauli paramagnetism was found, and micromagnetic behaviour may be deduced from the Mössbauer, magnetic and neutron depolarization measurements [16].

Rh-Fe-Gd. The hyperfine interactions of the compound $Gd(Fe_xRh_{1-x})_2$ for $x \leqq 0.3$ were studied at 300, 80 and 5 K by Mössbauer ^{57}Fe measurements; a strong dependence of the magnetic hyperfine interaction on the local environment was shown [14]. In alloys of general composition $Gd(Rh_{1-x}Fe_x)_2$ for $x \leqq 0.15$ the crystal structure was found to change from the cubic $MgCu_2$-type to $MgZn_2$-type at 15 at% iron. The substitution of iron for rhodium in $GdRh_2$ results in a change in the band structure of the compound [17].

Rh-Fe-Tb. The quasibinary compound of the general formula $Tb(Fe_{1-x}Rh_x)_2$ for $x \approx 0.5$ has been investigated by magnetic and nuclear γ-resonance data; with $x \approx 0.5$, a partial reorientation of spins in the Fe ions occurs, resulting in a state in which a compensated AF (antiferromagnetic) structure is incompletely realized [18].

Rh-Fe-Dy. The hyperfine interactions of the compound $Dy(Fe_xRh_{1-x})_2$ for $x \leqq 0.3$ were studied at 300, 80 and 5 K by Mössbauer ^{57}Fe measurements; a strong dependence of the magnetic hyperfine interaction on the local environment was shown [14].

Rh-Fe-Ho. The hyperfine interactions of the compound $Ho(Fe_xRh_{1-x})_2$ for $x \leqq 0.3$ were studied at 300, 80 and 5 K by Mössbauer ^{57}Fe measurements; a strong dependence of the magnetic hyperfine interaction on the local environment was shown [14]. The compound $Ho(Rh_xFe_{1-x})_2$ for x = 0 to 1 has been examined by [19].

Rh-Fe-Ge. Mössbauer and magnetic moment measurements were used to study the hyperfine interaction in Heusler alloy of composition Rh_2FeGe; the alloy was prepared by repeated melting in an induction furnace followed by annealing for several days, and then examined by X-ray diffraction; the material was single-phased with an $L2_1$-type structure with slight distortion of the cubic phase. Moments were determined in a field of 3 kOe at 77 K. The results are given below: a = 5.764, c = 6.191 Å, Curie temperature T_c = 490 K, magnetic moment/formula = 2.19 μ_B in a 3 kOe field, hyperfine field at Fe and Sn atoms at Ge site, 287(3), 92(2) and 61(5) kOe, respectively. The implications are discussed [12, 13].

Rh-Fe-Sn. The crystal structure and magnetic properties of the compound Rh_2FeSn have been reported. A large tetragonal distortion of the Heusler structure was observed, which was attributed to an electronic instability of the band Jahn-Teller type.

Structure at 20 °C: tetragonal, a=4.150, c=6.912 Å, c/a=1.18, V_0=238 Å; θ=595 K, T=583 K; μ_{exp}=3.7, μ_{calc}=3.68 μ_B.

V_0 is unit cell volume; μ_{exp} and μ_{calc} are the moment per formula unit measured at 4.2 K and 66 Oe, and moment per formula unit calculated from $2x\{(\mu/2)(1+\mu)^{3.5}\}$ [6]. Low field A.C. susceptibility measurements made on the same compound (measured lattice parameters a=5.85, c=6.903 Å) gave a magnetic moment of μ=3.04 μ_B and a Curie temperature of 573 K [7]. The hyperfine fields of tin and iron atoms in the Heusler alloy Rh_2FeSn has been determined by Mössbauer and NMR techniques; the results are given below.

T in K	HF(Sn) in kOe	HF(Fe) in kOe	Ref.
77	55.1±0.4	−288.5±1	[9]
88	59.1±1.5	−296.0±1.5	[8]
293	51.5±1.5	—	[8]
300	48 ±1	−257 ±1	[9]

HF=hyperfine field

The effect of pressure on the hyperfine magnetic fields at ^{119}Sn and ^{57}Fe nuclei in ferromagnetic Rh_2FeSn has been investigated using Mössbauer γ-ray spectroscopy; the ratio $\Delta H^{Sn}/H^{Sn}\cdot\Delta p$ was measured at 12.5 kbar at room temperature and found to be $-2.5\pm0.8\times10^{-3}$ $kbar^{-1}$. The corresponding room temperature figure for iron was $\Delta H^{Fe}/\Delta H^{Fe}\cdot\Delta p=+0.5\pm0.2\times10^{-3}$ $kbar^{-1}$ [10]. Mössbauer spectra were determined using a ^{119}Sn concentration of 0.2 at% in iron containing 2.3 and 6 at% rhodium; the interaction potential between Sn and Rh atoms at the first neighbour distance was found to be −32 meV, and the change in hyperfine field at ^{119}Sn due to the presence of a rhodium atom at first neighbour distance was +20 kOe. For concentrations of Rh>3 at%, and of Sn>0.2 at%, at 400 °C an intermetallic compound was formed in which the hyperfine field at the ^{119}Sn nucleus was 57 kOe with an isomer shift of −0.1 mm/s relative to tin in iron. A corner of the ternary compound at 673 K is given [11].

Rh-Fe-V. The transition temperature in RhFe was decreased by a 2 at% addition of vanadium [1].

Rh-Fe-(Nb, Ta, Mo, W). It was found that the elements listed in parentheses in amounts as small as 2 at% eliminated the exchange-inversion transition in RhFe [1].

Rh-Fe-Mn. Small additions of 2 at% Mn were found to decrease the exchange-inversion transition in FeRh [1]. The temperature and concentration dependence of the magnetic susceptibility of $(Fe_{55}Mn_{45})_{1-x}Rh_x$ alloys have been examined at concentrations up to 20 at% Rh. The temperature dependence was similar to that of the γ-$(Fe_{55}Mn_{45})$ alloy whilst above 20 at%, this changed and the dissolved solute had a drastic influence on the Néel temperature. It is suggested that this is caused by the combined effect of impurity scattering and modification of the density of states [5].

Rh-Fe-Ni. Ni in amounts as low as 2 at% eliminated the exchange-inversion transition in RhFe [1]. The magnetic behaviour of Ni_xRh_{1-x} (x=0.42 and 0.55) with additions of ~20 ppm ^{57}Fe between 11 and 0.05 K has been studied by Mössbauer technique; below 4.2 K deviations from a free spin behaviour were found [2]. The electric resistance of $(Fe_{1-x}Ni_x)_{0.49}Rh_{0.51}$ $(0\leqq x\leqq0.1)$ at 4.2 to 300 K in fields ≦7.5 T. The magnetic diagram is described. A field-induced AF → F transition was observed [3]. The effect of rhodium additions on the thermal expansion of iron-nickel alloys is described by [4].

Rh-Fe-Co. Co in amounts as low as 2 at% eliminated the exchange-inversion transition in RhFe [1].

References:

[1] Walter, P.H.L. (J. Appl. Phys. **35** [1964] 938/9).
[2] Seidel, E.R.; Litterst, F.J.; Gierisch, W.; Kalvius, G.M. (J. Magn. Magn. Mater. **1** [1975] 19/22).
[3] Baranov, N.V.; Barabanova, E.A. (Metallofizika **10** No. 4 [1988] 84/9).
[4] Solov'eva, N.A.; Yudevich, M.I.; Pasternak, I.I.; Pogosov, V.Z. (Metalloved. Term. Obrab. Met. **1968** No. 4, pp. 45/6; Metal Sci. Heat Treat. Metals [USSR] **1968** 293/4).
[5] Petrisor, T.; Giurgiu, A.; Pop, I.; Dadarlat, N. (J. Magn. Magn. Mater. **53** [1985] 185/8).
[6] Suits, J.C. (Solid State Commun. **18** [1976] 423/5).
[7] Dhar, S.K.; Grover, A.K.; Malik, S.K.; Vijayaraghavan, R. (Proc. Nucl. Phys. Solid State Phys. Symp. C **21** [1978] 549/51).
[8] Campbell, C.C.M.; Birchall, T.; Suits, J.C. (J. Phys. F **7** [1977] 727/43).
[9] Grover, A.K.; Pillay, R.G.; Nagarajan, V.; Tandon, P.N. (Phys. Status Solidi B **98** [1980] 495/505).
[10] Nikolaev, I.N.; Potapov, V.P.; Bezotosnii, I.Yu.; Mar'in, V.P. (Soviet Phys.-Solid State **20** [1978] 2135/6).

[11] Cranshaw, T.E. (J. Phys. Condens. Matter **1** [1989] 829/46).
[12] Patil, V.S.; Pillay, R.G.; Tandon, P.N.; Devare, H.G. (Proc. Nucl. Phys. Solid State Phys. Symp. C **25** [1982] 157/8).
[13] Patil, V.S.; Pillay, R.G.; Tandon, P.N.; Devare, H.G. (Phys. Status Solidi B **118** [1983] 57/61).
[14] Hrubec, J.; Steiner, W.; Reissner, M. (J. Magn. Magn. Mater. **29** [1982] 100/4).
[15] Hrubec, J.; Steiner, W.; Reissner, M. (J. Magn. Magn. Mater. **37** [1983] 93/100).
[16] Steiner, W.; Reissner, M. (Radex Rundschau **1986** 80/92).
[17] Tari, A.; Larica, C. (J. Phys. [Paris] **41** [1980] 35/40).
[18] Stetsenko, N.N.; Antipov, S.D.; Mostafa, M.A. (Pis'ma Zh. Eksperim. Teor. Fiz. **29** [1979] 684/7; JETP Letters **29** [1979] 627/30).
[19] Statsenko, P.N.; Avksent'ev, I.Yu.; Byrmii, Z.P. (Deposited Doc. VINITI-3249-85 [1984] 1/20).
[20] Deyris, B.; Roy-Montreuil, J.; Rouault, A.; Krumbügel-Nylund, M.M.; Sénateur, J-P.; Fruchart, R.; Michel, A. (Compt. Rend. C **278** [1974] 237/9).

[21] Tari, A. (Phys. Status Solidi **46** [1978] 173/7).

35.3 Quaternary Alloys

Rh-Fe-Gd-Ho. The quenching effect of iron substitution on the holmium moment in compounds of the type $Ho_{0.01}Gd_{0.99}(Rh_{1-x}Fe_x)_2$ have been made by spin-echo measurements on ^{165}Ho at the rhodium-rich end of these intermetallic compounds; with no iron present, the measured holmium moment was 9 μ_B but with the substitution of iron for rhodium weakens the exchange interaction resulting in further quenching of the holmium moment. This also produces an apparent increase in the total hyperfine field.

Tari, A. (Phys. Status Solidi **46** [1978] 173/7).

36 Alloys with Copper

36.1 The Rh-Cu System

Diffusion. The coefficient of bulk diffusion of Rh in Cu was measured by X-ray diffraction methods described in [8, 13], at temperatures between 750 and 1075 °C. Constants: $D_0 = 3.3\ cm^2/s$, $Q = 58000$ cal/mol [14].

Phase Diagram. A limited investigation of the system using thermal analysis, X-ray diffraction, optical microscopy and hardness data was made by [1]. It is suggested in [2] that the specimen annealing times used were insufficiently long to ensure equilibrium; however, their limits of solid solubility of 20 at% rhodium in copper and 10 at% copper in rhodium agree reasonably well with those of later investigations [3, 4]. The data given by [3] was obtained by X-ray diffraction measurements made on flakes produced by quenching from the molten state. Other findings by [1] suggested that ordered phases occurred at 50 and 75 at% and possibly 25 at% rhodium; the structure of all alloys was reported as face-centred cubic. A more detailed investigation by [4] using similar methods provided the main data for the diagram shown in **Fig. 164**, showing the two terminal solid solutions with a miscibility gap: The diagram has been modified slightly from that originally determined by slight changes in the miscibility gap boundary to include compositions for which the X-ray showed a two-phase structure [5].

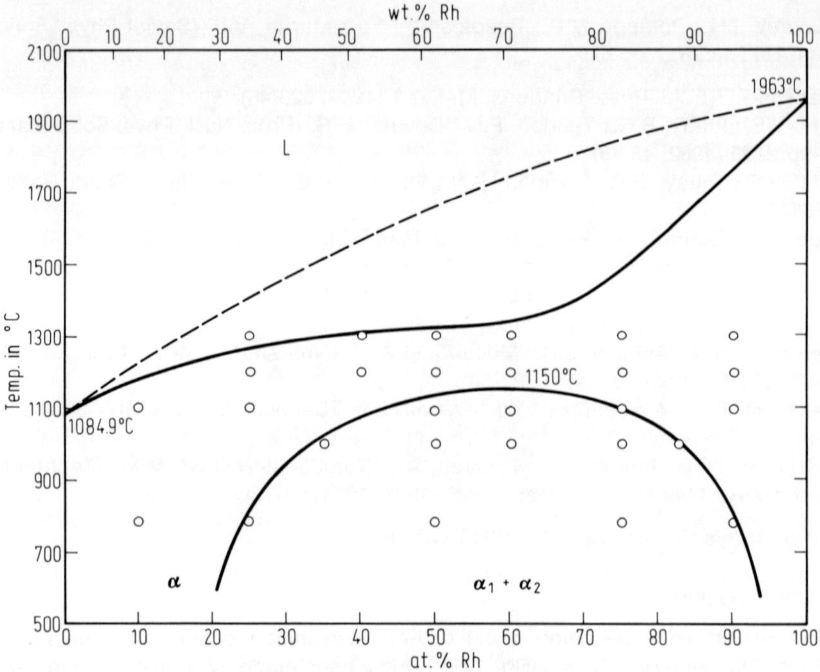

Fig. 164. Phase diagram Rh-Cu.

Preparation. Rhodium of 99.9% purity was melted with electrolytic copper in a high-frequency furnace in ceramic crucibles under a protective atmosphere; a check on the extent of contamination from the crucible material was made by comparison with alloys made by argon-arc melting. The latter were found to be very porous because of the high copper vapour pressure. The alloys were cold- or warm-worked. Annealing was carried out under hydrogen at high temperatures and at lower temperatures in evacuated quartz tubes [3]. The material was made from constituents of >99.9% purity melted by induction under hydrogen in alumina crucibles. Small ingots weighing about 3 g were cast and checked for weight loss; the uncertainties in composition were estimated as ±1.0 at%. Well mixed powders of 99.9% purity were vacuum-melted using an induction furnace. Homogenization treatments varied depending on the alloy composition; all were carried out in a vacuum of 10^{-5} Torr in quartz or wrapped in thin tungsten sheet [6, 7]. Alloys for surface tension measurements were prepared from 99.99% rhodium and 99.999% copper in zirconia crucibles under purified helium [8]. Some comments on the thermodynamic assessment of the system are given in [9].

Crystallographic Properties. The lattice constant drops almost linearly from 3.80 to 3.61 Å between 0 and 100 at% Cu content [4]. These results agree well with measurements made on melt-quenched material, the authors attributing a slight positive deviation to strain induced in the melt-quenching [3]. Surface segregation in a 5 at% copper alloy homogenized at different temperatures between 833 and 918 K has been studied by Auger electron spectroscopy; preferential segregation of copper to the surface was observed at all the temperatures used. The heat of segregation of copper was found to be 14±2 kcal/mol [10]. The solidification structures found in single-crystal alloys containing <15 at% rhodium grown unidirectionally from the melt have been studied microscopically and by electron microprobe analysis; it is shown that the microsegregation found could be reduced by high temperature annealing. The time required for homogenization is shorter than predicted by the Fleming model [11]. The microstructure and stability of rapidly solidified $Rh_{58}Cu_{42}$ has been studied by X-ray diffraction, transmission electron microscopy and continuous heating differential scanning calorimetry; this alloy did not form a complete extended solid solution, but the rapid solidification did increase the solubility limits beyond the equilibrium phase boundaries. During isothermal ageing at 900 °C the alloy underwent increased segregation to rhodium-rich and copper-rich phases. The quenched alloy was stable up to 375 °C under continuous heating at 10 °C/min; at temperatures up to 300 °C the alloy remained metastable [12].

References:

[1] Svyaginstsev, O.E.; Brunovskii, B.K. (Izv. Sekt. Platiny No. 12 [1935] 37/66).
[2] Hansen, M.; Anderko, K. (Constitution of Binary Alloys, 2nd Ed., New York-Toronto-London 1958).
[3] Luo, H.-L.; Duwez, P. (J. Less-Common Metals **6** [1964] 248/9).
[4] Raub, C.; Röschel, E.; Menzel, D.; Gadhof, M. (Metall [Berlin] **25** [1971] 761/2).
[5] Chakrabarti, D.J.; Laughlin, D.E. (Bull. Alloy Phase Diagrams **2** [1982] 460/3).
[6] Khan, H.R.; Raub, C.J.; Harmsen, N. (Mater. Res. Bull. **8** [1973] 1131/6).
[7] Negodayeva, N.Yu.; Sergin, B.I.; Timofeyev, A.I.; Dubinin, E.L. (Izv. Akad. Nauk SSSR Met. **1977** No. 6, pp. 52/6; Russ. Met. **1977** No. 6, pp. 43/6).
[8] Fogel'son, R.L.; Ugay, Ya.A.; Pokoyev, A.V.; Akimova, I.A. (Fiz. Tverd. Tela [Leningrad] **13** [1971] 1028/31; Soviet Phys.-Solid State **13** [1971] 856/8).
[9] Taylor, J.R. (Platinum Metals Rev. **29** [1985] 74/80).
[10] Sundaram, V.S.; Landers, R. (Appl. Surf. Sci. **10** [1982] 567/70).
[11] Tselentis, C.; Jardinier-Offerceld, M.; Bouillon, F. (Ann. Chim. [Paris] [15] **9** [1984] 141/4).
[12] Irons, L.; Mini, S.; Brower, W.E. (Mater. Sci. Eng. **98** [1988] 309/12).
[13] Fogel'son, R.L.; Ugay, Ya.A.; Pokoyev, A.V. (Fiz. Met. Metalloved. **33** [1972] 1102/4; Phys. Metals Metallog. [USSR] **33** No. 5 [1972] 194/6).
[14] Fogel'son, R.L.; Ugay, Ya.A.; Pokoyev, A.V. (Fiz. Met. Metalloved. **34** [1972] 1104/5; Phys. Metals Metallog. [USSR] **34** No. 5 [1972] 198/9).

Mechanical and Thermal Properties

Hardness. This rises with the solution of the alloying constituent and reaches a maximum of 450 kp/mm^2 at 60 at% rhodium [1]. Microhardness determinations showed a maximum of 554 kp/mm^2 at ~65 at% rhodium [2].

Surface Tension, Molar Volume. Fig. 165 shows experimental and calculated results obtained by [3]; the authors consider that the results suggest cluster grouping of the type $CuRh_5$.

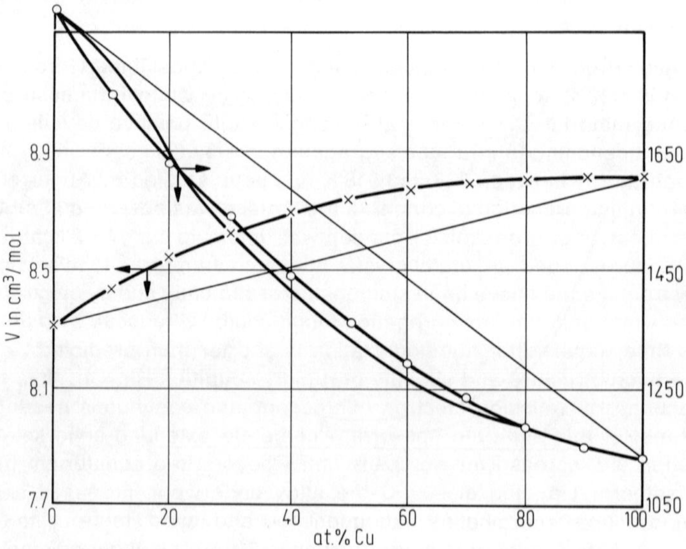

Fig. 165. Molar-volume and surface-tension isotherms for Rh-Cu alloys: solid lines show calculated values, dots show experimental values.

Low-Temperature Heat Capacity, Debye Temperature, Electronic Heat

Capacity. Low-temperature heat capacity measurements on a series of single-phase rhodium-copper alloys were made and θ_D and γ derived from C/T versus T^2 plots [2]; details of the single-phase alloy preparation are given in [4]. The C/T plots for two of the alloys are shown in a figure, and the derived θ_D and γ figures are given in the following table, together with microhardness figures.

Electronic heat capacity coefficient γ, Debye temperature θ_D and microhardness data for Rh-Cu alloys.

sample	γ (mJ/g-atom·K^2)	θ_D in K	microhardness $H_{V0.02}$
Rh	4.7	500	132
$Rh_{90}Cu_{10}$	3.1	339	373
$Rh_{80}Cu_{20}$	2.7	269	484
$Rh_{65}Cu_{35}$ a)	2.7	348	—
$Rh_{65}Cu_{35}$ b)	2.8	346	554
$Rh_{38}Cu_{62}$	2.5	324	304
$Rh_{17}Cu_{83}$	1.2	342	202
Cu	0.7	344	50

a) Rh not removed from surface, b) Rh removed from surface.

The initial drop in γ with increase in copper up to 20 at% is attributed to the filling of the d band, but above 20 at% to ~60 at% the copper conduction electron enters the s band resulting in no change in γ; this agreed with magnetic data given in [2]. The peaks in θ_D and microhardness at 40 at% copper are not considered to be due to crystallographic ordering [2]. Determinations of γ low-temperature specific heat and Debye temperature have also been made by [5].

Free Energy of Formation. Curves for solid and liquid alloys are shown in **Fig. 166** [6].

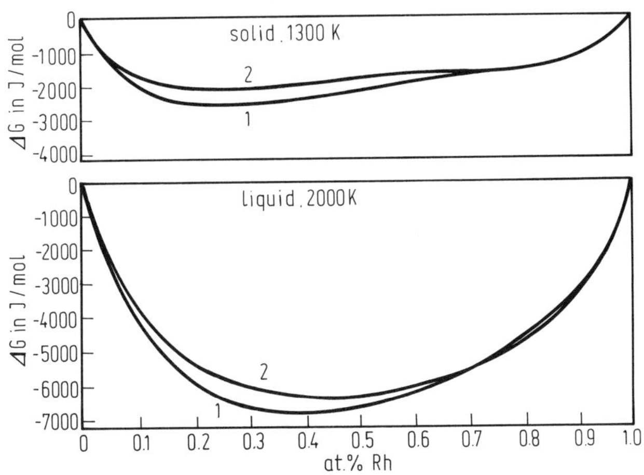

Fig. 166. Free energy of formation for solid and liquid Rh-Cu alloys.

References:

[1] Raub, C.; Röschel, E.; Menzel, D.; Gadhof, M. (Metall [Berlin] **25** [1971] 761/2).
[2] Vishwanathan, R.; Khan, H.R.; Raub, C.J. (J. Phys. Chem. Solids **37** [1976] 431/2).
[3] Negodayeva, N.Yu.; Sergin, B.I.; Timofeyev, A.I.; Dubinin, E.L. (Izv. Akad. Nauk SSSR Met. **1977** No. 6, pp. 52/6; Russ. Met. **1977** No. 6, pp. 43/6).
[4] Khan, H.R.; Raub, C.J.; Harmsen, N. (Mater. Res. Bull. **8** [1973] 1131/6).
[5] Kuentzler, R.; Ebert, H.; Winter, H.; Abart, J.; Voitlander, J. (Solid State Commun. **62** [1987] 145/8).
[6] Taylor, J.R. (Platinum Metals Rev. **29** [1985] 74/80).

Electrical and Magnetic Properties

de Haas-van Alphen Effect. This has been used to study the electronic structure of dilute alloys containing a few hundred ppm rhodium; the work showed the dominant d scattering of the conduction electrons expected from a Friedel virtual-bound state [1]. These results have been reanalysed which included measurements of residual resistivity, thermoelectric power and photoelectron spectra by [2].

Electronic Structure. For alloys of the composition Cu_xRh_{1-x} the total and component densities of state are calculated and compared with available experimental data [3]. Starting from the Mott formula, the thermoelectric power for rhodium in copper is calculated [4]. Alloys of composition Cu_xRh_{1-x} have been analysed in compositional steps of $x=0.25$ by a modified augmented spherical wave (ASW) method and the results checked against experimental magnetic susceptibility measurements [5]. Calculations using charge self-consistent

Korringa-Kohn-Rostoker coherent potential approximation methods (KKR-CPA) have been made and the results discussed in the light of recent photoemission, susceptibility and nuclear magnetic resonance data [6]. Calculations based on the tight-binding muffin-tin orbital method and single-site approximation have been made by [7].

Photoelectron Spectra. The photoelectron spectra of the valence bands of rhodium and rhodium-copper alloys have been measured by [8]; some of the results at the excitation energy of 40 eV are shown in **Fig. 167**. UV photoelectron spectra were determined on alloys up to 5.5 at% by [2]. Soft X-ray emission spectra on Cu_xRh_{1-x} with x=0, 0.05, 0.10, 0.80 and 1.0 were measured and the results compared with KKR-CPA (Korringa-Kohn-Rostoker coherent potential approximation) calculations [9].

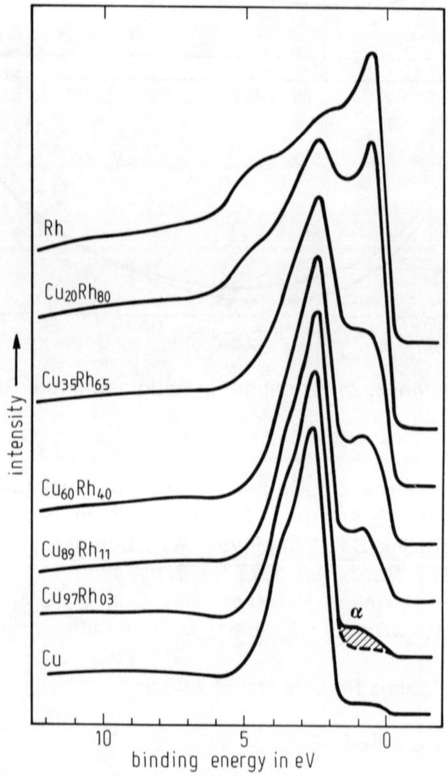

Fig. 167. The spectra of five alloys with different compositions as well as pure metals at the excitation energy of 40 eV.

Electric Resistance. The atomic raising of resistance Δϱ/atom for up to 20 at% copper in rhodium is given as 4.4 [10]. Resistivity measurements on alloys containing up to 10 at% rhodium measured in the temperature range 4 to 77 K show deviations from Matthiessen's law [11]. Residual resistivity measurements on dilute alloys are given by [1, 12, 13]. The change in resistivity Δϱ is plotted against at% rhodium for alloys up to 6 at% by [2].

Thermoelectric Force. This has been calculated by [4]. Measurements on dilute alloys at low temperatures have been made by [2].

Miscellaneous Magnetic Properties. The mass magnetic susceptibility as a function of composition at 135 and 290 K is shown in **Fig. 168**, whilst **Fig. 169** shows susceptibilities as a function of temperature [15]. The spin susceptibility calculated by the fixed-spin moment method is plotted against experimental bulk susceptibility obtained by [14] in [5].

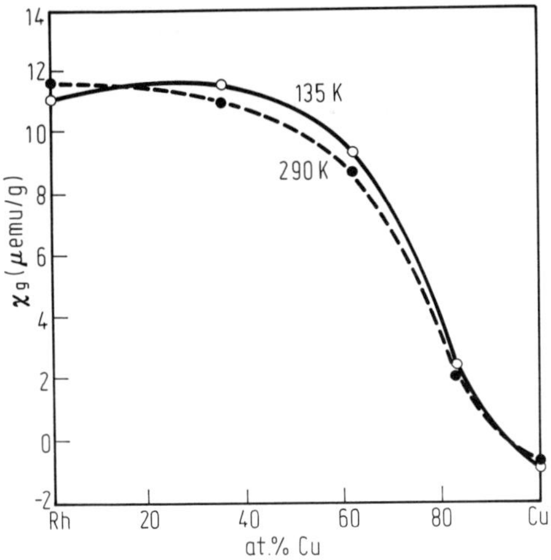

Fig. 168. Mass magnetic susceptibility of Cu-Rh alloys as a function of Cu at 135 and 290 K.

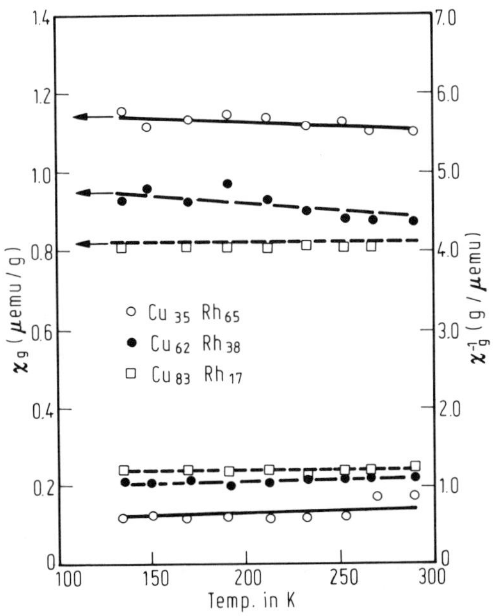

Fig. 169. Susceptibilities of three Rh-Cu alloys as a function of temperature.

Nuclear Magnetic Resonance. By means of rapid quenching, single-phased Cu_xRh_{1-x} alloys ($0 \leqq x \leqq 0.8$) were obtained and the Knight shift of ^{63}Cu and ^{103}Rh was determined using pulsed NMR at low temperatures; the magnetic susceptibility was also measured between 4.2 and 300 K. The composition dependence of the Knight shift for rhodium and copper is shown in **Fig. 170** and **171** [14]. These results are further discussed by [6].

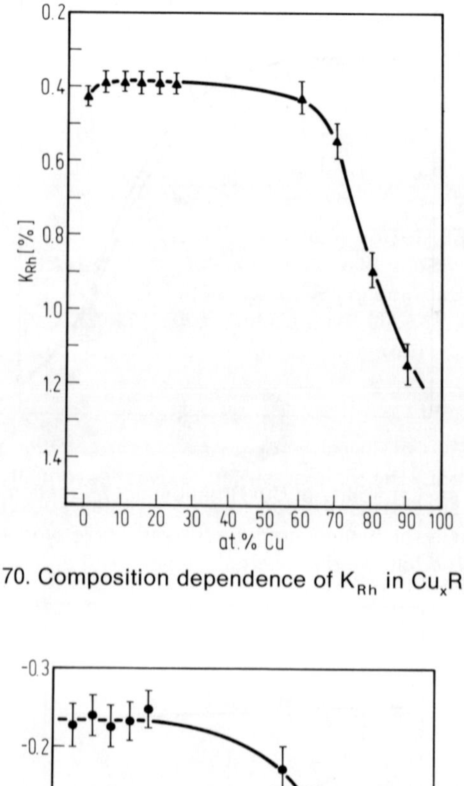

Fig. 170. Composition dependence of K_{Rh} in Cu_xRh_{1-x}.

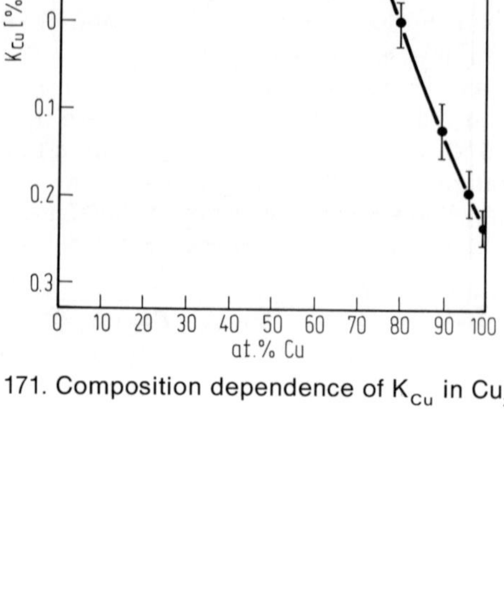

Fig. 171. Composition dependence of K_{Cu} in Cu_xRh_{1-x}.

References:

[1] Coleridge, P.T.; Templeton, I.M.; Toyoda, T. (J. Phys. F **11** [1981] 2345/7).
[2] Julianus, J.A.; Myers, A.; Bekker, F.F.; van der Marel, D.; Allen, E.F. (J. Phys. F**15** [1985] 111/39).
[3] Masek, J.; Kudrnovsky, J. (Solid State Commun. **54** [1985] 981/4).
[4] Mertig, I.; Mrosan, E.; Zeller, R.; Dederichs, P.H.; Ziesche, P. (Phys. Status Solidi B**129** [1985] 407/14).
[5] Mohn, P.; Schwarz, K. (Solid State Commun. **57** [1986] 103/7).
[6] Ebert, H.; Winter, H.; Abart, J.; Voitländer, J. (J. Phys. F**17** [1987] 1457/69).
[7] Kudrnovsky, J.; Drchal, V. (Solid State Commun. **65** [1988] 613/6).
[8] Ishii, H.; Miyahara, T.; Hanyu, T. (J. Phys. Soc. Japan **53** [1984] 2151/6).
[9] Franz, W.; Steeb, S.; Ebert, H.; Winter, H.; Voitländer, J. (Z. Physik B**69** [1987/88] 257/61).
[10] Pawlek, F.; Reichel, K. (Z. Metallk. **47** [1956] 347/56).

[11] Loegel, B. (J. Phys. F**3** [1973] L106/L109).
[12] Toyoda, T.; Kume, K. (Solid State Commun. **15** [1974] 1889/90).
[13] Toyoda, T. (J. Phys. Soc. Japan **39** [1975] 76/83).
[14] Abart, J.; Voitländer, J. (Solid State Commun. **40** [1981] 277/80).
[15] Khan, H.R.; Raub, C.J.; Harmsen, N. (Mater. Res. Bull. **8** [1973] 1131/6).

36.2 Ternary Alloys

Rh-Cu-Al. Single crystals of $Rh_{50}Al_{50-x}Cu_x$ were prepared by the solution growth method using copper as a solvent. The crystal structure was essentially the same as that of RhAl(CsCl-type B2). No superstructure due to an orientated substitution by copper atoms was observed. The stability of these compounds was correlated to the atomic size and the electronic structure. The following lattice parameters and densities were found [3]:

composition	a in Å	D in g/cm³
$Rh_{50}Al_{47}Cu_3$	2.973(4)	8.1
$Rh_{50}Al_{48}Cu_2$	2.974(7)	8.1
$Rh_{50}Al_{46}Cu_4$	2.973(9)	8.4
$Rh_{50}Al_{43}Cu_7$	2.970(7)	8.5
$Rh_{50}Al_{44}Cu_6$	2.970(3)	8.4
$Rh_{50}Al_{42}Cu_8$	2.969(5)	8.5
$Rh_{50}Al_{41}Cu_9$	2.968(5)	8.6

Rh-Cu-Sn. The crystal structure and magnetic properties were determined for a series of compounds of the formula Rh_2TSn where T = Mn, Ni, Cu, V, Cr, Fe or Co. The alloys were made by mixing powders and annealing in evacuated quartz ampoules, firstly at 250 °C for 7 h, followed by 6 days at 700 °C; finally, the samples were pulverized and reannealed in vacuum for 5 days at 700 °C. Where T = Cu, the fully ordered $L2_1$ Heusler structure was observed at room temperature. The lattice constants were a = 6.146, c = 6.146 Å [1]. The interaction of ^{119}Sn has been studied by the Mössbauer effect in $Rh_xCu_{0.998-x}Sn_{0.002}$; it was found that there was an attractive interaction between the Sn and Rh impurities in the copper host [2].

Rh-Cu-Fe. An abrupt change in magnetization of $FeRhCu_{0.0833}$ occurred at a critical temperature (T_c); σ was 143.2 emu/g at T_c = 486 °C. The lattice parameter a was reported as 2.989 Å (CsCl-type) [1].

References:

[1] Suits, J.C. (Solid State Commun. **18** [1976] 423/5).
[2] Kmiec, R.; Hrynkiewicz, A.Z.; Królas, K.; Tomala, K. (J. Phys. F **17** [1987] 1349/55).
[3] Shishido, T.; Takei, H. (J. Less-Common Metals **119** [1986] 75/82).

36.3 Quinary Alloys Rh-Cu-Fe-Co-Ni

The effect of the addition of rhodium to an alloy with 0.4% Cu, 13.7% Co, 30% Ni, rest Fe was to increase the coefficient of expansion without altering the Curie temperature.

Solov'eva, N.A.; Yudkevich, M.I.; Pasternak, I.I.; Pogosov, V.Z. (Metalloved. Term. Obrab. Met. **1968** No. 4, pp. 45/6; Metal. Sci. Heat Treat. Metals [USSR] **1968** 293/4).

37 Alloys with Silver

37.1 The Rh-Ag System

Phase Diagram. Crystallography. Early work on the system suggested that rhodium was insoluble in molten silver [1]. Later work using X-ray methods showed that all alloys consisted of practically pure silver and a solid solution of silver in rhodium containing at most 0.1 at% of silver [2]. A comprehensive study carried out more recently paid particular attention to the rhodium-rich end of the compositional range and indicated a solid solubility of about 5 wt% of silver in rhodium; this increases to 10 wt% at 1400 °C. The solubility of rhodium in silver is extremely small. Liquid immiscibility was noted between 25 and 99.9 wt% silver. The authors made use of metallography, differential thermal analysis, hardness, electric resistivity and thermoelectric measurements in establishing their diagram. The equilibrium diagram is shown in the original paper [3]. An assessed phase diagram based largely on the experimental results of [3], but with the addition of some thermodynamically calculated results is shown in **Fig. 172**; some further diagrams indicating the effect of hydrostatic pressure on the vapour-phase boundaries are given by [4]. In an investigation of binary systems with positive heats of formation and likely to have small miscibility in the solid state, the silver-rhodium alloys were found to show an extension of the terminal solid solubilities in the metastable state obtained by ion-beam mixing. Alloys were formed from multilayered samples bombarded at 77 K by 400 keV Kr^+ ions. The following lattice parameters are given [5]:

composition . .	$Ag_{70}Rh_{30}$	$Ag_{60}Rh_{40}$	$Ag_{50}Rh_{50}$	$Ag_{25}Rh_{75}$
a in Å	4.052(5)Ag fcc	4.050(15)Ag fcc 3.960(16)Rh fcc	3.926(5)Rh fcc	3.874(5)Rh fcc

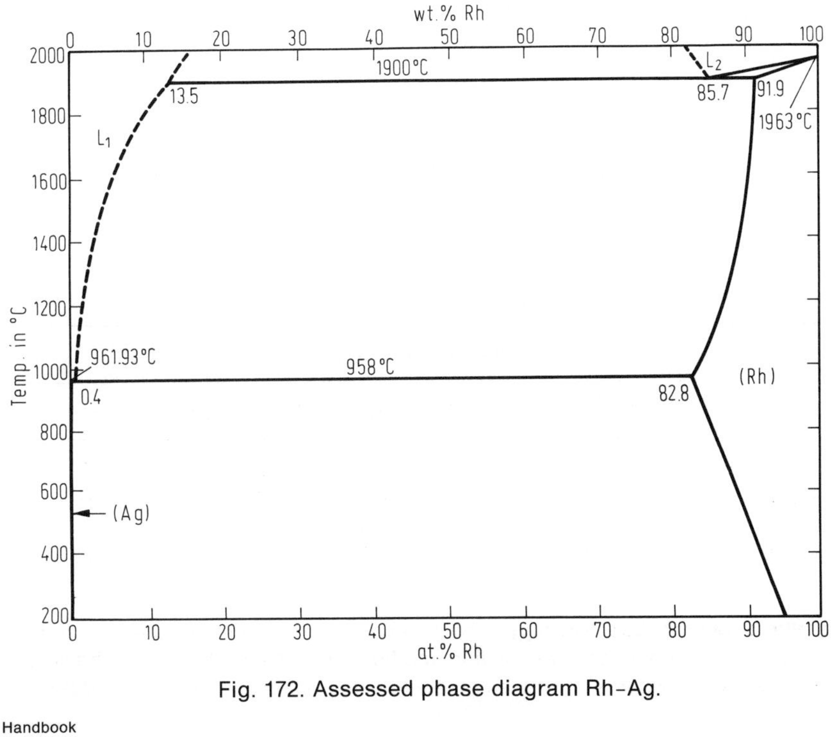

Fig. 172. Assessed phase diagram Rh-Ag.

The **heat of formation** of the $Ag_{70}Rh_{30}$ alloy was 14.1 kJ/mol [5].

Preparation. Alloys were made from refined rhodium and silver produced by zinc reduction of the chloride; they were melted in alumina crucibles using a high-frequency furnace. Silver-rich alloys were kept molten for 4 h to ensure maximum solution of the rhodium. The electrical properties were measured on 1 and 0.5 mm wires where possible, otherwise a regular section was obtained by sucking into 2 to 3 mm porcelain tubes. Annealing was carried out for 125 h in a resistance furnace, the rhodium-rich alloys at 1200 °C and the silver-rich ones at 850 °C, cooling to room temperature at 100 °C/d. Quenched samples of rhodium-rich alloys were prepared from 1400 °C and silver-rich ones from 850 °C [3].

Mechanical and Electrical Properties. These are summarized in the following tables [3].

Properties of silver-rhodium alloys.

wt% Ag	Brinell hardness in kg/mm^2	microhardness of annealed alloys		microhardness of quenched alloys		electric resistivity, in $\Omega \cdot cm$		temperature coefficient of resistivity	Brinell hardness for quenched alloys
		Rh phase	Ag phase	Rh phase	Ag phase	ϱ_{25}	ϱ_{100}		
5	175	154.2	—	80.0	—	14.63	15.63	9.53	196.4
10	137	217.2	—	85.0	—	11.5	13.75	28.0	195.1
15	131.7	247.2	—	118.0	—	7.14	8.82	34.0	107.2
20	108.0	283.8	—	150.0	—	5.86	7.17	34.0	79.3
30	104.1	283.8	—	175.0	—	5.43	6.4	23.4	—
40	—	283.8	—	175.0	—	—	—	—	—
90	27.5	—	20.2	—	23.2	—	—	—	—
95	27.0	—	20.2	—	23.5	1.54	1.94	42.0	—
97	26.7	—	20.4	—	23.2	1.52	1.93	43.5	—
98	26.5	—	20.0	—	23.0	1.5	1.92	41	—
99	26.7	—	20.45	—	22.5	1.46	1.87	43	—
99.5	26.0	—	20.2	—	22.0	1.48	1.90	40.5	—
99.9	26.5	—	20.5	—	22.5	—	—	—	—

Thermoelectric power of silver-rhodium alloys.

wt% Ag	integral thermal emf									
	100	200	300	400	500	600	700	800	900	950
99.5	+0.8	+1.75	+3.05	+4.6	+6.45	+8.4	+10.6	+13.3	+16.0	+17.45
99	+0.8	+1.75	+3.06	+4.6	+6.45	+8.4	+10.6	+13.3	+16.0	+17.45
98	+0.79	+1.73	+3.02	+4.5	+6.41	+8.35	+10.6	+13.3	+16.3	+17.45
97	+0.79	+1.73	+3.03	+4.5	+6.4	+8.35	+10.6	+13.3	+16.03	+17.45

X-ray Photoelectron Spectroscopy. The valence band spectra of silver-rhodium films prepared by the simultaneous evaporation method have been determined. The rhodium spectra in silver were observed at 1.0 eV; the width was estimated to be 1.4 eV [6].

References:

[1] Rössler, H. (Chemiker-Ztg. **24** [1900] 733/5).
[2] Drier, R.W.; Walker, H.L. (Phil. Mag. [7] **16** [1933] 294/8).
[3] Rudnitskii, A.A.; Khotinskaya, A.N. (Zh. Neorg. Khim. **4** [1959] 2308/12; Russ. J. Inorg. Chem. **4** [1959] 1053/6).
[4] Karakaya, I.; Thompson, W.T. (Bull. Alloy Phase Diagrams **7** [1986] 362/5).
[5] Peiner, E.; Kopitzki, K. (Nucl. Instrum. Methods Phys. Res. B**34** [1988] 173/80).
[6] Ishii, H.; Hanyu, T.; Yamaguchi, S. (J. Phys. Soc. Japan **44** [1978] 1395/6).

37.2 Ternary Alloys

Rh-Ag-As. The substitution of silver into $RhAs_3$ to obtain compounds of the type $Ag_xRh_{1-x}As_3$ has been investigated up to temperatures of 750 °C. No subsitution with silver was found, the phases present being $RhAs_3$ with no change in lattice parameter, Ag_8As and As [3].

Rh-Ag-Dy. A section of the ternary diagram at 600 °C was determined by [2].

Rh-Ag-Sn. The interaction between two soluble impurities in a metal host, in this case of composition $Ag_{0.997}Rh_{0.001}Sn_{0.002}$, has been studied using the Mössbauer effect on ^{119}Sn nuclei. It was found that there is an attractive interaction between the tin and rhodium atoms. The binding energy of the rhodium-tin pairs was found to be −194 25 meV [1].

References:

[1] Kmiec, R.; Hrynkiewicz, A.Z.; Królas, K.; Tomala, K. (J. Phys. F**17** [1987] 1349/55).
[2] Slavev, A.G.; Portnoi, V.K.; Sokolovskya, E.M. (Deposited Doc. VINITI-906-77 [1977] 10 pp.; C.A. **90** [1979] No. 91055).
[3] Bennett, S.L.; Heyding, R.D. (Can. J. Chem. **44** [1966] 3017/30).

38 Alloys with Gold

38.1 The Rh–Au System

Diffusion. The bulk diffusion of gold into rhodium in thin films has been studied by [13]. D_0 was found to be 4.2×10^{-11} cm^2/s and $Q = 0.93 \pm 0.28$ eV. The bulk diffusion coefficients are plotted against reciprocal temperature in **Fig. 173**. The authors point out that these values of Q and D_0 are lower than would be expected for lattice diffusion in face-centred cubic metals, and may not be applicable to bulk material.

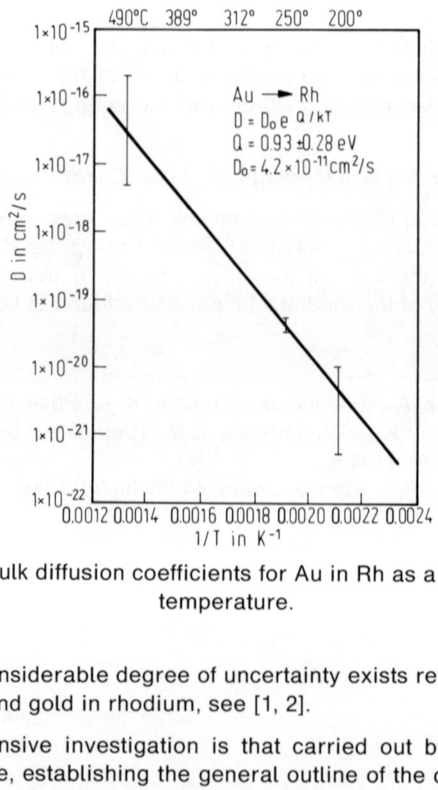

Fig. 173. Thin-film bulk diffusion coefficients for Au in Rh as a function of reciprocal temperature.

Phase Diagram. A considerable degree of uncertainty exists regarding the solid solubilities of rhodium in gold and gold in rhodium, see [1, 2].

The most comprehensive investigation is that carried out by [1], which covered the whole composition range, establishing the general outline of the diagram, showing a liquid immiscibility gap, terminal solid solutions, and monotectic and peritectic reactions. The diagram was established from thermal analysis and microscopical studies; however, hardness, electric resistivity, tensile and thermoelectric measurements were also made. There were some obvious inconsistencies in the paper which are pointed out by [3, 4]. The terminal solubilities and peritectic reaction were investigated by thermal analysis, microscopy, and lattice parameter measurements by [2]; **Fig. 174** shows the diagram due to [1] with these results added. The available data have been assessed by [3], and supplemented by the authors' own thermodynamical models of the probable diagram.

Preparation. The alloys were melted in alumina crucibles using a high-frequency furnace. The gold and rhodium were refined and reduced by ethyl alcohol, the losses in melting not exceeding 1.3%; electrical properties were measured on 0.5 to 1.0 mm wires and more brittle alloys were produced by sucking the melt into 2 to 3 mm porcelain tubes. Details of annealing procedure are given [1].

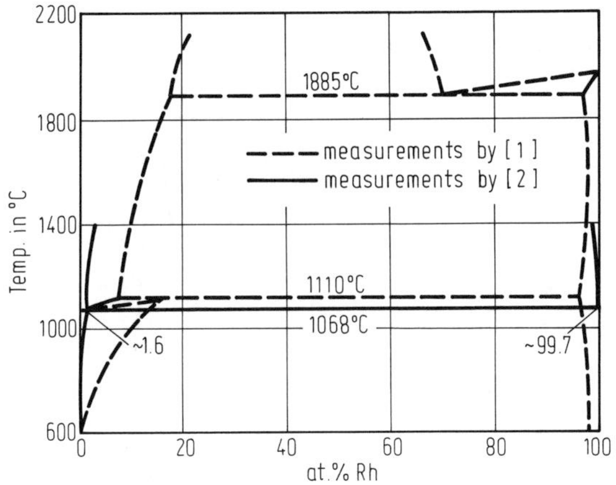

Fig. 174. Phase diagram Rh-Au.

Crystallographic Properties. The lattice constants of gold-rhodium alloys are given in the table below with details of the annealing treatments [2].

Rh in at%	wt%	annealing t in °C	time in h	lattice constant in kX Au	Rh
3	1.59	1000	17	4.06_7	3.78_7
		800	22	4.06_6	—
6.4	3.45	800	326	4.06_6	3.78_7
		600	34	4.07_0	3.79_5
9.2	5.03	1000	55	4.06_0	3.79_5
		600	470	4.07_0	3.79_5
12.9	7.18	1000	55	4.06_5	3.79_6
		600	470	4.07_0	3.79_4
15	8.44	1000	17	4.06_7	3.79_6
		800	86	4.06_7	3.79_6
		600	34	4.06_7	3.79_5
50	34.31	1000	17	4.06_7	3.79_7
92	85.71	800	126	—	3.79_6
97	94.39	600	327	—	3.79_6
97.3	94.95	800	326	—	3.79_6
		600	34	—	3.79_7
98	96.24	800	86	—	3.79_6
99	98.08	600	158	—	3.79_6
99.5	99.02	1000	55	—	3.79_6
99.7	99.41	600	470	—	3.79_4

A number of alloys showing positive heats of formation and hence the possibility, as in Rh-Au alloys of a liquid or solid state miscibility gap, were bombarded in multilayered form at 77 K by 400 keV Kr^+ ions. In the case of Rh-Au a continuous series of single-phase metastable fcc solid solutions were produced. The lattice parameters are given: for $Au_{75}Rh_{25}$, $Au_{50}Rh_{50}$ and $Au_{25}Rh_{75}$, a = 4.033(6), 3.98(5) and 3.914(6) Å [5].

Hardness. The Brinell hardness and the microhardness of gold and rhodium phases of annealed and quenched alloys are given in **Fig. 175** [1].

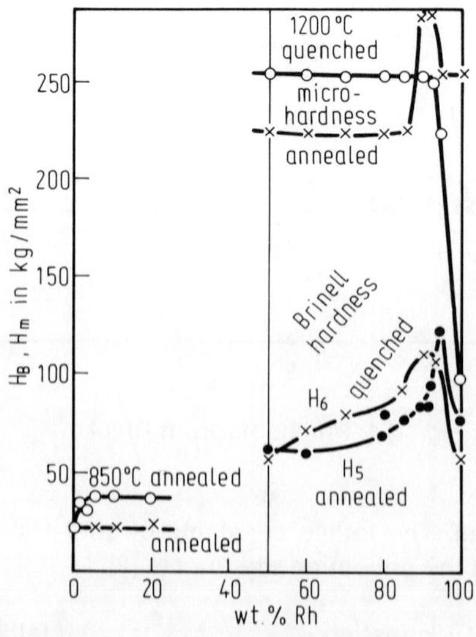

Fig. 175. Brinell hardness H_B and microhardness H_m for annealed and quenched Rh-Au alloys.

Tensile Properties. These are shown in **Fig. 176** [1].

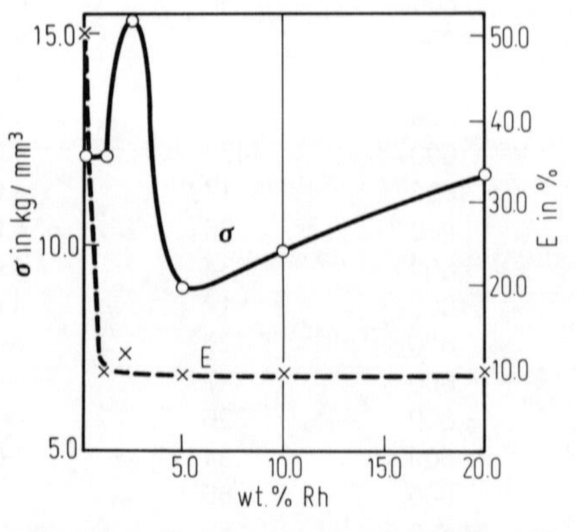

Fig. 176. Tensile strength σ and extensibility E of Rh-Au alloys.

Dissociation Energy of Gaseous Alloys. By mass spectrometric Knudsen effusion technique, for RhAu(g) D (298 °C) was found to be 55.9±8 kcal/mol (233.9±33.5 kJ/mol) [6]. The heat of formation of $Au_{75}Rh_{25}$ was said to be 10.7 kJ/mol [13].

Electric Resistivity. Determinations on annealed and quenched alloys are shown in **Fig. 177** [1]. The resistivity of thin films of rhodium-gold alloys on glass, produced from resinated media is given by [7]. The resistivity of gold alloys containing up to 1% rhodium has been measured at low temperatures; the resistance was found to remain constant below 4.2 K [8]. The residual resistivities of dilute solutions of 4 d transition metal impurities were measured at 1.5 to 4.2 K by [9, 10]. Residual resistivities of a number of dilute 4 d element alloys have been measured at 4.2 K; results for rhodium are compared with those of [10, 11]. The following low-temperature results on dilute alloys were made by [12].

at% Rh	0.29	0.3	0.63	1.05
ϱ in $\mu\Omega\cdot$cm . .	0.96	0.97	2.12	2.82
T in K	7	7	10	12

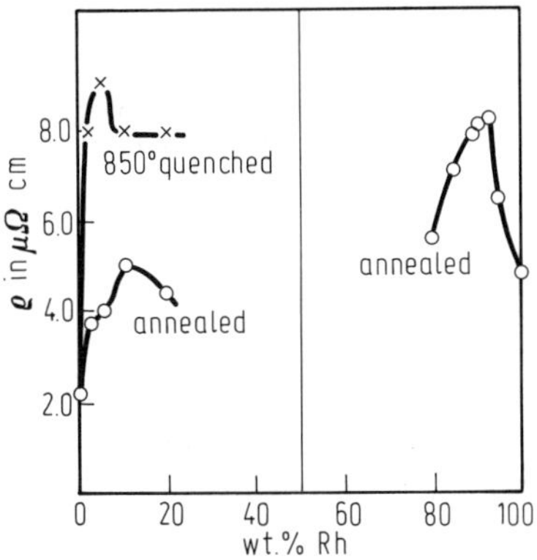

Fig. 177. Electric resistivity of annealed and quenched Rh-Au alloys.

Magnetoresistance. This was measured on gold containing up to 1 at% rhodium; $\Delta R/R_{H=0}$ was found to be positive [8].

Thermoelectric Properties. The thermal emf of the alloys against platinum was determined by [1]. Results are shown in the following table.

Thermoelectric properties of gold-rhodium alloys [1].

wt% Rh	integral thermo emf in mV										
	t in °C										
	100	200	300	400	500	600	700	800	900	1000	1100
0	+0.7	+2.8	+3.05	+4.5	+6.05	+7.9	+9.9	+12.05	+14.3	+16.8	–
1	+0.1	+0.4	+0.75	+1.5	+2.0	+2.8	+3.75	+ 4.8	+ 5.8	+ 9.6	+10.75
2.5	−0.15	−0.1	+0.05	+0.3	+0.65	+1.1	+1.7	+ 2.45	+ 3.35	+ 4.3	+ 5.25
10	−0.1	−0.05	−0.05	+0.4	+0.8	+1.35	+2.0	+ 2.8	+ 3.7	+ 4.8	+ 5.9
20	−0.05	+0.05	+0.25	+0.55	+0.95	+1.45	+2.10	+ 2.9	+ 3.65	+ 4.4	+ 4.9

The low temperature (1.5 to 15 K) absolute thermoelectric power has been measured on dilute alloys containing 0.29, 0.3, 0.63 and 1.05 at% rhodium; the results are shown in a figure in the original paper [12].

References:

[1] Rudnitskii, A.A.; Khotinskaya, A.N. (Russ. J. Inorg. Chem. **4** [1959] 1160/3).
[2] Raub, E.; Falkenberg, G. (Z. Metallk. **55** [1964] 392/7).
[3] Okamoto, H.; Massalski, T.B. (Bull. Alloy Phase Diagrams **5** [1984] 384/7).
[4] Elliott, R.P. (Constitution of Binary Alloys, 1st Suppl., New York – Toronto – London 1965, pp. 1/877).
[5] Peiner, E.; Kopitzki, K. (Nucl. Instrum. Methods Phys. Res. B**34** [1988] 173/80).
[6] Cocke, D.L.; Gingerich, K.A.; Kordis, J. (High Temp. Sci. **7** [1975] 61/73).
[7] Prinsen, P. (Electrocomponent Sci. Technol. **5** [1978] 41/3).
[8] Knook, B.; van den Berg, G.J. (Proc. 7th Intern. Conf. Low Temp. Phys., Toronto 1960 [1961], pp. 257/8).
[9] Toyoda, T. (J. Phys. Soc. Japan **39** [1975] 76/83).
[10] Toyoda, T.; Kume, K. (Solid State Commun. **15** [1974] 1889/90).

[11] Myers, A.; Bekker, F.F.; van Nassou, H. (J. Phys. F**10** [1980] 461/9).
[12] Julianus, J.A.; Bekker, F.F.; de Châtel, P.F. (J. Phys. F**14** [1984] 2061/76).
[13] de Bonte, W.J.; Poate, J.M.; Melliar-Smith, C.M.; Levesque, R.A. (J. Appl. Phys. **46** [1975] 4284/90).

38.2 Ternary Alloys

Rh-Au-Ti. Interdiffusion between Rh/Au/Ti thin film systems was studied in detail by [1].

Rh-Au-Nb. Superconductivity was investigated of alloys of the type $Nb_3Rh_{1-x}B_x$ where B = Au; all these alloys have the Cr_3Si-type structure. The lattice parameters and transition temperatures are given below [2].

compound	lattice parameter in Å	T_c in K
$Rh_{0.98}Au_{0.02}Nb_3$	5.133	2.53
$Rh_{0.95}Au_{0.05}Nb_3$	5.137	2.52
$Rh_{0.90}A_{0.10}Nb_3$	5.1412	2.70
$Rh_{0.70}Au_{0.30}Nb_3$	5.1573	4.6
$Rh_{0.5}Au_{0.50}Nb_3$	5.1688	6.6
$Rh_{0.30}Au_{0.70}Nb_3$	5.1827	9.5
$Rh_{0.10}Au_{0.90}Nb_3$	5.1960	10.8
$Rh_{0.05}Au_{0.95}Nb_3$	5.200	11.0
$Rh_{0.02}Au_{0.98}Ab_3$	5.203	10.9

References:

[1] DeBonte, W.J.; Poate, J.M.; Melliar-Smith, C.M.; Levesque, R.A. (J. Appl. Phys. **46** [1975] 4284/90).
[2] Zegler, S.T. (Phys. Rev. [2] **137** [1965] A 1438/A 1440).

39 Alloys with Ruthenium

39.1 The Rh-Ru System

Phase Diagram. It was suggested that the solid solubility of rhodium in ruthenium is >20 wt% (19.7 at%) and that of ruthenium in rhodium is >20 wt% (20.3 at%) [2]. A comprehensive investigation of the whole composition range between 900 and 2300 °C was carried out, using metallography, X-ray diffraction, differential thermal analysis, electron microprobe and microhardness measurements. The system was found to be of the simple peritectic type, and intermetallic phases were not found. **Fig. 178** shows the phase diagram due to [3].

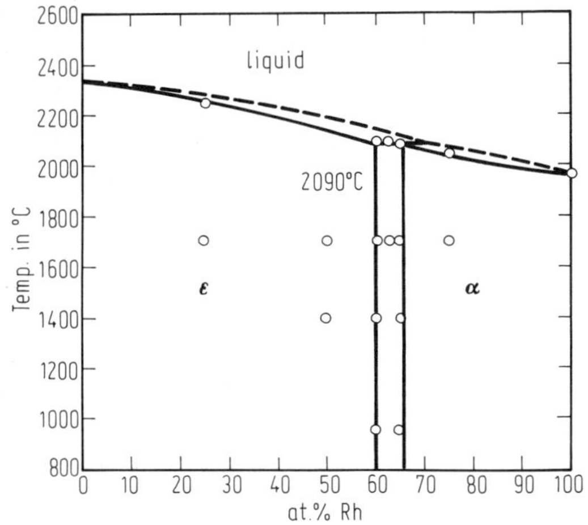

Fig. 178. Phase diagram Rh-Ru.

Preparation. The alloys were prepared from purified sponges in the form of 5 to 10 g buttons which were melted two or three times in an argon-arc furnace and then partly homogenized by heating in the arc for about 5 min as near the melting point as possible. The resulting material was ground to powder in a tungsten carbide percussion mortar and then annealed for 12 h at 1050 °C in sealed evacuated silica tubes [2]. Powdered 99.9% sponges were used; the powders were hand-mixed and pressed in a steel die to form pellets, which were melted under reduced argon pressure and homogenized under a high vacuum at various temperatures which are shown in the lattice parameter data. The rate of cooling from the annealing temperature was at 250 K/min. Checks were made of any weight losses [3].

Crystallography. The lattice parameters of pure ruthenium, and the dilute solid solutions of rhodium in it were determined by [1]. The alloy results are given below; the lattice constants given in kX units at 25 °C are uncorrected for refractive index.

at% Rh	a in kX	c in kX	c/a
1.93	2.7003(2)	4.2736(8)	1.5826(5)
3.12	2.7000(5)	4.2736(4)	1.5828
5.30	2.6999(3)	4.2739(9)	1.5830

The following table gives details of heat treatments, composition, and lattice parameters of sample used [3]:

Heat treatment, composition and lattice parameters of samples from the Ru–Rh system.

heat treatment	nominal compo-sition (at% Rh)	composition [a] (at% Rh)		phases	lattice parameters in pm			
		ε	α		ε			α
					a	c	c/a	a
	0				270.6	428.1	1.582	
1700 °C, 24 h	25	26.1±0.1		ε	270.5	429.1	1.586	
	50	50.3±0.3		ε	270.6	430.4	1.591	
	60	60.8±0.3	—	ε, α	270.5	431.5	1.595	380.2
	62.5	—	65.4±0.1	ε, α	270.8	432.0	1.595	380.7
	65	—	65.5±0.1	(ε), α	(270.9)	—	—	380.6
	75		74.9±0.1	α				380.7
1400 °C, 80 h	50	49.7±0.3		ε	271.2	431.2	1.590	
	60	59.5±0.3		ε	271.2	432.3	1.594	
	65	60.2±0.1	64.4±0.5	ε, α	271.0	432.0	1.594	380.6
950 °C, 450 h	60	60.2±0.4		ε	271.2	433.6	1.598	
	65	—	65.3±0.1	ε, α	271.2	433.9	1.599	381.5
	100							380.4

[a] Measured by electron microprobe analysis, normalized to 100% in the binary Ru–Rh system.

Hardness. Fig. 179 shows microhardness results at room temperature carried out under a load of 1 N [3].

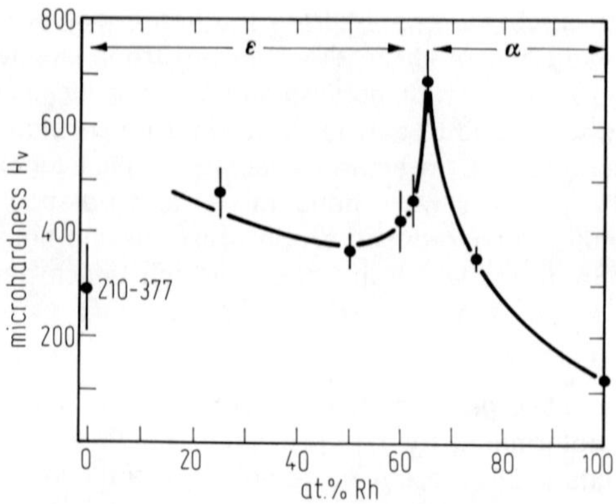

Fig. 179. Microhardness of Rh–Ru alloys annealed at 1700 °C.

Specific Heat, Debye Temperature. The low-temperature (1.4 to 4.2 K) specific heat has been determined for a series of alloys containing up to 40 at% Ru; cT versus T^2 plots showed straight line behaviour. The following figures for the electronic specific heat coefficient γ and the Debye temperature θ_D were determined; the electronic implications are discussed [4].

at% Ru	$\gamma \cdot 10^{-1}$ cal·mol^{-1}·deg^{-2}	θ_D K
40	11.00	527.9
35	10.40	444.3
25	10.48	396.9
20	10.97	454.7
15	10.89	427.5
10	11.29	429.8
5	11.65	456.6
0	12.03	528.6

References:

[1] Hellawell, A.; Hume-Rothery, W. (Phil. Mag. [7] **45** [1954] 797/806).
[2] Jaffee, R.I.; Maykuth, D.J.; Douglass, R.W. (CNR-039-067 [1960]; Metall. Soc. Conf. **11** [1960] 383/463; C.A. **56** [1962] 3246).
[3] Paschoal, J.O.A.; Kleykamp, H.; Thümmler, F. (J. Less-Common Metals **98** [1984] 279/84).
[4] Tsang, P.J.M.; Wei, C.T. (NBS-SP-323 [1971] 579/85).

39.2 Ternary Alloys

Rh-Ru-As. The substitution of ruthenium in $RhAs_3$ at 750 and 1050 °C has been attempted; there was, however, no change in the lattice parameter of $RhAs_3$ and therefore no solid solution [10].

Rh-Ru-Ce. Compositions of the formula $Ce(RH_{1-x}Ru_x)_2$ have been examined for some electrical and magnetic properties. Resistivity, susceptibility, and superconduction have been measured; striking violations of Nordheim's and Matthiessen's rules have been observed for the resistivity and it is suggested that large changes in the d-electron spin fluctuations may be occurring. Large variations in the susceptibility are also observed [8]. The valence of Ce in $Ce(Rh_{1-x}Ru_x)_2$ has been measured by L_{III} absorption. It was found to vary between 3.13 at $CeRh_2$ to 3.20 at $CeRu_2$. Lattice constants are also given [9].

Rh-Ru-Ti. In a study of CsCl-type phases and related distorted structures in the alloys of transition metals, the isothermal section of the system at 1000 °C was determined [7]. This is shown in **Fig. 180**, p. 270.

Rh-Ru-Zr. The superconducting transition temperature of $Ru_{0.5}Rh_{3.5}Zr_8$ was determined as ~11 K by [6].

Rh-Ru-Nb. Alloys were prepared in the form of 4 g buttons by arc-melting on a water-cooled copper hearth under helium. The niobium used was 99.95% pure. Melting losses were <0.3% of the total charge weight. The alloys were annealed initially for one week at 1400 °C in a dynamic vacuum of 2×10^{-5} Torr and slowly cooled to room temperature. They were then reannealed in evacuated quartz tubes for 3 to 6 weeks at 900 °C and water-quenched. Alloys having the Cr_3Si-type structure are of the general composition $Nb_3Rh_{1-x}B_x$ where B is Co, Ru, or Au. Lattice constants and superconducting temperatures for x = 0.02, 0.05 and 0.10 are according to [5].

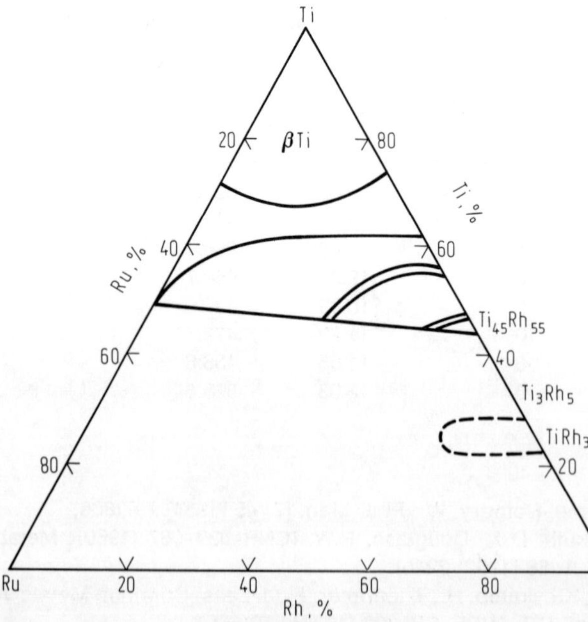

Fig. 180. Isothermal section of the Rh-Ru-Ti system at 1000 °C (compositions in at%).

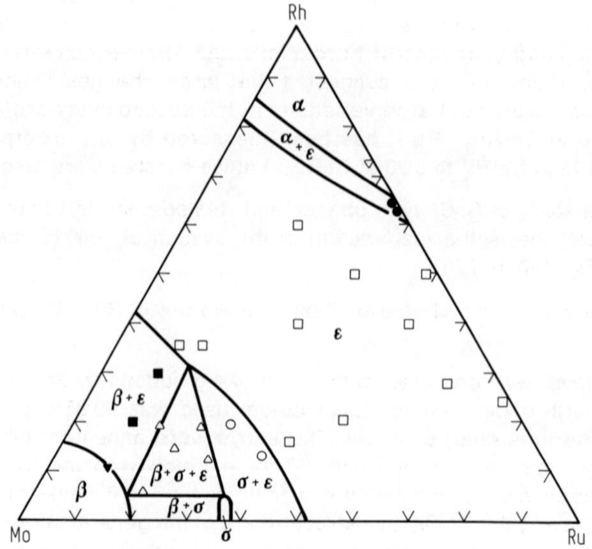

Fig. 181. Isothermal section of the Rh-Ru-Mo system at 1700 °C (concentrations in at%).

Rh–Ru–Mo. Fig. 181 shows the isothermal section at 1700 °C. Samples for phase studies were made from powders of ≧99.7% purity, compacted in a steel die, and melted in an arc-furnace under argon. Homogenization was carried out at 1700 °C in a vacuum of approximately 10^{-8} bar for 20 to 100 h. Samples with a weight loss >10% were discarded. Samples

Compositions and lattice parameters of phases in the Mo–Ru–Rh system.

nominal composition in at%			phases	lattice parameters in pm		
Mo	Ru	Rh		a	c	c/a
10	63	27	ε	272.0	430.8	1.584
10	50	40	ε	272.0	431.6	1.587
15	35	50	ε	272.8	433.7	1.590
20	20	60	ε	272.5	434.2	1.594
30	30	40	ε	274.2	436.6	1.592
33.3	46.7	20	ε	274.5	437.8	1.595
45	38.5	16.5	ε	275.2	441.5	1.604
50	35	15	ε	276.1	443.6	1.607
			σ	956.7	493.4	0.516
50	15	35	ε	276.1	443.7	1.607
52	28.5	19.5	ε	276.2	443.8	1.607
			σ	956.4	494.8	0.517
55	10	35	ε	276.3	444.0	1.607
57	24	19	ε	276.6	443.4	1.603
			σ	956.3	495.2	0.518
			β	314.2		
60	10	30	ε	276.6	444.9	1.609
			β	314.1		
60.7	27.8	11.5	ε	276.2	443.2	1.605
			β	313.4		
			σ [1]	960.0	492.0	0.512
65.9	20.1	14.0	ε	276.1	443.4	1.606
			β	313.5		
			σ [2]	—	—	—
65	15	20	ε	275.9	443.8	1.609
			β	313.7		
			σ	956.5	496.0	0.519
69.5	11.0	19.5	ε	276.3	444.0	1.607
			β	313.8		
75	20	5	ε	275.9	442.9	1.605
			β	313.4		
			σ	959.1	494.6	0.516
78.5	11	10.5	ε	276.8	444.0	1.604
			β	313.8		

[1] lattice parameters calculated from two reflections, [2] only one reflection

were analysed metallographically and by X-ray diffraction, electron probe microanalysis, and microhardness measurements. The following table shows details of composition, phases present, and lattice parameters [4].

x	a in Å	T_C in K
0.02	5.132	2.42
0.05	5.135	2.42
0.10	5.1346	2.44

Rh-Ru-Fe. The alloy $(Ru_{0.75}Rh_{0.25})_{0.99}Fe_{0.01}$ shows a local magnetic moment between 1.4 K and room temperature because of a Curie-Weiss dependence of χ on T. The following table shows the magnetic moment for a number of compositions containing 1 at% Fe, and the associated Curie temperatures; the quantum-mechanical implications are discussed [1].

Magnetic moment and Curie temperature for 1% solutions of Fe in Ru-Rh alloys [1].

alloy	structure	N	μ in μ_B (Curie-Weiss)	θ in K
Ru	hcp	8.0	0.0	–
$Ru_{0.75}Rh_{0.25}$	hcp	8.25	0.0	–
$Ru_{0.63}Rh_{0.37}$	hcp	8.37	0.8	-21 ± 2
$Ru_{0.5}Rh_{0.5}$	hcp	8.5	1.3	-13 ± 2
$Ru_{0.25}Rh_{0.75}$	fcc	8.75	1.7	-17 ± 2
Rh	fcc	9.0	2.2	-14 ± 2

The composition $FeRhRu_{0.083}$ was examined in a study of various transiton elements on the magnetic properties of iron-rhodium alloys; the alloys containing between 50 and 60 at% Rh only show an abrupt change in magnetization at a critical temperature which was a function of composition; small alloy additions considerably alter the critical temperature [2]. The magnetic susceptibility and the electric resistivity were measured for several rhodium-ruthenium alloys containing iron [3]; an earlier paper had shown that the magnetic susceptibility of binary rhodium-iron alloys followed an approximate Curie-Weiss law, and that χ and ϱ followed an equation of the type $\varrho(T)=\varrho_0+AT\chi(T)$.

Alloys containing ruthenium, however, showed Curie-Weiss temperatures which were independent of iron concentration, and which changed only slightly as the composition changed from $(Rh_{0.85}Ru_{0.15})_{0.999}Fe_{0.001}$ to $(Rh_{0.75}Ru_{0.25})_{0.99}Fe_{0.01}$. It was also found that the linear relationship between ϱ and χ no longer holds in the ternary alloys; these findings disprove a two-band model put forward earlier to explain the electric resistivity [3].

Excess susceptibility at 1.3 K, the host susceptibility, and the Curie-Weiss θ's for the alloys measured [3].

sample	χ (1.3 K $\times10^{-2}$ emu/mol Fe)	χ_{host} ($\times10^{-6}$ emu/g)	θ in K
$Rh_{0.995}Fe_{0.005}$	8.64	0.97	10.0 ± 1
$(Rh_{0.85}Ru_{0.15})_{0.999}Fe_{0.001}$	7.9	0.77	10.4 ± 1
$(Rh_{0.85}Ru_{0.15})_{0.995}Fe_{0.005}$	8.19	0.77	10.0 ± 1
$(Rh_{0.85}Ru_{0.15})_{0.99}Fe_{0.01}$	6.99	0.77	10.5 ± 1
$(Rh_{0.75}Ru_{0.25})_{0.99}Fe_{0.01}$	5.38	0.80	17 ± 1

References:

[1] Clogston, A.M.; Matthias, B.T.; Peter, M.; Williams, H.J.; Corenzwit, E.; Sherwood, R.C. (Phys. Rev. [2] **125** [1962] 541/52).
[2] Walter, P.H.L. (J. Appl. Phys. **35** [1964] 938/9).
[3] Knapp, G.S.; Sarachik, M.P. (J. Appl. Phys. **40** [1969] 1474/5).
[4] Paschoal, J.O.A.; Kleykamp, H.; Thümmler, F. (Z. Metallk. **74** [1983] 652/64).
[5] Zegler, S.T. (Phys. Rev. [2] **137** [1965] A 1438/A 1440).
[6] Havinga, E.E.; Damsma, H.; Kanis, J.M. (J. Less-Common Metals **27** [1972] 281/91).
[7] Dwight, A.E.; Beck, P.A. (Trans. Metall. Soc. AIME **245** [1969] 389/90).
[8] Harrus, A.; Timlin, J.; Mihalisin, T.; Batlogg, B. (J. Appl. Phys. **55** [1984] 1990/2).
[9] Mihalisin, T.; Harrus, A.; Raaen, S.; Parks, R.D. (J. Appl. Phys. **55** [1984] 1966/8).
[10] Bennett, S.L.; Heyding, R.D. (Can. J. Chem. **44** [1966] 3017/30).

39.3 Quaternary Alloys

Rh-Ru-Mo-Zr. A new superconducting phase with an α-Mn-type structure has been discovered in an alloy of composition $Ru_{23}Rh_6Mo_{19}Zr_{10}$. Measurements of low-temperature specific heat have been carried out and the principles underlying the stability of this peculiar structure type are discussed. The following results are given: $\gamma=3.75\ mJ\cdot g-atom^{-1}\cdot K^{-1}$, $\theta_D=336$ K, $T_c=4.58$ K [1].

Rh-Ru-U-Al. The role of hybridization in alloys of general formula $URu_{1-x}Rh_xAl$ has been studied with measurements of susceptibility, magnetization and specific heat. The development of magnetism occurring in compounds with rhodium concentrations >20 at% could be attributed mainly to a varying strength of the hybridization of the U 5f orbitals with the transition metal d orbitals [2].

References:

[1] Waterstrat, R.M.; Kuentzler, R. (J. Less-Common Metals **142** [1988] 163/8).
[2] Veenhuizen, P.A.; Klaasse, J.C.P.; de Boer, F.R.; Sechovsky, V.; Havela, L. (J. Appl. Phys. **63** [1988] 3064/6).

Physical Constants and Conversion Factors

Avogadro constant N_A (or L) $= 6.02214 \times 10^{23}$ mol^{-1}
Faraday constant $F = 9.64853 \times 10^4$ C/mol
molar gas constant $R = 8.31451$ J·mol^{-1}·K^{-1}
molar volume (ideal gas) $V_m = 2.24141 \times 10^1$ L/mol
(273.15 K, 101325 Pa)

Planck constant $h = 6.62608 \times 10^{-34}$ J·s
elementary charge $e = 1.60218 \times 10^{-19}$ C
electron mass $m_e = 9.10939 \times 10^{-31}$ kg
proton mass $m_p = 1.67262 \times 10^{-27}$ kg

1 kg = 2.205 pounds
1 m $= 3.937 \times 10^1$ inches = 3.281 feet
1 m^3 $= 2.642 \times 10^2$ gallons (U.S.)
1 m^3 $= 2.200 \times 10^2$ gallons (Imperial)

Force	N	dyn	kp
1 N	1	10^5	1.019716×10^{-1}
1 dyn	10^{-5}	1	1.019716×10^{-6}
1 kp	9.80665	9.80665×10^5	1

Pressure	Pa	bar	kp/m^2	at	atm	Torr	lb/in^2
1 Pa = 1 N/m^2	1	10^{-5}	1.019716×10^{-1}	1.019716×10^{-5}	9.86923×10^{-6}	7.50062×10^{-3}	1.450378×10^{-4}
1 bar = 10^6 dyn/cm^2	10^5	1	1.019716×10^4	1.019716	9.86923×10^{-1}	7.50062×10^2	1.450378×10^1
1 kp/m^2 = 1 mm H_2O	9.80665	9.80665×10^{-5}	1	10^{-4}	9.67841×10^{-5}	7.35559×10^{-2}	1.422335×10^{-3}
1 at (technical)	9.80665×10^4	9.80665×10^{-1}	10^4	1	9.67841×10^{-1}	7.35559×10^2	1.422335×10^1
1 atm = 760 Torr	1.01325×10^5	1.01325	1.033227×10^4	1.033227	1	7.60×10^2	1.469595×10^1
1 Torr = 1 mmHg	1.333224×10^2	1.333224×10^{-3}	1.359510×10^1	1.359510×10^{-3}	1.315789×10^{-3}	1	1.933678×10^{-2}
1 lb/in^2 = 1 psi	6.89476×10^3	6.89476×10^{-2}	7.03069×10^2	7.03069×10^{-2}	6.80460×10^{-2}	5.17149×10^1	1

Work, Energy, Heat	J	kW·h	kcal	Btu	eV
1 J = 1 W·s = 1 N·m = 10^7 erg	1	2.778×10^{-7}	2.39006×10^{-4}	9.4781×10^{-4}	6.242×10^{18}
1 kW·h	3.6×10^{6}	1	8.604×10^{2}	3.41214×10^{3}	2.247×10^{25}
1 kcal	4.1840×10^{3}	1.1622×10^{-3}	1	3.96566	2.6117×10^{22}
1 Btu (British thermal unit)	1.05506×10^{3}	2.93071×10^{-4}	2.5164×10^{-1}	1	6.5858×10^{21}
1 eV	1.602×10^{-19}	4.450×10^{-26}	3.8289×10^{-23}	1.51840×10^{-22}	1

1 cm^{-1} = 1.239842×10^{-4} eV   1 Hz = 4.135669×10^{-15} eV

1 hartree = 27.2114 eV   1 eV ≙ 23.0578 kcal/mol

Power	kW	hp	$kp \cdot m \cdot s^{-1}$	kcal/s
1 kW = 10^3 J/s	1	1.35962	1.01972×10^{2}	2.39006×10^{-1}
1 hp (horsepower, metric)	7.3550×10^{-1}	1	7.5×10^{1}	1.7579×10^{-1}
1 $kp \cdot m \cdot s^{-1}$	9.80665×10^{-3}	1.333×10^{-2}	1	2.34384×10^{-3}
1 kcal/s	4.1840	5.6886	4.26650×10^{2}	1

References:

Mills, I. (Ed.), International Union of Pure and Applied Chemistry, Quantities, Units and Symbols in Physical Chemistry, Blackwell Scientific Publications, Oxford 1988.

The International System of Units (SI), National Bureau of Standards Spec. Publ. 330 [1972].

Landolt-Börnstein, 6th Ed., Vol. II, Pt. 1, 1971, pp. 1/14.

ISO Standards Handbook 2, Units of Measurement, 2nd Ed., Geneva 1982.

Cohen, E. R., Taylor, B. N., Codata Bulletin No. 63, Pergamon, Oxford 1986.